Microscale and Nanoscale Heat Transfer

Fundamentals and Engineering Applications

T0139171

Microscale and Nanoscale Heat Transfer

Fundamentals and Engineering Applications

C. B. Sobhan

National Institute of Technology
Calicut, India

G. P. Peterson

University of Colorado
Boulder, U.S.A.

CRC Press
Taylor & Francis Group
Boca Raton London New York

CRC Press is an imprint of the
Taylor & Francis Group, an **informa** business

CRC Press
Taylor & Francis Group
6000 Broken Sound Parkway NW, Suite 300
Boca Raton, FL 33487-2742

First issued in paperback 2019

© 2008 by Taylor & Francis Group, LLC
CRC Press is an imprint of Taylor & Francis Group, an Informa business

No claim to original U.S. Government works

ISBN-13: 978-0-8493-7307-7 (hbk)
ISBN-13: 978-0-367-40350-8 (pbk)

Library of Congress Cataloging-in-Publication Data

Sobhan, Choondal B.
    Microscale and nanoscale heat transfer : fundamentals and engineering applications / Choondal B. Sobhan, G.P."Bud" Peterson.
        p. cm.
    Includes bibliographical references and index.
    ISBN 978-0-8493-7307-7 (hardback : alk. paper)
    1. Heat--Transmission. I. Peterson, G.P. (George P.) II. Title.

TJ260.S63 2008
621.402'2--dc22   2008012305

Visit the Taylor & Francis Web site at
http://www.taylorandfrancis.com

and the CRC Press Web site at
http://www.crcpress.com

CV 09.19.2019 2223

Contents

Preface

Heat transfer in size-affected domains has become one of the most widely studied areas in thermal science and engineering in recent times. The ever increasing quest for miniaturization, especially in relation to microelectronics, has made this a topic of considerable interest over the past decade. The possibility of including heat sinks as an integral part of individual components, and the operational- and fabrication-related challenges in ensuring effective heat dissipation, have inspired investigators to focus their attention on heat transfer problems at very short length scales. This has led to a tremendous growth of research and the resultant publications in this field, particularly with regard to experimental and computational analyses. The size effect, which becomes more and more pronounced when the domain size is reduced, offers special challenges to the researcher; whether the approach is based on microscale and nanoscale measurements using advanced (microelectrical systems) MEMS applications or on conventional, modified, or discrete computational methods.

The past few years have seen an enormous amount of literature being published in the area of microscale and nanoscale thermal phenomena. One of the objectives of this work is to compile and discuss the most relevant findings from these investigations, comparing and contrasting the methods and observations related to size-affected domains with conventional heat transfer analyses. Being active in research on microscale and nanoscale thermophysical phenomena in fluid domains, the authors have attempted to provide extensive discussions, based on work accomplished by their research groups, as well as those published in the open literature. Although not the authors' direct areas of research, microscale conduction and radiation are presented as additional areas, based on information obtained from excellent research material published by eminent researchers on these topics.

The opening chapter of the book provides an introduction to microscale heat transfer, where the general trends and observations in the technological area are discussed, focusing on both the early work and the more recent contributions. Chapter 2 deals with microscale conduction heat transfer, wherein application-oriented research, basic thermal conductivity models, microscale conduction measurement methods, and related topics are presented based on information taken from the most relevant publications and treatises. Chapters 3 and 4 present material on the fundamentals and engineering applications of microscale convective heat transfer, respectively. Several methods of special analysis pertaining to microscale domains have been discussed in Chapter 3. Analysis and design aspects pertaining to engineering applications, related to microchannels, slip flow domains, and micro heat pipes are included in Chapter 4. Experimental studies using optical techniques for flow and heat transfer in small channels are also included in this chapter. A comprehensive review of microscale convective heat transfer, related to engineering applications, is also presented. Chapter 5 deals with the fundamentals of microscale radiation, and some case studies from the literature on microscale radiation phenomena, which require special treatment. Important analysis methods and results from microscale radiation literature are reproduced and discussed in this chapter. Chapter 6 on nanoscale thermal phenomena emphasizes nanofluids, and their thermal behavior. A large fraction of the material presented in Chapters 4 and 6 is the result of the original efforts undertaken at the authors' laboratories in the United States and India.

Worked out numerical examples have been included in the last chapter (Chapter 7). It is expected that these examples will help the reader in the application of the results of microscale and nanoscale heat transfer analysis in practical design problems, where such results can be directly applied. However, it should be noted that a majority of the microscale heat transfer problems require special analysis utilizing particular formulation methods and solution approaches. Still, the problems presented in Chapter 7, covering all the modes of heat transfer discussed in the book, are meant to throw light on some of the differences and deviations to be taken into consideration while developing engineering designs related to microscale systems.

One of the limitations while dealing with dynamic emerging technologies is the difficulty in being comprehensive and inconclusive in the treatment of the subject in the form of a book. Notwithstanding this difficulty, and within the limited exposure to the vast resources of the subject matter, the authors do expect that the information provided would be useful and of interest to the graduate and research communities in this exciting new domain of knowledge.

C.B. Sobhan
G.P. Peterson

Acknowledgments

This book is an outcome of the ongoing academic and research collaboration between the authors, which began in 2003, when C.B. Sobhan spent a sabbatical year at the Two-Phase Heat Transfer Laboratory headed by G.P. Peterson at Rensselaer Polytechnic Institute, Troy, New York. G.P. Peterson has subsequently relocated to the University of Colorado at Boulder. The contents of the book include a large amount of original work performed in the authors' laboratories, in the United States and in India, and they would like to thank their institutions for providing the resources to accomplish this research.

The first major opportunity for C.B. Sobhan to enter the exciting area of microscale heat transfer was provided by Professor Suresh V. Garimella at Purdue University, West Lafayette, Indiana, in 1999, whose encouragement and motivation are gratefully acknowledged. G.P. Peterson first became involved in the field following an introductory course taught by A. Bar-Cohen and A. Kraus in 1981.

A number of the authors' students and research scholars have contributed to this book, through valuable research findings, discussions, and support in manuscript preparation. In particular, C.B. Sobhan would like to acknowledge the contributions of Sankar Narayanan, Nithin Mathew, V. Sajith, U.B. Jayadeep, and Shijo Thomas toward the material for the book. The tremendous help offered by K. Prabhul and Praveen P. Abraham, who spent many hours on manuscript editing, and the helping hands extended by Rahul Jayan, Raison Chacko, Jose Devachan, Satish Kumar Pappu, Yadu Vasudev, Divya Haridas, and P. Nithin at various stages of the work are gratefully acknowledged. The drawings were prepared by P. Beljith, and his efforts in this regard are acknowledged. Also of particular note is the work of J. Ochterbeck, A. Duncan, B. Babin, J.T. Dickey, J.M. Ha, B.H. Kim, H.B. Ma, Y. Wang, J. Li, C.H. Li, and C. Li, all of whom have taught GPP more than what he could have ever learned on his own.

Finally, the authors also would like to thank their respective families for their support, patience, and understanding, which have made this project a reality.

Authors

Choondal B. Sobhan is a professor in the Department of Mechanical Engineering, National Institute of Technology Calicut, Kozhikode, Kerala, India. He is also a research collaborator at the Two-Phase Heat Transfer Laboratory, Department of Mechanical Engineering, University of Colorado at Boulder.

Professor Sobhan received his bachelor of technology degree in 1984 from Regional Engineering College, Calicut, India. He obtained his master of technology and PhD from the Indian Institute of Technology, Madras, India in 1986 and 1990, respectively. Since then, he has been working as a faculty member in the Department of Mechanical Engineering at Regional Engineering College, Calicut, which is now National Institute of Technology, Calicut. He became a professor of mechanical engineering at the institute in 2006.

Professor Sobhan has been a visiting faculty member at Rensselaer Polytechnic Institute, Troy, New York (2003–2006) and a visiting scholar at the Cooling Technologies Research Consortium, School of Mechanical Engineering, Purdue University, West Lafayette, Indiana (1999–2000). He has also held postdoctoral research positions at Nanyang Technological University, Singapore (1997–1999), and the University of Wisconsin, Milwaukee (1999).

Professor Sobhan's background is on research topics such as interferometric measurement of heat transfer, heat pipes, heat exchangers, microchannels, and micro heat pipes. His current research interests are in optical measurements at the microscale, molecular dynamics modeling of nanoscale thermal phenomena, thermal management of electronics, and numerical modeling of phase change heat spreaders and heat pipes. He has been a reviewer for (American Society of Mechanical Engineers) ASME and (American Institute of Aeronautics and Astronautics) AIAA journals, the *International Journal of Heat and Mass Transfer, Nanoscale and Microscale Thermophysical Engineering, Elsevier Journal of Aerospace Science and Technology, Journal of Microfluidics and Nanofluidics*, and *Heat Transfer Engineering*.

Professor Sobhan performs collaborative research internationally with research groups at the Stokes Research Institute, University of Limerick, Ireland, and National Research Laboratory, Dong-A University, South Korea, in addition to the group at the Two-Phase Heat Transfer Laboratory, University of Colorado at Boulder. He has coauthored more than 50 papers in refereed international journals and conferences, and a chapter in the *MEMS Handbook*.

George P. Peterson is currently the chancellor at the University of Colorado at Boulder, Colorado. Before his appointment as chancellor in July 2006, Professor Peterson served for six years as provost at Rensselaer Polytechnic Institute in Troy, New York. He received his BS in mechanical engineering in 1975, a BS in mathematics in 1977, and an MS in engineering in 1980, all from the Kansas State University, and a PhD in mechanical engineering from Texas A&M University in

1985. In 1981 and 1982, Professor Peterson was a visiting research scientist at the NASA Johnson Space Center, and in 1985 he moved to a faculty position in the mechanical engineering department at Texas A&M University, where he conducted research and taught courses in thermodynamics and heat transfer. In 1990, he was named the Halliburton Professor of Mechanical Engineering and in 1991 was named the College of Engineering's Tenneco Professor. In 1993, Professor Peterson was invited to serve as the Program Director for the Thermal Transport and Thermal Processing Division of the National Science Foundation (NSF) where he received the NSF award for outstanding management. From June 1993 through July 1996, he served as head of the Department of Mechanical Engineering at Texas A&M University and in 1996 was appointed to the position of executive associate dean of the College of Engineering, where he also served as the associate vice chancellor for the Texas A&M University system. Before joining Texas A&M University, Professor Peterson was head of the General Engineering Technology Department at Kansas Technical Institute (now Kansas State University, Salina).

Throughout his career, Professor Peterson has played an active role in helping to establish the national education and research agendas, serving on numerous industry, government, and academic task forces and committees. In this capacity, he has served as a member of a number of congressional task forces, research councils, and advisory boards, most recently serving as a member of the board of directors and vice president for education for the AIAA. In addition, he has served in a variety of different roles for federal agencies, such as the Office of Naval Research (ONR), the National Aeronautics and Space Administration (NASA), and the Department of Energy (DOE), and for national task forces and committees appointed by the National Research Council (NRC) and the National Academy of Engineering (NAE). He is currently serving on a number of national groups whose focus is on postsecondary education, such as the American Association of Colleges and Universities, the Middle States Commission on Higher Education, and the New England Association of Schools and Colleges, where he is currently serving as accreditation team chair.

A fellow of both the ASME and AIAA, Professor Peterson is the author/coauthor of 12 books/book chapters, 160 refereed journal articles, more than 150 conference publications, and holds 8 patents. He has been an editor or associate editor for eight different journals and is currently serving on the editorial advisory board for two other journals. He is a registered professional engineer in the state of Texas, and a member of Pi Tau Sigma, Tau Beta Pi, Sigma Xi, and Phi Kappa Phi societies. He has received several professional society awards, which include the Ralph James and the O.L. Andy Lewis awards from ASME, the Dow Outstanding Young Faculty Award from ASEE, the Pi Tau Sigma Gustus L. Larson Memorial Award from ASME, the AIAA Thermophysics Award, the ASME Memorial Award, the AIAA Sustained Service Award, and the Frank J. Malina Award from the International Astronautical Society. While at Texas A&M, Professor Peterson was selected to receive the Pi Tau Sigma, J. George H. Thompson Award for excellence in undergraduate teaching and the Texas A&M University Association of Former Students Outstanding Teaching Award at both the college and university levels.

Nomenclature

A, A_c	cross-sectional area
a	lattice constant
a_λ	absorption coefficient
B	body force term
C, c	specific heat
C_p, c_p	specific heat at constant pressure
C_v, c_v	specific heat at constant volume
C_f	specific heat of the film
C_l	specific heat of the liquid
C_{sub}	specific heat of the substrate
C_f	skin friction coefficient
c_o	light velocity in vacuum
c_T	speed of temperature propagation
D	diffusion coefficient
D_h	hydraulic mean diameter
d	one side of a channel
dA	strip area
E	energy of a quantized energy packet
E_l	total energy of the liquid per unit volume $[E_l = \rho_l(C_lT + 1/2u_l^2)]$
E_v	total energy of the vapor per unit volume
E_o	original amplitude of electric field $[E_v = \rho_v(C_vT + 1/2u_v^2)]$
F	force vector
f	friction coefficient
f	particle distribution function
f_o	Fermi–Dirac distribution function
G	mass velocity
g	gravitational constant
g^*	imposed acceleration in reduced units
Gr	Grashof number
H	height
H_c	channel height
h	heat transfer coefficient
h	Planck's constant
h_{fg}	latent heat of vaporization
h_l	local heat transfer coefficient
h_o	heat transfer coefficient at the condenser
i	grid number in x-direction
j	grid number in y-direction

k	thermal conductivity
$k*$	thermal conductivity ratio with respect to copper
k_{conv}	convective component of thermal conductivity
k_{CP}	thermal conductivity according to Cahill–Pohl model
k_E	thermal conductivity according to Einstein's model
k_{eff}	effective thermal conductivity
k_f	thermal conductivity of the fluid
k_f	film component of thermal conductivity
k_g	lattice component of thermal conductivity
k_m	measured thermal conductivity of metallic crystal
k_p	thermal conductivity of particle
k_φ	molecular vibration component of thermal conductivity
k_n	thermal conductivity in direction n
k_s	thermal conductivity of substrate
k_{sub}	substrate component of thermal conductivity
k_{ph}	phonon wave vector
Kn	Knudsen number
L	characteristic length
L	length of the heat pipe
L	length of fin array
L_c	condenser section length
L_e	evaporator section length
L_e	entrance length
L_h	hydrodynamic entry length
L_t	thermal entry length
l	shortest length dimension
M	atomic mass
m	complex refractive index
N	number density of atoms
N	number density of molecules in Equation 2.48
n	number density of atoms
n	direction of heat flux
$\bar{n}$	average number of free electrons per unit volume
n	refractive index
n	real part of complex refractive index
Nu	Nusselt number
P	pressure
ΔP	pressure drop
P'	power dissipation per unit length
p	momentum
Pe	Peclet number
Pr	Prandtl number
P_{sat}	saturation pressure
Q	heat transfer rate

Q_{in}	heat input to the heat sink
$Q_{in}*$	heat input to the individual micro heat pipe
Q_m	mass flow rate in the microchannel
q, q''	heat flux
q_{in}	heat flux into the heat sink
$q''_{ONB,exp}$	onset heat flux from experiment
$q''_{ONB,mod}$	onset heat flux from model
q_n, q_n''	heat flux in direction n
q_{im}	imaginary part of the wave vector
R	gas constant
R	overall thermal resistance
R	radial coordinate
r	radius of curvature of the meniscus
r	dimensionless thermal resistance
r	position vector
r	radius of curvature of liquid–vapor meniscus
r_0	initial radius of curvature of the meniscus
Re	Reynolds number
Re_{cri}	critical Reynolds number
R_{th}	theoretical thermal resistance
S	fin spacing
S_x	component of Poynting vector in x direction
T, t	temperature
t	time
ΔT	temperature difference
ΔT_{sub}	substrate component of temperature oscillation
T_{amb}	ambient temperature
T_b	bulk temperature
T_w	wall temperature
T_α	free stream temperature
U	dimensionless velocity in x-direction
u	velocity in x-direction
V	dimensionless velocity in y-direction
V, ν	velocity
V_{il}	interfacial liquid velocity
V_{iv}	interfacial vapor velocity
V_{tv}	thermal voltage noise
V_z	velocity in the flow direction in Equation 3.12.
V	volume
ν	particle velocity
ν	sound velocity
$\underline{\nu}$	velocity in y-direction
ν_e	electron velocity
W	width of the metal

We	Weber number $(G^2 L/\rho\sigma)$
W_T	width of the substrate
X	dimensionless distance along the x-direction
ΔX	dimensionless grid size in the x-direction
x	distance coordinate
x	displacement of an atom in CP model
Δx	distance
Y	dimensionless distance along the y-direction
ΔY	dimensionless grid size in the y-direction

Greek Symbols

α	ratio of thermal conductivities of particle and base fluid
α	thermal diffusivity
β	coefficient of volumetric expansion
β_i, β_l, β_{lw}	geometric area coefficients
γ	coefficient in Equation 2.38
ζ	Fermi energy
ζ	zeta potential
η	joule heating frequency
θ	dimensionless temperature
θ	nondimensional temperature difference
κ	extinction coefficient
Λ	mean free path
λ	wavelength of the heat carriers
λ_o	wave length in vacuum
λ_{ph}	phonon wavelength
μ, μ_f	dynamic viscosity
μ, μ_1, μ_2	coefficients of viscosity
μ	magnetic permeability
ν	frequency
ν	kinematic viscosity
γ	wave number
Ξ	total energy of phonons
Ξ_{ph}	energy in any phonon mode
ξ	dimensionless vorticity
ρ	density
ρ_f	density of fluid
ρ	electrical resistivity
σ	surface tension
σ	accommodation coefficient
σ	characteristic distance
τ	relaxation time
τ	dimensionless time
$\Delta\tau$	dimensionless time step
Φ	viscous dissipation term
φ	heat generation per unit volume
Ψ	coefficient for heat spreading in the film
ψ	dimensionless stream function
ω	angular frequency
ω_{ph}	phonon angular frequency

Subscripts

0	base
∞	ambient
a	atmosphere
av	average
c	condenser
cav	center line average up to the fin height
e	evaporator
f	fin
fav	fin average
i	interface
j	temperature jump at gas flow boundary
l	liquid
li	liquid interface
lw	liquid wall
max	maximum
s	slip at gas flow boundary
sc	scattering
t	energy
u	momentum
v	vapor
vi	vapor interface
vw	vapor wall
w	wall
x	local

1 Introduction to Microscale Heat Transfer

1.1 MICROSCALE HEAT TRANSFER: A RECENT AVENUE IN ENERGY TRANSPORT

The influence of the physical size of the domain under consideration, or the extent of the medium, on heat transfer has been of interest to scientists for over 400 years. Whether it is heat conduction in solids, where the surface boundaries or interfaces can be of significance; convective heat transfer in fluids, where the basic mechanism is conduction into a fluid layer adjacent to a surface; or radiative heat transfer, where the heat transport mechanism is electromagnetic in nature, size effects are important, particularly when dealing with small-length-scale systems. While the fundamental physics within the physical domain of microscale dimensions are typically the same, the small-length-scale effects often make conventional approaches to the analysis of heat transport at small-length scales inappropriate or, in some cases, inaccurate.

A number of explanations have been developed and proposed to explain some of the differences and deviations that occur between the observed behavior of large-scale and microscale systems. In the case of conduction-aided thermal transport, the most popular and widely accepted of these is the order of magnitude similarity between the reduced characteristic length of the physical domain and the scattering mean free path of the energy carriers, especially near the surfaces, due to termination and/or interaction at the boundaries. In convective heat transport, the impact and relative importance of the boundary layer can be of increased significance in microscale systems, while in radiative heat transfer, the microscale size effects are believed to be of increased importance when the characteristic length scales of surfaces or structures are of the same order of magnitude as the wavelength. These effects have gained considerable importance for both research and practice, particularly in applications involving conduction and radiation in thin films, layers and microscale structures in electronics and other micromanufacturing processes, and microscale cooling options for the thermal management of microelectronic devices and systems.

Though fluid flow and heat transfer in channels with very small hydraulic mean diameters have been of interest for quite some time, it has only been recently that the study of microscale thermal phenomena in engineering applications has gained relevance and importance. In particular, microscale conduction and radiative heat transfer have become an important area of study, due to the numerous applications involving the fabrication processes and performance monitoring of semiconductor devices. Focused studies have been reported over a broad range of microscale thermal transport phenomena, and the depth and breadth of knowledge has increased dramatically. More recently, due to the significant potential of applications in micromachining, microelectronics, and microelectromechanical systems (MEMS), the field has expanded into the domain of nanoscale heat transfer, which calls for an altogether different approach and analysis methodology to clearly understand and apply the physical phenomena associated with thermal transport at these length scales.

There has been immense progress in the level of sophistication of the experimental techniques and theoretical analyses of microscale heat transfer over the past two decades. While earlier studies were conducted using conventional measurement techniques, probes and instrumentation, with the advent of MEMS-based technology, have facilitated the development of system-integrated measurement techniques. These have helped in revolutionizing the level of understanding, as well as serving to correct many of the inaccurate hypotheses associated with microscale heat transfer, most of which were the result of the limitations in measurement techniques as well as computational capabilities. Through improvements in these areas, there is now much more information available on the size limitations that govern the continuum assumption used as the basis for conventional flux laws for heat transfer, and the domain range over which these laws can be applied.

1.2   STATE OF THE ART: SOME INTRODUCTORY REMARKS

Research and investigations focused on microscale heat transfer have progressed through two relatively distinct phases. The first of these occurred before the advent and introduction of sophisticated MEMS-based probes and measurement systems, and the second occurred after the introduction of these devices and techniques. Broadly speaking, the research in the 1990s and thus far in the present century have pointed toward significant disparities in the reported data and the interpretation of these data, in large part because of the level of resolution possible and the interference resulting from the mere presence of the probe or measurement devices. Because investigators had to depend on the available instrumentation in the early stages of microscale experimentation, these data indicated wide disparities for similar physical situations. Moreover, the fabrication techniques were also not as well advanced as they are today. As a result, the sophistication level applicable to the manufacture of microscale structures, especially microchannel arrays and heat sinks, was comparatively limited.

For instance, much of the early experimental findings reported during this period suggested that convective heat transfer in microchannels deviated from that in conventional channels, and that the heat transfer and fluid friction correlations in conventional channels could not accurately predict the phenomena observed in microchannels. While many investigators agreed that there might be some fundamental differences between the observed behavior in the microscale and that observed for larger conventional channels, it was apparent that in addition to not agreeing with the values predicted by the conventional approaches, much of the experimental heat transfer data reported by various investigators were not in agreement, making it difficult to draw any conclusions about microscale heat transfer in general and microscale convection in particular. For microchannel flow phenomena, early evidence indicated that the deviations from conventional channels could be due to poor instrumentation and fabrication, flow maldistribution, and calculation errors that neglected the entrance and exit effects in the channel arrays.

Extensive comparative analysis of studies on microchannels, as presented in the literature (Sobhan and Garimella 2001, Garimella and Sobhan 2003), indicated that additional investigation and data were required to understand the phenomenon, and that these data should be obtained using the most sophisticated instrumentation and data reduction methods possible, before any overarching statements could be made regarding the applicability of conventional analyses to microscale heat transfer or the generalized deviation from the existing correlations. Overall, however, there is sufficient evidence to support the premise that continuum modeling can be used to analyze most of the work reported in the literature thus far on the heat transfer and liquid flow characteristics in microchannels.

While most of the experimental research in microchannels focused on investigations of the similarities and differences of microscale and conventional heat transfer, a number of theoretical investigations have attempted to use the continuum governing equations for microchannel fluid flow and heat transfer problems, and to apply established numerical methods in solving the computational domains for systems with very small physical dimensions. This led to comparisons of the experimental results with new and varied numerical results. Some of these investigations were aimed at the optimization of microchanneled heat sinks for the best heat transfer performance in electronic packages and systems.

As stated earlier, recent microscale heat transfer research has benefited greatly from the use of sophisticated instruments and system-integrated probes to measure fluid flow and heat transfer characteristics, designed and manufactured using MEMS technology. Interestingly, almost all of the recent research indicates that microchannels are "fairly large channels" for continuum modeling to fail, and the conventional correlations and predictive tools should be quite sufficient to understand and predict the performance of microchannels unless the physical dimensions are too small. Experimental evidence indicates that the continuum assumptions can be comfortably used, down to hydraulic diameters of around 50 μm. Other than microprobes, sophisticated optical measurement methods such as micro-PIV (Particle Image Velocimetry) and microscopy have been used to better understand single-phase flow, phase change, and two-phase flow in microchannels, leading to results which precisely matched conventional predictions. The experimental and theoretical research dealing with flow and heat transfer in microchannels in the recent past has been reviewed extensively in the literature (Sobhan and Peterson 2006).

Size effects are an important consideration while studying and computing thermal transport phenomena as the bulk thermophysical properties such as thermal conductivity, used in such computations, result from macroscopic flux laws. A thorough investigation into the size effects of transport in small domains is required to analyze these types of problems. This leads to the analysis of thin films of solid materials, as well as the liquid domains for small geometries, using space and timescales compatible with physical transport phenomena.

Significant advances in the measurement of the thermal conductivity of thin films using highly accurate methods, including optical and electrical techniques, have been made in the more recent past. It is still not clear, however, at what physical dimensions the continuum assumption may break down in the analysis of microscale heat transfer, as well as at what physical sizes does the use of bulk thermophysical properties, such as thermal conductivity, which are defined based on flux laws, start to become invalid. Under such cases, it is necessary to employ an altogether different modeling methodology, such as modeling based on ballistic transport equations or a molecular dynamics modeling approach, for microscale heat transfer computations.

1.3   OVERVIEW OF MICROSCALE TRANSPORT PHENOMENA

While extensive discussions of the various modes of microscale heat transfer are presented in the succeeding chapters, a very brief overview of the studies is presented here. Investigators have enthusiastically pursued the study of all modes of heat transfer, the fundamental modes of conduction in microstructures, convection in microchannels, and microscale radiation phenomena, as all of these find applications in many of the modern engineering problems. Apart from these, thermal transport phenomena such as phase change, single and two-phase flows, and combined mode heat transfer are also important in many of the engineering applications, and have been studied.

Among the many areas of application of microscale heat transfer, the most demanding at present is in the electronics industry, where concepts, analysis, and data are essential to identify and resolve manufacturing and operational issues for both systems and packages. Fabrication of microelectronics and the thermal management of electronic components and systems pose a variety of challenging problems with respect to microscale heat transfer. A number of other areas such as biomedical engineering, lasers, microelectromechanical systems, and nanotechnology, among others, must also employ the analysis and design of microscale heat transfer applications, which are extremely important in dealing with problems specific to these industries.

While it is not clear "how small is small" in deciding whether conventional theory can be applied to the microscale, it is certain that deviations from the conventional approaches, due to the size effect, become important at some level of physical dimensions. This is expected to be so in microchannels, or while dealing with conduction and radiation in thin films and microscale structures. All of these situations will lead to modeling and solutions based on special techniques, while approaching the problems theoretically. As mentioned before, ballistic heat transport models and discrete computation including molecular dynamics modeling are some of the techniques that have been applied in the mathematical modeling of these physical situations. From an experimental perspective, it is important to have sophisticated techniques and instrumentation with physical dimensions suitable for measurements, as non-intrusively as possible, and this has led to an interesting coexistence of MEMS and microscale heat transfer in the recent past.

1.3.1 MICROCHANNEL FLOW AND CONVECTIVE HEAT TRANSFER

The classification of channels into large, mini, and micro is quite arbitrary in nature. Over the past few years, because of a large number of research publications appearing in the literature on fluid flow and heat transfer in channels of varying size, a general method of classification has evolved. This has resulted in a terminology or classification such that channels for which the largest dimension of the polygonal cross section is below 1 mm are classified as microchannels. Channels with cross-sectional dimensions in the range of a few millimeters are called minichannels (typically 1–5 mm) and those above this are considered as conventional or large channels. In the case of irregular shapes, or very deep or broad channels, the hydraulic mean diameter is often used as the criterion for the classification as mentioned above.

Of the fundamental modes of heat transfer, the one most studied in relation to microscale heat transfer is convective heat transfer in microchannels. The pioneering work in the application of small channels for heat dissipation from silicon substrates was reported by Tuckerman and Pease (1981). This investigation focused on the design and testing of very compact water-cooled heat sinks integrated into silicon substrates. The channels used in these heat sinks were as small as 50 μm wide and 300 μm deep, which are typical of microchannels in use today and are characterized by dimensions less than a millimeter. This first effort held the promise of using fluid flow through microchannels as an effective means of heat dissipation from silicon-integrated circuits. Heat dissipation of the order of 790 W/cm^2 was achieved, with a substrate temperature rise of 71°C above the inlet water temperature. Problems associated with coolant selection, packaging and the development of effective headers, microstructure selection, fabrication and bonding, and some successful strategies for forced convection cooling using microchannels were all discussed by Tuckerman and Pease (1982). The use of microchannels provided a means of achieving two of the most sought for solutions in electronics cooling—a high heat removal rate, and the possibility of integrating the heat sink into silicon substrates.

Though the available instrumentation facilities at the early stages of microchannel research were limited in terms of precision, it is worthwhile and interesting to look at the results of these early investigations to understand the premises that led to the development of microchannel convection as a distinct area of study. With the numerical and experimental methods available at the time, a number of investigators studied the flow and heat transfer characteristics of single microchannels and arrays of microchannels, most frequently, deducing results from observations of effects external to the channel, in the substrates. It was noted early on that the results presented considerable variations and deviations from what would normally be expected from an analysis of the problem using conventional models and correlations. This led to more careful observations of similar channels. The primary aim of these studies was the quest to identify the conditions under which flow in a channel would deviate from conventional theory based on the continuum hypothesis. Early results were somewhat convincing and some investigators believed that this might be possible even at dimensions of the order of a few hundreds of microns, which are classified as "fairly large" microscale dimensions at present.

While extensive discussions on the literature on convective transport in micro-channels will be presented later, a very brief overview of the early studies mentioned above is given here. As discussed, studies on microchannel convection had resulted in findings that were mutually contradictory in both the values and the trends. A comparison of some of these investigations, particularly those reported in the last decade of the twentieth century, is presented in Figures 1.1 and 1.2. As illustrated, the results presented by various investigators, both on fluid friction and on heat transfer, show large variations, and also indicate significant deviations from the values predicted using well-accepted theoretical models and algorithms. The salient features of these investigations are discussed below.

According to conventional predictions, the product of the friction factor and the Reynolds number ($f \times$ Re) should be constant for laminar flow (64 for circular and other constant values for other cross sections). For turbulent flow, $f \times$ Re should increase linearly while plotted on a log–log scale against the Reynolds number, for conventional turbulent flow as described by:

$$f = 0.014 \, \mathrm{Re}^{-0.182} \qquad (1.1)$$

Deviations from these expected results were reported for microchannels as is apparent in the findings presented in Figure 1.1. No conclusive remarks can be made on

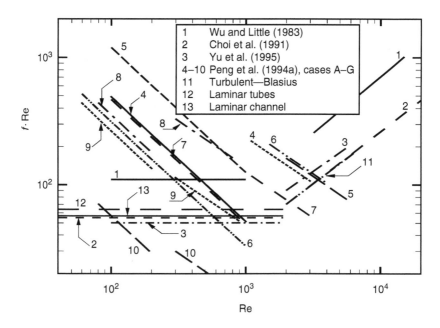

FIGURE 1.1 Comparison of the fluid friction characteristics from various investigations on convective flow in microchannels. (From Sobhan, C.B. and Garimella, S.V., *Microscale Thermophys. Eng.*, 5, 293, 2001. With permission.)

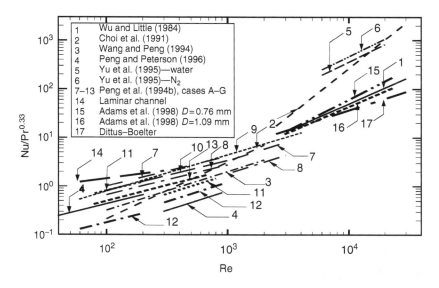

FIGURE 1.2 Comparison of the heat transfer characteristics from various investigations on convective flow in microchannels. (From Sobhan, C.B. and Garimella, S.V., *Microscale Thermophys. Eng.*, 5, 293, 2001. With permission.)

the variations of $f \times Re$ depicted in the plots, which, as indicated, exhibit increasing and decreasing trends, both in the presumably laminar and the turbulent flow regimes. These effects cannot be explained in relation to the physical dimensions or flow transitions encountered in conventional channels.

Deviations and variations were also observed in the heat transfer characteristics, as illustrated in Figure 1.2. In the laminar flow regime, it was observed that the dependence of the Nusselt number on the Reynolds number is generally stronger for the case of microchannels, compared to conventional channels, as indicated by the steeper slopes of the plots. Further, the general trends reported by various investigations differed significantly, some of them lying below and some above the values predicted by conventional correlations, both in the laminar and the turbulent flow regimes.

Another interesting observation from these early experiments was that it appeared that the transition from laminar flow was not characterized by the Reynolds number (based on the hydraulic diameter) criterion of 2000–2300, as would be expected from previous theory and empirical correlations. The critical Reynolds number (characterized by the change in trends of the plots in Figure 1.1), which determines the upper boundary of the laminar flow regime, was found to deviate from this theoretically predicted range of values, and variations depended on the hydraulic diameters of the channels used. Though it was not clear whether the deviations and differences were due to actual physical effects or limitations in fabrication and instrumentation, these reported results paved the way for more serious thinking and closer observations of the problem of fluid flow and heat transfer in microchannels.

1.3.2 PHASE CHANGE AND TWO-PHASE FLOW

A majority of the investigations of phase change and two-phase flow in microchannels has been directed toward the cooling of electronic equipment. The objectives of these studies were to determine and minimize the overall thermal resistance, as well as to obtain and correlate critical heat flux data. Friction models have been developed for two-phase liquid–vapor flows. Visualization studies on flow regimes have been presented by a number of investigators. Comparisons between the effectiveness of single-phase and two-phase cooling methods have also been presented in the context of electronics cooling.

Similar to the observations in single-phase flow and heat transfer in microchannels presented above, deviations from conventional theory were also reported for phase change heat transfer and two-phase flows in microchannels. Some of the early investigations (Bowers and Mudawar 1994) were on circular mini- and microchannels, and were designed to obtain the critical heat flux values and optimize the channel thickness to diameter ratio as a compromise between the structural and heat transfer considerations. Pressure drop models were also developed for mini- and microchannels, and the theoretical predictions were compared with the experimental results. These investigations were intended to help determine and explain the heat transfer and pressure drop characteristics in relation to the flow acceleration, due to the phase change, and the channel erosion, due to flow boiling. Studies directed at the exploration of the effects of parameters such as the flow velocity, subcooling, property variations, and channel configurations on two-phase flow behavior were also undertaken.

Observations indicated that some of the characteristics differed from those anticipated in conventional channels. For instance, some of the early experiments (Peng et al. 1995) on nucleate flow boiling indicated that the liquid velocity and subcooling did not affect fully developed nucleate boiling, but greater subcooling increased the velocity and suppressed the initiation of flow boiling. Experiments on the flow boiling of binary water/methanol mixtures in microchannels (Peng et al. 1996) indicated that the heat transfer coefficient at the onset of flow boiling and in the partial nucleate boiling region was greatly influenced by liquid concentration, microchannel and plate configuration, flow velocity, and subcooling, but these parameters had little effect in the fully nucleate boiling regime. Similar investigations of V-shaped microchannels (Peng et al. 1998) gave an impression that in contrast to conventional channels, no bubbles were observed in these microchannels during flow boiling, even with heat fluxes as high as 10^6 W/m^2. Investigations using single- and two-phase flow experiments on refrigerant flow in microchannel heat exchangers (Cuta et al. 1996) inferred that a substantial improvement in thermal performance could be achieved in microscale heat exchangers without a large increase in pressure drop, though conclusive explanations were not offered for this observation. As was the case for single-phase flow, the two-phase flow analyses mentioned above were limited in their sophistication, due to constraints on the instrumentation. In many of these cases, inferences were drawn based only on external temperature measurements and overall measurements of the pressure drop, and not based on nonintrusive local measurements. More recent experimentations obtained using flow visualization

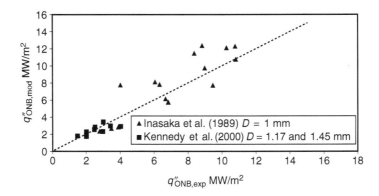

FIGURE 1.3 Comparison of microtube ONB data with predictions from theoretical model. (Adapted from Ghiaasiaan, S.M. and Chedester, R.C., *Int. J. Heat Mass Transfer*, 45, 4599, 2002. With permission.)

studies, however, have resulted in a dramatic modification in the understanding of the phase change and two-phase flow phenomena in these microscale channels. Various new mathematical modeling techniques were also utilized and predictions very close to the experimental findings were obtained, as illustrated in Figure 1.3, particularly on the onset of nucleate boiling in small channels (Ghiaasiaan and Chedester 2002).

Separate and distinct from boiling, condensation and the associated two-phase flows have also been studied extensively more recently (Garimella et al. 2005). Hybrid models combining experimental data and theoretical modeling have been developed to address the progression of the condensation process from the vapor phase to the liquid phase, including the overlap and transition between flow regimes. The relevant results from the early and recent investigations of phase change and two-phase flow in small channels are dealt with more extensively in Chapters 3 and 4.

1.3.3 CONDUCTION AND RADIATION IN THE MICROSCALE

Size effects in electron transport phenomena, which had been observed when the characteristic dimensions (for instance, in thin metallic films and wires) become small, have recently led to the study of microscale conduction in greater detail. The applications of thin films have been extensive in semiconductor technology, as microelectronics utilizing silicon-on-insulator (SOI) devices have become more prevalent as these devices are being more widely utilized. The manufacture, as well as the operation of these devices, makes the thermal management problem one of the key challenges. Analysis of conduction heat transfer in microstructures is essentially required for the prediction and design of the thermal effects in these types of systems. In addition, microscale conduction with pronounced size effects is encountered in structures where the characteristic length dimensions are of the same order as the scattering mean free path of the electrons. In this region, reductions in the transport coefficients, such as the thermal conductivity, are expected as a result of the shortening of the mean free path near the surface, due to the presence of the boundary. One of the primary goals of a majority of the investigations in microscale conduction has been the measurement, or estimation through various microscopic mathematical models, of the behavior of the thermal conductivity when the physical dimensions of the analyzed structures are in the microscale range.

A majority of the past studies have reported a significant reduction in the thermal conductivity of thin films, compared to those associated with the bulk material (Flik and Tien 1990). However, the physical and chemical nature and composition of the thin film material is found to significantly affect the degree of deviation from the bulk material thermal property behavior. This information is of particular importance to the semiconductor industry, as films of similar chemical composition can have wide variations in microstructure, based on the manufacturing method. This implies that thermal analysis and design of such structures cannot be generalized, as it is impossible to work based on a unique thermal property database. In reality, the analysis of each component or structure requires an understanding of the structural influence of the transport properties. For the case of optical elements, the nature of the material microstructure, voids, and cracks can all significantly affect the life of the device, as can laser-induced damage.

In addition to experimental measurements, the theoretical analysis of microscale conduction is of critical importance in understanding the behavior of these systems. As indicated above, comprehensive analysis of microscale conduction systems should take into account the thermal behavior associated with the film material used for the component or the structure. Past attempts (Kumar and Vradis 1991) to use fundamental transport equations such as the Boltzmann equation have indicated that the reduction in thermal and electrical conductivities in microscales was virtually identical in most of the practical cases for thin films. Fundamental approaches such as the use of the Boltzmann equation or molecular dynamics simulation were found to provide useful insight into the transport property behavior of thin films and microscale physical structures. Theoretical maps defining the macroscale and microscale regimes have proven useful for the thermal analysis of certain materials and take into account factors, such as material purity and defect characteristics. Attempts have also been made to approach the problem of microscale conduction theoretically using modified (hyperbolic) heat conduction equations, in place of the conventional Fourier heat conduction analysis. Some representative results from these theoretical analyses (Ju and Goodson 1999) are shown in Figure 1.4. Detailed discussions of microscale conduction models and measurement are presented in Chapter 2.

Radiation-length-scale effects occur when the characteristic length scales of surfaces or structures undergoing radiative exchange are of the same order of magnitude as the wavelength. Microscale radiative transfer has become increasingly significant, as micromachining technology has become capable of producing many structures and solid films with these characteristic dimensions. As in the case of microscale conduction, theoretical and experimental approaches have been utilized to study microscale radiative behavior, as well as to use electromagnetic radiation for making measurements in the microscale. These have involved investigations of phenomena such as emissions from microscale structures and the measurement

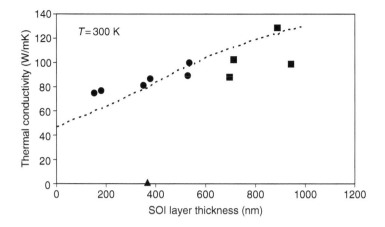

FIGURE 1.4 Variation of the in-plane thermal conductivity of silicon layers with respect to the layer thickness at 300 K, from various investigations. (Adapted from Ju, Y.S. and Goodson, K.E., *Microscale Heat Conduction in Integrated Circuits and Their Constituent Films*, Kluwer Academic Publishers, Boston, MA, 1999. With permission.)

of emissivity; studies on scattering, reflection, absorption, and transmission in thin films and structures; and the estimation of the related radiation properties. The use of many of these phenomena for the measurement of local temperatures and radiative fluxes in thin films and microscale structures and devices has also gained importance.

As the size of structures has continued to decrease to the order of nanometers using current nanotechnology and MEMS technologies, the radiative characteristics of these properties have been investigated using electromagnetic theory and experimental methods. Phenomena such as the departure of spectral emittance behavior of highly doped silicon microgrooves from that normally expected have to be well understood before using the radiative properties of such structures in design calculations. As a result, the quest to understand similar deviations and differences has been the motivation for many of the advances in microscale and nanoscale radiation.

Microscale radiation studies have focused on applications such as that of the superconducting states, where prediction of the radiative properties of thin superconducting films with respect to different wavelengths of electromagnetic radiation has been one of the primary objectives. Studies on radiative properties have also found applications in microsensor technology using optical sensors. Chapter 5 of this book is devoted to a review of the fundamentals and advances in microscale radiative heat transfer and its applications.

1.4 DISCUSSIONS ON SIZE-EFFECT BEHAVIOR

The starting point for any analysis of small-scale systems is to identify whether or not the approach could be based on continuum theory. That is, an investigation to find out whether the physical size of the system demands analysis incorporating size effects is to be done, before approaching the problem for analysis. A comparison of the characteristic length dimensions with the carrier mean free path is often recommended to determine what kind of analysis is required. It should also be noted that the timescale of the transport processes is also of importance, in determining whether an approach different from conventional analyses is required. More extensive discussions on the length and timescales are presented in later chapters.

Due to the assumption of the size independence of the thermal conductivity, analysis based on Fourier's law tends not to be adequate in the analysis of size-effect dominated problems. Further, Fourier's law does not consider the propagation speed of the heat carriers, or time variations in the heat flux. Conduction heat transfer analysis in the microscale, therefore, should consider the temperature-dependent motion of the heat carriers, and proceed utilizing appropriate space and timescales, without continuum assumptions. The analysis method should also be based on considerations of the temperature levels and heating conditions encountered. The mean free path has a temperature dependence, which shows an increase with decreasing temperature; the analysis at low and cryogenic temperatures should consider size effects as very significant aspects in the modeling of these situations.

New hypotheses, other than lattice vibrations, have been put forward to describe energy transport processes, such as a wave-type transport occurring in the conducting medium. Wave-type energy transport has been detected in metals such as gold films, while the heat transfer process takes place under length scales as low as 10^{-9} m (nanoscale) and timescales of the order of 10^{-12} s.

As indicated previously, reduced values of thermal conductivities are expected in the size-effect domain, because the length of the mean free path will shorten near the surfaces owing to the termination at the boundary. Similarly, a pronounced size effect on the electron transport phenomena in metallic films and wires exists, with sizes (characteristic lengths) of the order of the electron mean free paths. Thermal conductivity measurements in thin copper films, in a thickness range of 400–8000 Å and a temperature range of 100–500 K, and calculation of thermal conductivity of films using electrical analogies (Nath and Chopra 1974), have clearly indicated the reduction of thermal conductivity with a reduction in the film thickness, while the temperature dependence agreed with variations in bulk copper.

One of the major questions in microscale heat transfer analysis is, "what is the range of parameters where results based on Fourier's law are no longer valid?" Deviations from the macroscale theory can be understood from thermal conductivity measurements, but then the most challenging task becomes the accurate measurement of the thermal conductivity at the physical sizes under consideration. This is especially true, as non-Fourier effects become significant, i.e., when the length scales get smaller and smaller.

The effort taken by investigators (Flik et al. 1991) to study thermal transport in domains with various length scales to provide information to the designer as to whether conventional theory could be applied to a given microstructure is noteworthy in this regard. These investigations led to regime maps, which identify boundaries between microscale and macroscale heat transfer regimes, relating the smallest geometric dimension to the temperature. Additional maps have been developed for conduction, convection, and radiation modes of heat transfer; a typical microscale regime map for conduction heat transfer is shown in Figure 1.5 to demonstrate its application. These can be used to determine whether microscale

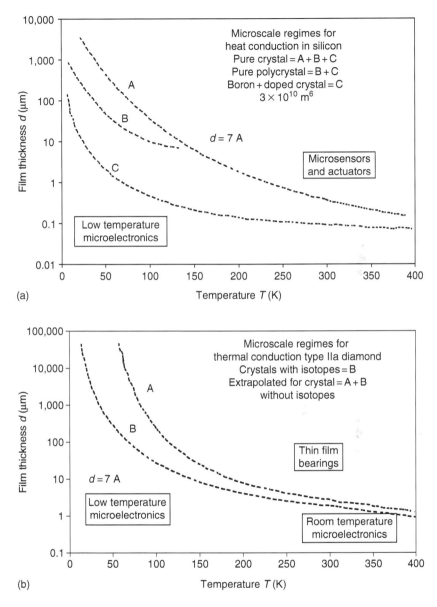

FIGURE 1.5 (a) Microscale regime maps for thermal conduction in silicon films. (Adapted from Duncan, A.B. and Peterson, G.P., *Appl. Mech. Rev.*, 47, 397, 1994. With permission.) (b) Microscale regime maps for thermal conduction in diamond films. (Adapted from Duncan, A.B. and Peterson, G.P., *Appl. Mech. Rev.*, 47, 397, 1994. With permission.)

heat transfer consideration is required to analyze and design microstructures pertaining to a given technology, by observing and positioning it appropriately on the map.

An interesting finding has been reported in the literature (Flik et al. 1991), which could serve as a guide in assessing whether conduction in thin films should be analyzed as a microscale conduction problem or not. This was because thermal conduction in a film is microscale, if the layer thickness is less than seven times the mean free path (approximately) for conduction across the film, and less than 4.5 times the mean free path for conduction along the film. A method commonly recommended for analysis of microscale conduction is the use of the Boltzmann transport equation (BTE), though conventionally this was used for the analysis of gases.

1.4.1 COMMENTS ON CONTRADICTORY OBSERVATIONS IN MICROSCALE CONVECTION

In addition to the differences in observed behavior and required analysis discussed above, the problem of microscale convective flow and heat transfer presents a few more additional points of concern. One of the major reasons put forward by early investigators on observed differences in the performance of microchannels from expected large channel predictions was the dominating influence of the surface roughness, whose effect could become considerably large due to the small size of the channels themselves. However, this is a problem related to the fabrication or manufacturing methods used to produce the channels, rather than a basic fluid flow problem. Still, there is no conclusive methodology to incorporate the effect of the roughness parameter into a mathematical model of the problem. In a one-dimensional approach, a friction factor modified for the microchannel flow could be used, but this has to depend on experiments whose accuracy and repeatability cannot be guaranteed. There have been attempts to correlate the data using dimensionless parameters such as the Brinkman number (Tso and Mahulikar 1998) in place of the Reynolds number, so that conclusive interpretations can be realized and highlighted, based upon the observed differences and unusual behavior in microchannels. In addition to these, the effects of flow maldistribution in the channels studied in the early investigations when coupled with the lack of sophisticated or nonintrusive instrumentation also remain as two possible reasons for the reported differences.

Various arguments and hypotheses have been put forward to explain the observations, within the limitations of the available instrumentation facilities, particularly in the early stages when many of the pioneering ideas in microscale convective heat transfer were being developed. As explained above, one of the reasons suggested for the differences in flow and convective heat transfer was the influence of surface roughness. There have been attempts to explain the effects in terms of property variations, due to temperature distributions in microchannels, the presence of effects such as the electrical double layer (Mala et al. 1997), and the possible absence of nucleate boiling during phase change in microchannels. Almost all of these were based on overall observations rather than local measurements in the channels, which were not possible during the early stages of microchannel research. More careful observations in the later stages led to the opinion that the unusual effects observed were possibly due to maldistribution of flow in channel arrays or due to inaccuracies in experimentation in data reduction, which have been supported by results from experiments using MEMS probes and nonintrusive visualization techniques. Regardless, understanding the heat dissipation levels and characteristics of heat transfer in the microscale where the size effects become predominant, as well as understanding the size limits where continuum modeling may become insufficient for analyzing heat transfer problems based on the flux laws, are challenges for the heat transfer engineer who deals with transport phenomena in physical dimensions of small scale.

1.4.2 RECENT TRENDS

Experimental research on microscale heat-transfer-related issues has progressed considerably over the past several years, due to the application of better fabrication methods, sensors, and data acquisition techniques. Most of the reported studies indicate that conventional solutions and correlations for fluid flow and heat transfer in channels and tubes predict the behavior of microchannel flows quite adequately. The deviations, variations, and discrepancies observed in the early stages of investigations on microscale heat transfer appear to be the result of inadequacies in the sensors or the various measurement techniques utilized, or imperfections in the fabrication techniques used at these small dimensions. Modern fabrication methods and more sophisticated instrumentation have made it possible to generalize the results to a much larger extent than what was possible in the early studies, indicating that in the vast majority of microscale applications conventional correlations are still valid.

1.5 FUNDAMENTAL APPROACH FOR MICROSCALE HEAT TRANSFER

As discussed in the beginning of this chapter, the continuum modeling approach for transport phenomena starts to become inadequate only when the physical dimensions of the domain under analysis are reduced to a level comparable with the mean free path of the molecules and the distances associated with the molecular motion. The question in most of such cases is how then to determine the critical physical dimensions. Though, at the early stages of investigations, there was a tendency to believe that the differences between the theoretical and experimental values observed in the microscale were due to deviations from continuum conditions, it is clear now that such dimensions are still large enough for the continuum assumptions to be valid. The observed differences, however, were in all likelihood attributable to difficulties associated with the fabrication and measurement of microscale structures and systems.

Still the question of determining the lower limits of microscale dimensions where continuum analyses become inadequate remains. Particularly, as advances in microscale transport phenomena have led to the design and invention of nanoscale structures and processes, the necessity of applying fundamental molecular level approaches for the analysis of problems has become all the more important. The basic approach in all of the methods of this kind is to attribute the physical processes of energy transport to the dynamics of the fundamental particles of matter. Various modeling methodologies have been developed utilizing the well-known principles of fundamental physics, but the solution of these within the dimensions of the domains still entail a number of problems that must be addressed. For instance, numerical computational methods would become extremely difficult and time consuming, particularly when applied to analyze energy transport problems through the molecular dynamics approach, as it takes numerous computational steps to perform the calculations for a physical domain, which can be compared with the results from macroscopic investigations on bulk material behavior.

Fundamental approaches have been utilized to analyze the thermal behavior of materials and to estimate the thermophysical properties and coefficients (such as thermal conductivity and viscosity) defined by classical macroscopic theories, with a view to explore whether such properties differ in the size-effect domains. Some of these are introduced in the subsequent sections.

1.5.1 MICROSCOPIC VIEW POINT AND ENERGY CARRIERS

As mentioned earlier, in the size-affected domain, the conventional approach based on phenomenological laws fails to predict the heat transfer behavior and properly model the associated thermophysical properties of various media. In all modes of heat transfer, a microscopic approach that considers the heat transfer process as a manifestation of the interactions of the basic energy carriers is required.

The fundamental carriers of energy are the phonon, the electron, and the photon. Phonons are quantized lattice vibrations, which have a wave nature of propagation. Energy transfer in crystalline substances is well explained by phonon transport, whereas the conduction process in metals is postulated to occur because of the interactions of the free electrons or "Fermi gas." Photons are energy carriers that transfer radiative energy until they fall on a surface, transferring energy to the electrons exciting them, which subsequently give up energy, producing phonons.

Discussions of the transport phenomena associated with the fundamental energy carriers (phonons, electrons, and photons) are presented in subsequent chapters in a number of occasions. An overview of the basics of phonons as energy carriers is dealt with in Chapter 2 under conduction heat transfer. The free-electron model for conduction in metals is also discussed in some detail. Photons and their transition to thermal energy is discussed in Chapter 5, which deals with microscale radiation heat transfer.

The major concept useful in the understanding of energy transport phenomena from a microscopic perspective is referred to as particle–wave duality. This approach provides a common basis for relating the perspectives of matter and energy to describe and analyze energy transport. The BTE provides a mechanism by which the process of carrier transport can be described from a particle point of view. The governing equations used in the study of electromagnetic fields are the Maxwell equations or modifications of these. As these two are the fundamental approaches from which many of the analysis methods for microscale energy transport in various applications have evolved, they are introduced here.

1.5.2 BOLTZMANN TRANSPORT EQUATION

The BTE uses a combination of classical mechanics and quantum approach (Cercignani 1988). Newtonian mechanics is used to describe the classical motion of particles, while the dissipative effects are accounted by quantum mechanical (statistical) considerations, including scattering rates and probabilistic distributions of particles. Thus the equation becomes especially useful while analyzing energy transport from a microscopic viewpoint, in systems where gradients of thermodynamic properties such as density and temperature exist, producing bulk effects. Analysis based on this method can be used to study physical phenomena, as well as to obtain transport coefficients. Further, the formulation is based on a nonequilibrium approach, and the equilibrium state is a special case of the general formulation.

The BTE is given by (Cercignani 1988):

$$\frac{\partial f}{\partial t} + \mathbf{v} \cdot \nabla_r f + \mathbf{F} \cdot \nabla_p f = \left(\frac{\partial f}{\partial t} \right)_{sc} \tag{1.2}$$

where f is a distribution function of particles, expressed as $f = f(r,p,t)$, which indicates the probability of occupation of particles with momentum p at location r and time t. To simplify the equation a relaxation time approximation is often applied to the right-hand side of the equation, which describes the scattering process as follows:

$$\left(\frac{\partial f}{\partial t} \right)_{sc} = \frac{f_0 - f}{\tau(r,p)} \tag{1.3}$$

Here f_0 is the equilibrium distribution and τ is the relaxation time. For instance, appropriate distributions could be the Fermi–Dirac distribution for electrons and Bose–Einstein distribution for phonons.

Various methods have been suggested for the solution of the Boltzmann equation, other than by use of the relaxation time approximation mentioned above. For some transport problems, iterative integration methods have been utilized. Other effective methods suggested include the Monte Carlo method (which is a random sampling method), conversion of the BTEs to a cellular automata problem (this method assumes cells or lattices in the analysis domain and proceeds with computations by updating the states of such cells simultaneously, while marching in time), and the use of a spherical harmonic series expansion of the distribution (spherical harmonic method). A detailed treatment of the Boltzmann equation, its solution, and applications is given by Cercignani (1988) and the reader is recommended to refer to the book for an in-depth understanding of this powerful analytical approach.

1.5.3 ELECTROMAGNETIC WAVES AND MAXWELL'S EQUATIONS

The theory of propagation of energy as electromagnetic waves is also useful in analyzing microscale heat transport. The analysis of radiative transfer, to a large extent, is based on electromagnetic wave propagation. Combinations of the theory with particle theory for energy transport in media by simultaneous solution of the basic equations for the electromagnetic field along with the Boltzmann equation have also been used to analyze microscale energy transport related to radiation heating in thin films, for instance (Kumar and Mitra 1999). While discussions on such applications are presented later in Chapter 5, it is useful to introduce the governing equations of the electromagnetic field here.

The Maxwell equations describe the propagation of electromagnetic waves, in a mathematical form, in the following relationship between the electric and magnetic fields (Kong 1990, Born and Wolf 1999, Jackson 1999, Zhang et al. 2003):

$$\nabla \times E = -\frac{\partial(\mu H)}{\partial t} \tag{1.4}$$

$$\nabla \times H = \sigma E + \frac{\partial(\varepsilon E)}{\partial t} \tag{1.5}$$

$$\nabla \cdot (\varepsilon E) = \rho \tag{1.6}$$

$$\nabla \cdot (\mu H) = 0 \tag{1.7}$$

where
  H is the magnetic field vector
  E is the electric field vector
  ε is the permittivity
  μ is the magnetic permeability
  ρ is the electric charge density
  σ is the electrical conductivity (Zhang et al. 2003)

The following general form can be obtained from this set of equations:

$$\nabla^2 E = \mu\sigma\frac{\partial E}{\partial t} + \mu\varepsilon\frac{\partial^2 E}{\partial t^2} \tag{1.8}$$

and for a dielectric medium (medium with no electrical conductivity), the equation takes the form (Zhang et al. 2003):

$$\nabla^2 E = \mu\varepsilon\frac{\partial^2 E}{\partial t^2} \tag{1.9}$$

The solution of the equation for a monochromatic wave can be shown to be

$$E = E_+ e^{-i(\omega t - q \cdot r)} \tag{1.10}$$

Here, E_+ is a complex vector, ω is the angular frequency, and q is the wave vector giving the direction of propagation (Zhang et al. 2003). The wave vector is a function of the complex refractive index of the medium given by $(n + i\kappa)$, where κ, the imaginary part, is called the extinction coefficient.

The vector product $(E \times H)$, called the Poynting vector, is related to the power flux of the electromagnetic filed as follows:

$$S = \frac{1}{2}\mathrm{Re}\left(E \times H^*\right) \tag{1.11}$$

In the above equation, the right-hand side is half the real part of the Poynting vector. H^* is the complex conjugate of the magnetic field vector H (Zhang et al. 2003).

The electric and magnetic fields can be assumed to attenuate exponentially in an absorbing medium. An absorption coefficient is derived by combining these exponential attenuations with the Maxwell equations, such that

$$S_x = \frac{n}{2\mu c_0} E_0^2 e^{-2q_{im}x} = \frac{n}{2\mu c_0} E_0^2 e^{-a_\lambda x} \tag{1.12}$$

where the absorption coefficient is given by $a_\lambda = 4\pi\kappa/\lambda_0$ (Zhang et al. 2003).

The reciprocal of the absorption coefficient is called the radiation penetration depth. Physically, this is the depth, traveling through which the radiation is attenuated by $1/e$ from the original power (Zhang et al. 2003).

Another theory which is used to describe radiation phenomenon is the particle theory, where radiation is considered as a collection of quantized energy packets called photons, each having an energy given by

$$E = h\nu = \frac{h}{2\pi}\omega \tag{1.13}$$

where h is the Planck's constant.

For a more detailed treatment of the topic presented above, readers are directed to the extensive discussions in the literature (Kumar and Mitra 1999, Zhang et al. 2003).

1.5.4 BASICS OF MOLECULAR DYNAMICS MODELING

Molecular dynamics simulation is a discrete computational method which finds use in analyzing transport phenomena and estimating transport properties of physical systems (Allen and Tildesely 1989, Maruyama 2000). The method approaches problems from a molecular perspective, and, essentially, all physical phenomena are brought down to the framework of dynamics, applied to the molecules. Thus, the primary result of a molecular dynamics computation is the information on the positions and velocities of molecules (or particles or atoms, as appropriate) in the computation domain, which in turn quantifies the energies associated with them. These can then be used to analyze the bulk physical phenomena, and obtain thermophysical properties using physical principles.

Essentially, in molecular dynamics simulations, the dynamic equilibrium of a many-body system is used as the governing condition. The time evolution of a set of interacting molecules or atoms, as appropriate, is simulated using the equations of motion. The classical Newton's law is followed:

$$F_i = m_i a_i \tag{1.14}$$

for each molecule or atom i in a system,

where
  m_i is the mass
  a_i is the acceleration
  F_i is the force due to the interaction with neighbors

The simulation procedure consists of an equilibration process, where time-evolution computations in a particle system are continued until the properties become invariant with respect to time. The equilibration process is followed by determining the values of interest (such as positions, velocities, and the related energies), which is termed "measurement," and then by calculation of the required system properties where required, using correlations where these are written as functions of the position and momentum of the discrete computational particles (molecules or atoms). Applying principles of statistical mechanics, the physical quantities (thermodynamic or thermophysical properties) are obtained as averages of the values when the system has reached equilibrium. Application and results of molecular dynamics simulations, designed and performed to analyze thermal phenomena in fluids, and the thermophysical properties and behavior of nanofluids, are further described in Chapter 6 under nanoscale heat transfer.

1.6 INTRODUCTION TO ENGINEERING APPLICATIONS OF MICROSCALE HEAT TRANSFER

A majority of the developments in microscale heat transfer have been focused on the electronics industry, though the applications of electronics are numerous and varied. Conduction and radiation analyses and experimental methods in this field find ample use in the design and testing of integrated circuits and other electronic components, which deal with films of semiconductors and insulators. The application of heat transfer could be in the manufacture of these or in the thermal management of the resulting devices or systems. Microscale conduction and radiation heat transfer are also useful in determining the variations of the thermophysical properties and the related operational behavior of such devices at different heating levels. Applications also occur in laser devices and optical components, which in most cases are related to electronics. Convective heat transfer and phase change processes in microchannels have found tremendous application as effective cooling options for electronics, as microchannel arrays can form integrated heat exchangers within microelectronic devices and silicon substrates. This is an area where extensive studies have been carried out, both theoretically and experimentally, particularly over the past few years. Some discussion on the important applications are given below, and some of these will be dealt with more extensively in the chapters to follow.

1.6.1 THIN FILMS

As mentioned previously, one of the potential areas of application of microscale conduction analysis is the semiconductor industry, where a variety of thin film structures are used. The manufacture as well as operation of such devices encompasses a number of heat transfer problems. Silicon-on-insulator (SOI) devices essentially consist of films of semiconductors and insulator layers and interconnects, similar to the typical construction shown in Figure 1.6. The heat conduction in such devices is focused on the crystalline silicon device layers (films), amorphous silicon dioxide layers (used as insulators), and metallic interconnects.

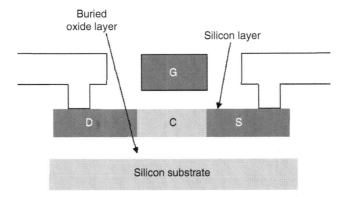

FIGURE 1.6 A typical arrangement of silicon and passivation layers and interconnects in an SOI device. (Adapted from Ju, Y.S. and Goodson. K.E., *Microscale Heat Conduction in Integrated Circuits and Their Constituent Film*, Kluwer Academic Publishers, Boston, MA, 1999. With permission.)

The major thermal problem in SOI devices is the failure of the silicon layers, the passivation (insulation) layers, and the interconnects. The main reason for this is the differences in the heat conduction and the resulting associated thermal gradients in these films, which are closely related to the differences and directional nature of the thermal conductivities of the constituent films. Thus, a precise knowledge of the thermal conductivities of the films, which are normally in the size-effect domain, is required. This makes the accurate measurement of thermal conductivity of these films an important problem (Ju and Goodson 1999).

To design thin-film-based electronic devices, prediction of the temperature patterns and estimation of local heat dissipation rates in the films are also required. This requires that a theoretical analysis of conduction in the size-effect domains be performed. The temperature fields in SOI device layers often produce large gradients, and have to be analyzed at extremely small timescales, due to the highly transient nature of the processes. A good summary of the thermal problems associated with SOI devices is presented in the literature by Ju and Goodson (1999). Also included is an extensive treatment of the measurement and analytical methods used for microscale conduction in the size-effect domain of interest. Advances and techniques for investigating conduction at very small length scales are presented and discussed in Chapter 2.

1.6.2 MICROCHANNEL HEAT EXCHANGERS

Fluid flow and convective heat transfer in microchannels are useful in a number of applications in the cooling of electronics, because such channels can be used in arrays to form microchannel heat exchangers. These miniature heat exchangers can be integrated by fabricating them directly in silicon substrates, for effective heat dissipation using a flowing/circulating fluid in the system. This approach helps to avoid the problem of thermal contact resistance encountered while attaching

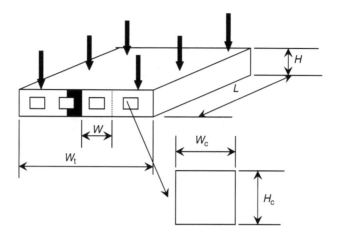

FIGURE 1.7 Schematic arrangement of integrated microchannels to form a heat exchanger in a solid substrate. The channels, fabricated in the substrate, transport the coolant through them. The shaded portion represents a typical domain for theoretical analysis of the substrate.

conventional metallic heat sinks to the dissipating surfaces. A typical design of an integrated microchannel heat exchanger in a silicon substrate is shown in Figure 1.7. Microchannel heat sinks are found to efficiently achieve the two most important objectives in electronics cooling, namely the reduction of the device maximum temperature to optimal operating levels and the minimization of temperature gradients on the device surface i.e., the elimination of localized hot spots by heat spreading.

To date, the majority of studies of micro heat exchangers have focused on theoretical modeling and experimentation leading to optimal designs of the heat exchanger structure. There have been studies based on rigorous numerical solutions of the governing flow and heat transfer equations for the fluids in the passages, as well as calculations based on thermal resistance models, taking into consideration the overall temperature gradients. Optimization of the overall size and weight of the compact or integrated exchangers, as well as the arrangement and dimensions of the passages, has been of interest. Guidelines and suggestions for the various fabrication processes of these microchannel structures have been developed for use in heat exchanger applications (Hoopman 1990), and comparisons of the merits of the various types of microchannel structures have been presented. The application of microchannel heat exchangers is not limited to electronics cooling; this has also been used in systems such as compact heat exchangers, heat shields, and fluid distribution systems. A review of literature on the optimum design of microchannel heat sinks has been presented by Goodling and Knight (1994).

1.6.3 MICRO HEAT PIPES AND MICRO HEAT SPREADERS

A heat pipe is a device of very-high effective thermal conductivity, capable of transferring large quantities of heat over considerable distances, without an appreciable temperature gradient. The high effective thermal conductance of the heat pipe maintains the vapor core temperature almost uniform while transferring heat, making it capable of being used also as a "heat spreader." As described by Peterson (1994), a heat pipe operates in a closed two-phase cycle in which heat added to the evaporator region causes the working fluid to vaporize and move to the cooler condenser region, where the vapor condenses, giving up its latent heat of vaporization. In traditional heat pipes, the capillary forces existing in a wicking structure pump the liquid back to the evaporator.

While the concept of utilizing a wicking structure as part of a device capable of transferring large quantities of heat with a minimal temperature drop was first introduced by Gaugler (1944), it was not until much more recently that the concept of combining phase change heat transfer and microscale fabrication techniques for the dissipation and removal of heat was first proposed by Cotter (1984). This initial introduction envisioned a series of very small micro heat pipes incorporated as an integral part of semiconductor devices. While no experimental results or prototype designs were presented, the term "micro heat pipe" was first defined as one "so small that the mean curvature of the liquid–vapor interface is necessarily comparable in magnitude to the reciprocal of the hydraulic radius of the total flow channel" (Babin et al. 1990). In the micro heat pipe, liquid is transported along the heat pipe channel due to the capillary action of the corners of the channel themselves, and so, a separate wicking structure is not required. Early proposed applications of these devices included the removal of heat from laser diodes (Mrácek 1988) and other small localized heat-generating devices (Peterson 1988a,b), the thermal control of photovoltaic cells (Peterson 1987a,b), the removal or dissipation of heat from the leading edge of hypersonic aircraft (Camarda et al. 1997), applications involving the nonsurgical treatment of cancerous tissue through either hyper- or hypothermia (Fletcher and Peterson 1993), and space applications in which heat pipes are embedded in silicon radiator panels to dissipate the large amounts of waste heat generated (Badran et al. 1993).

While not all of these applications have been implemented, micro heat pipes ranging in size from 30 μm to 1 mm in characteristic cross-sectional dimensions and from 10 mm to 60 mm in length have been analyzed, modeled, and fabricated, and the larger of these are currently commonplace in commercially available products such as laptop computers or high precision equipment where precise temperature control is essential. Reported studies include those on individual micro heat pipes and micro heat pipe arrays made as an integral part of silicon substrates. Theoretical and experimental analysis has led to the characterization of the influence of geometrical and operational parameters on the performance of these devices. Determination of the operating limitations of micro heat pipes also has been an objective of the research. The schematic arrangement of a channel in a micro heat pipe is shown in Figure 1.8, which shows the operation of a triangular micro heat pipe channel with a typical evaporator and condenser section.

Various polygonal cross sections have been used in micro heat pipes. A new concept of a wire-sandwiched structure that can be used as a micro heat pipe array has also been proposed and experimented with. Micro heat pipes made of flexible and (biologically) implantable materials are also being developed.

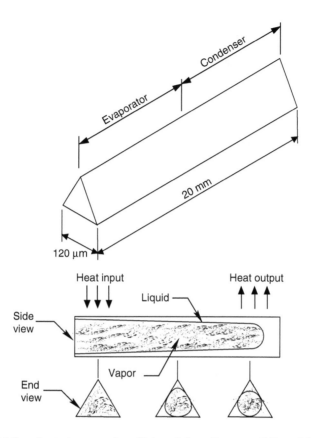

FIGURE 1.8 Micro heat pipe operation. (Adapted from Peterson, G.P. and Sobhan, C.B., *The MEMS Handbook, 2nd ed. Applications*, Mohammed Gad-el Hak (Ed.), Taylor & Francis, Boca Raton, 2006. With permission.)

More recently, the work on micro heat pipes has expanded to include micro heat spreaders fabricated in silicon or in new metallized polymeric materials, which can be used to produce highly conductive, flexible heat spreaders capable of dissipating extremely high heat fluxes over large areas, thereby reducing the source heat flux by several orders of magnitude. Flat-plate heat spreaders, capable of distributing heat over a large two-dimensional surface, have been proposed by Peterson (1992), in which, a wicking structure fabricated in silicon multichip module substrates promote the distribution of the fluid and the vaporization of the working fluid (Figure 1.9). Several methods for wick manufacture have been developed (Peterson et al. 1998). More detailed discussions on the working principle, construction, analysis, and optimization of micro heat pipes and heat spreaders are presented in Chapter 4.

1.6.4 OTHER APPLICATIONS

Though most of the developments in microscale heat transfer have been directed toward applications in the thermal management of electronics, there have been many

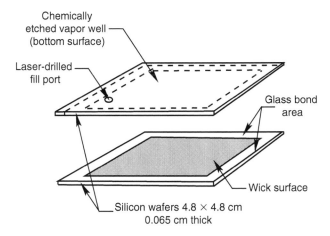

FIGURE 1.9 Flat-plate micro heat spreader. (Adapted from Peterson, G.P. and Sobhan, C.B., *The MEMS Handbook. 2nd ed. Applications*, Mohammed Gad-el Hak (Ed.), Taylor & Francis, Boca Raton, 2006. With permission.)

applications in other areas. Thin films have been used extensively in optical systems and the manufacture and operation of many of the optical components using thin films are associated with heat dissipation problems. The behavior of solid films in the presence of electromagnetic radiation results in a number of applications, such as the characterization of materials, the measurement of microscale physical dimensions, and the optical determination of temperature and heat flux patterns.

Coated surfaces have been utilized effectively to reduce the contact thermal resistance. Various methods have been developed for coating metallic and other surfaces to obtain good thermal contact. With modern techniques, it is possible to achieve coating thicknesses at the micro- and nanoscale, but the analysis of coated surfaces to optimize the process is still complex, as when the coating layer thicknesses become extremely small, conventional conduction analysis and assumption of continuum modeling appear to be inadequate, due to size effects. Better modeling methods for thin layers of coating materials are of interest, and so are experimental methods to estimate contact thermal resistances across these layers.

Analysis of microscale fluid flow and convective heat transfer has also found applications in engineering fields other than electronics. A good example of the use of microducts is in refrigeration systems. Phase change phenomena in micro-tubes have been studied extensively in connection with miniaturized cooling and refrigerating systems. Biological systems also present a number of applications of microscale fluid flow. Analysis of blood and other body fluids in arteries has been effectively utilized in designing various processes of drug dosing and delivery applications. With the advent of microencapsulation techniques for drug delivery and implantable devices of micro- and nanoscales, the thermal problems associated with micro- and nanoscale are gaining importance in the biomedical arena.

Diverse uses for the micro heat pipe and micro heat spreader can also be found in biomedical applications. One such application is in catheters that provide a hyperthermia or hypothermia source, which can be used in the treatment of tumors and cancers (Fletcher and Peterson 1993). In one of the designs, the micro heat pipe catheter enables the hypo- or hyperthermic treatment of cancerous tumors or other diseased tissue. The heat pipe is about the size of a hypodermic needle and is thermally insulated along a substantial portion of its length. The heat pipe includes a channel partially charged with an appropriate working fluid. The device provides the delivery or removal of thermal energy directly or from the tumor or diseased tissue site. In a second design, the catheter uses a variety of passive heat pipe structures alone or in combination with feedback devices. This catheter is particularly useful in treating diseased tissue that cannot be removed by surgery, such as a brain tumor. Another biomedical application under development is the polymer-based micro heat pipe heat spreader, which is being proposed for the treatment of neocortical seizures by implanting a device that can provide localized cooling. More discussions on applications of microchannel heat transfer in micro heat spreaders is presented in greater detail in Chapter 4.

A wide range of new applications have recently been developed related to investigations of polymer-based flexible micro heat pipes for use in spacecraft radiators (McDaniels and Peterson 2001). Theoretical and experimental investigations have been reported on micro heat pipe arrays designed for this application. The focus of this investigation consisted of micro heat pipe arrays that were made from a composite of two layers: an ungrooved metal foil and a grooved polymer film.

1.7 CONCLUDING REMARKS

While the present trends of microscale heat transfer studies tend to lead to the analysis and design of heat and fluid flow systems at even smaller length scales such as the nanoscale, the validity of continuum models and results applying flux laws have yet to be investigated on, to determine the dimensional limits of application of such results. Under such conditions, nonconventional modeling based on molecular dynamics considerations would have to be utilized to better incorporate the problem physics at small length and timescales.

In situ local measurements in the channels using flow visualization studies have thrown more light on the physical processes taking place inside small channels, compared to the overall measurements that were possible in the early stages of microchannel research. Single-phase flow in microchannels has been investigated more extensively than two-phase flow and the conclusions indicate that conventional modeling is adequate. However, as phase change processes provide a much more effective mechanism of heat dissipation, such processes at the microscale will continue to be investigated to obtain more conclusive and definitive results. Measurement of nucleation and two-phase flow associated with boiling in microchannels requires optical instrumentation based on microscopy and high-speed photography. This is an area that presents a number of considerable challenges in instrumentation. Other optical, nonintrusive methods could also be utilized in these measurements. Extension of microchannel research into areas such as biomedical engineering, MEMS, and nanotechnology also seems to be of increasing importance and relevance.

REFERENCES

Adams, T.M., S.I. Abdel-Khalik, S.M. Jeter, and Z.M. Qureshi. 1998. An experimental investigation of single-phase forced convection in microchannels. *Int. J Heat Mass Transfer* 41: 851–857.

Allen, M.P. and D.J. Tildesely. 1989. *Computer Simulation of Liquids*. Oxford: Clarendon Press.

Babin, B.R., G.P. Peterson, and D. Wu. 1990. Steady state modeling and testing of a micro heat pipe. *Journal of Heat Transfer* 112: 595–601.

Badran, B., J.M. Albayyari, F.M. Gerner, P. Ramadas, H.T. Henderson, and K. Baker. 1993. Liquid metal micro heat pipes. *Proceedings of the 29th National Heat Transfer Conference HTD*, Atlanta, Georgia 236: 71–85.

Born, L. and E. Wolf. 1999. *Principles of Optics*. 7th ed. Cambridge: Cambridge University Press.

Bowers, M.B. and I. Mudawar. 1994. High flux boiling in low flow rate, low pressure drop mini-channel and micro-channel heat sinks. *International Journal of Heat and Mass Transfer* 37: 321–332.

Camarda, C.J., D.R. Rummler, and G.P. Peterson. 1997. Multi heat pipe panels. NASA Tech Briefs. LAR-14150.

Cercignani, C. 1988. *The Boltzmann Equation and Its Applications*. 1st ed. New York: Springer Verlag.

Choi, S.B., R.F. Barron, and R.O., Warrington. 1991. Fluid flow and heat transfer in microtubes. *Micromechanical Sensors, Actuators and Systems ASME DSC* 32: 123–134.

Cotter, T.P. 1984. Principles and prospects of micro heat pipes. *Proceedings of the 5th International Heat Pipe Conference*, Tsukuba, Japan: 382.

Cuta, J.M., C.E. McDonald, and A. Shekarriz. 1996. Forced convection heat transfer in parallel channel array microchannel heat exchanger. *Advances in Energy Efficiency, Heat/Mass Transfer Enhancement ASME* HTD 38: 17–23.

Duncan, A.B. and G.P. Peterson. 1994. Review of microscale heat transfer. *Applied Mechanics Reviews* 47: 397–428.

Fletcher, L.S. and G.P. Peterson. 1993. A micro heat pipe catheter for local tumor hyperthermia. U.S. Patent No. 5,190,539.

Flik, M.I. and C.L. Tien. 1990. Size effect on thermal conductivity of high T_c thin film superconductors. *Journal of Heat Transfer* 112: 872–880.

Flik, M.I., B.I. Choi, and K.E. Goodson. 1991. Heat transfer regimes in microstructures. *American Society of Mechanical Engineers Dynamic Systems and Control* 32: 31–47.

Garimella, S., A. Agarwal, and J.D. Killion. 2005. Condensation pressure drop in circular microchannels. *Heat Transfer Engineering* 35: 26–28.

Garimella, S.V. and C.B. Sobhan. 2003. Transport in microchannels—A critical review. *Annual Review of Heat Transfer* 13: 1–50.

Gaugler, R.S. 1944. Heat Transfer Device. U.S. Patent No. 2: 348–350.

Ghiaasiaan, S.M. and R.C. Chedester. 2002. Boiling incipience in microchannels. *International Journal of Heat and Mass Transfer* 45: 4599–4606.

Goodling, J.S. and R.W. Knight. 1994. Optimal design of microchannel heat sinks: A review. *Optimal Design of Thermal Systems and Components ASME HTD* 279: 65–78.

Hoopman, T.L. 1990. Microchanneled structures, microstructures, sensors and actuators. *American Society of Mechanical Engineers Dynamic Systems and Control* 108: 171–174.

Inasaka, F., H. Nariai, and T. Shimura. 1989. Pressure drop in subcooled boiling in narrow tubes, *Heat Transfer-Japanese Research* 18: 70–82.

Jackson, J.D. 1999. *Classical Electrodynamics*. 3rd ed. New York: John Wiley & Sons.

Ju, Y.S. and K.E. Goodson. 1999. *Microscale Heat Conduction in Integrated Circuits and Their Constituent Films*. Boston, MA: Kluwer Academic Publishers.

Kennedy, J.E., G.M. Roach, Jr., M.F. Dowling, S.I. Abdel-Khalik, S.M. Ghiaasiaan, and S.M. Jeter. 2000. The onset of flow instability in uniformly heated horizontal microchannels. *Journal of Heat Transfer* 122: 118–125.

Kong, J.A. 1990. *Electromagnetic Wave Theory*. 2nd ed. New York: John Wiley & Sons.

Kumar, S. and G.C. Vradis. 1991. Thermal conduction by electrons along thin films, effects of thickness according to Boltzmann transport theory. *American Society of Mechanical Engineers Dynamic Systems and Control* 32: 89–101.

Kumar, S. and K. Mitra. 1999. Microscale aspects of thermal radiation transport and laser applications. *Advances in Heat Transfer* 33: 187–294.

Mala, G.M., D. Li, and J.D. Dale. 1997. Heat transfer and fluid flow in microchannels. *International Journal of Heat and Mass Transfer* 40: 3079–3088.

Maruyama, S. 2000. Molecular dynamics method for microscale heat transfer. *Advances in Numerical Heat Transfer* 2: 189–226.

McDaniels, D. and G.P. Peterson. 2001. Investigation of polymer based micro heat pipes for flexible spacecraft radiators. *American Society of Mechanical Engineers HTD* 142.

Mrácek, P. 1988. Application of micro heat pipes to laser diode cooling. Annual Report of the VÚMS, Prague, Czechoslovakia.

Nath, P. and K.L. Chopra. 1974. Thermal conductivity of copper films. *Thin Solid Films* 20: 53–63.

Peng, X.F. and G.P. Peterson. 1996. Convective heat transfer and flow friction for water flow in microchannel structures. *International Journal of Heat Mass Transfer* 39: 2599–2608.

Peng, X.F., G.P. Peterson, and B.X. Wang. 1994a. Frictional flow characteristics of water flowing through microchannels. *Experimental Heat Transfer* 7: 249–264.

Peng, X.F., G.P. Peterson, and B.X. Wang. 1994b. Heat transfer characteristics of water flowing through microchannels. *Experimental Heat Transfer* 7: 265–283.

Peng, X.F., G.P. Peterson, and B.X. Wang. 1996. Flow boiling of binary mixtures in microchannel plates. *International Journal of Heat and Mass Transfer* 39: 1257–1264.

Peng, X.F., H.Y. Hu, and B.X. Wang. 1998. Flow boiling through V-shape microchannels. *Experimental Heat Transfer* 11: 87–90.

Peng, X.F., B.X Wang, G.P. Peterson, and H.B. Ma. 1995. Experimental investigation of heat transfer in flat plates with rectangular microchannels. *International Journal of Heat and Mass Transfer* 38: 127–137.

Peterson, G.P. 1987a. Analysis of a heat pipe thermal switch. *Proceedings of the 6th International Heat Pipe Conference*, Grenoble, France: 177–183.

Peterson, G.P. 1987b. Heat removal key to shrinking avionics. *Aerospace America* 8: 20–22.

Peterson, G.P. 1988a. Investigation of miniature heat pipes. Final report, Wright Patterson AFB, Contract No. F33615 86 C 2733, Task 9.

Peterson, G.P. 1988b. Heat pipes in the thermal control of electronic components. *Proceedings of the 3rd International Heat Pipe Symposium*, Tsukuba, Japan: 2–12.

Peterson, G.P. 1992. An overview of micro heat pipe research. *Applied Mechanics Review* 45: 175–189.

Peterson, G.P. 1994. *An Introduction to Heat Pipes: Modeling, Testing and Applications*. New York: John Wiley & Sons.

Peterson, G.P. and C.B. Sobhan. 2006. Micro heat pipes and micro heat spreaders. In *The MEMS Handbook. 2nd ed. Applications*, Mohammed Gad-el Hak (Ed.), 11:1–11:37. Boca Raton: Taylor & Francis.

Peterson, G.P., L.W. Swanson, and F.M. Gerner. 1998. Micro heat pipes. In *Microscale Energy Transport*, Tien, C.L., Majumdar, A., and Gerner, F.M. (Eds.), 295–337. Washington DC: Taylor & Francis.

Sobhan, C.B. and G.P. Peterson. 2006. A review of convective heat transfer in micro channels. *Journal of Engineering and Machinery* 47: 10–65.

Sobhan, C.B. and S.V. Garimella. 2001. A comparative analysis of studies on heat transfer and fluid flow in microchannels. *Microscale Thermophysical Engineering* 5: 293–311.

Tso, C.P. and S.P. Mahulikar. 1998. Use of the Brinkman number for single phase forced convective heat transfer in microchannels. *International Journal of Heat and Mass Transfer* 41: 1759–1769.

Tuckerman, D.B. and R.F.W. Pease. 1981. High-performance heat sinking for VLSI. *IEEE Electron Device Letters* EDL-2: 126–129.

Tuckerman, D.B. and R.F.W. Pease. 1982. Ultrahigh thermal conductance microstructures for cooling integrated circuits. *Proceedings of the 32nd Electronic Components Conference IEEE*, EIA, CHMT, San Diego, California: 145–149.

Wang, B.X. and X.F. Peng. 1994. Experimental investigation on liquid forced convection heat transfer through microchannels. *International Journal of Heat Mass Transfer* 37: 73–82.

Wu, P.Y. and W.A. Little. 1983. Measurement of friction factor for the flow of gases in very fine channels used for micro miniature Joule Thompson refrigerators. *Cryogenics* 23: 273–277.

Yu, D., R. Warrington, R. Barron, and T. Ameel. 1995. An experimental and theoretical investigation of fluid flow and heat transfer in microtubes. *Proc. ASME/JSME Thermal Engineering Conference* 1: 523–530.

Zhang, Z.M., C.J. Fu, and Q.Z. Zhu. 2003. Optical and thermal radiative properties of semiconductors related to micro/nanotechnology. *Advances in Heat Transfer* 37: 179–296.

2 Microscale Heat Conduction

2.1 REVIEW OF CONDUCTION HEAT TRANSFER

Because conduction relates to solid materials, it is the easiest mode of heat transfer to observe experimentally and the equation that describes it, Fourier's law, represents the macroscopic expression of the observed heat transfer phenomena associated with the solid media. Fourier's Law gives a mathematical expression of the observed physical fact that the heat transfer (per unit time) by conduction across a plane is proportional to the area of the cross section of the plane, normal to which the heat flows and the gradient of temperature in the direction of heat flow. It also confirms mathematically, the second law of thermodynamics, which ensures that heat flows in the direction of the negative temperature gradient. As such, within a macroscopic domain, Fourier's law can be expressed as

$$q = -k\frac{\Delta T}{\Delta x} \tag{2.1}$$

This equation can be further expressed in differential form as

$$q_n = -k\frac{\partial T}{\partial n} \tag{2.2}$$

This differential form of the Fourier's law of heat conduction, when combined with the principle of conservation of energy, can be used to derive the general heat conduction equation for a macroscopic continuum. The general heat conduction equation with heat generation in Cartesian coordinates can be expressed as

$$\nabla(k_n \nabla T) + \phi = \rho c\frac{\partial T}{\partial t} \tag{2.3}$$

For an isotropic material with no internal heat sources

$$\nabla^2 T = \frac{1}{\alpha}\frac{\partial T}{\partial t} \tag{2.4}$$

where $\alpha = \dfrac{k}{\rho c}$, the thermal diffusivity of the medium.

At steady state, with internal heat generation as (Poisson equation)

$$\nabla^2 T + \frac{\phi}{k} = 0 \qquad (2.5)$$

And with no internal heat generation, and steady state as (Laplace equation)

$$\nabla^2 T = 0 \qquad (2.6)$$

Depending on the physical problem being analyzed, these expressions can be transformed into equivalent equations using either cylindrical or polar coordinate systems.

The solution of these heat conduction equations yields temperature distributions in the continuum in which conduction heat transfer occurs. The resulting temperature distributions can then be used to calculate the direction of the heat fluxes by applying Fourier's law in the differential form at any given point of interest. The derivation of the differential equation of heat conduction, and its solution under various conditions are treated extensively in the literature, and are more fully developed in the many fundamental text books that address conduction heat transfer, such as the book referenced here by Holman (1989). In the standard form, the general equation derived from the energy conservation principles presents a parabolic governing differential equation for the heat conduction problem.

For transient problems, the parabolic heat conduction equation given above does not account for the propagation speed of the temperature front in the medium. However, there are methods by which this can be introduced through the incorporation of appropriate terms. One model developed by Cattaneo (Tamma and Zhou 1998) uses a relaxation time to obtain a modified version of Fourier's law. In this, the time-dependent expression that replaces Fourier's law of heat conduction is given as

$$\tau \frac{\partial q}{\partial t} = -q - k\nabla T \qquad (2.7)$$

where
τ is called the relaxation time
k is the thermal conductivity of the medium
q is the heat flux

This expression can be coupled with the transient energy equation such as the following equation for one-dimensional heat transfer in the absence of an internal heat source:

$$\rho c \frac{\partial T}{\partial t} + \frac{\partial q}{\partial x} = 0 \qquad (2.8)$$

where
ρ is the density
c is the specific heat capacity

Equations 2.7 and 2.8 lead to the following hyperbolic heat conduction equation (Tamma and Zhou 1998):

$$\tau \rho c \frac{\partial^2 q}{\partial t^2} - k \frac{\partial^2 q}{\partial x^2} + \rho c \frac{\partial q}{\partial t} = 0 \qquad (2.9)$$

The resulting governing equation in terms of temperature can be written as

$$\tau \rho c \frac{\partial^2 T}{\partial t^2} - k \frac{\partial^2 T}{\partial x^2} + \rho c \frac{\partial T}{\partial t} = 0 \qquad (2.10)$$

and the corresponding speed of temperature propagation is defined as (Tamma and Zhou 1998)

$$c_{\mathrm{T}} = \sqrt{\frac{k}{\rho c \tau}} = \sqrt{\frac{\alpha}{\tau}} \qquad (2.11)$$

Substituting, Equation 2.10 can be rewritten as

$$\frac{\partial^2 T}{\partial t^2} + \frac{1}{\tau} \frac{\partial T}{\partial t} - c_{\mathrm{T}}^2 \frac{\partial^2 T}{\partial x^2} = 0 \qquad (2.12)$$

Assuming infinite speed of propagation, the hyperbolic expression in Equation 2.9 and the parabolic heat conduction expression, Equation 2.4, assume the same form. This is also true for steady-state conditions.

Additional modifications to the standard heat flux approximation have been suggested, in particular for conduction problems associated with microscale heat transfer. These include the Jeffrey's type heat flux model involving a relaxation time and a retardation time, and two-step relaxation–retardation time models based on the microscopic approach (Tamma and Zhou 1998). In closing, it is important to recognize that even though these modifications may provide a practical method for analyzing heat conduction in situations where the Fourier law may be inappropriate, such as in the microscale domain, the hyperbolic heat conduction equation in the standard form is not entirely consistent with the second law of thermodynamics.

2.2 CONDUCTION AT THE MICROSCALE

As discussed above, the Fourier's law for a macroscopic domain is formulated based on a continuum assumption. The thermophysical property that relates the heat flux and the temperature gradient, the thermal conductivity, is assumed to be independent of both the length and timescales. The thermal conductivity of a material is independent of the size of the domain, and the rate at which heat is added to the material.

For problems in which both space and timescales are beyond the limitations of the continuum assumption, Fourier's law becomes invalid and while it can be applied, it may result in inaccurate results, the degree of inaccuracy depending upon the deviation from the initial assumptions. The domains for these types of applications are referred to as size-effect domains or size-affected domains, where the mean free paths of the molecules or atoms of the medium are comparable in magnitude, to the physical dimensions of the actual domain. The heat transfer and temperature relationships in these types of applications have been shown not to adhere to Fourier's law, due to the assumptions connected with the definition of the thermal conductivity, as a material property. To determine and understand the relationship between the size-effect domain and the macroscopic domain, a modified value of thermal conductivity can be determined experimentally, utilizing the same form of the definition and Fourier's law. It should be noted, however, that this definition carries with it no actual physical meaning as the value is dependent on both the physical size and timescale. Any accurate approach by which the equivalent thermal conductivity is computed for these cases must be based on the fundamental dynamic behavior of the molecules or atoms and the interactions occurring, due to the basic potentials of each.

It has been shown that the assumption of a continuum does not account for the fundamental dynamic behavior of the actual heat carriers or the molecular or atomic particles in the medium. This particular aspect becomes important when dealing with physical dimensions that are similar in size to the path distances. In addition, Fourier's law does not incorporate the length and timescales associated with the domain or the physical processes, which are of particular importance when dealing with energy transport at the micro- and nanoscale. In summary, Fourier's law, which is the most conventional and common law for conduction analysis, is only useful for applications in which macroscopic information is required, i.e., temperature fields and heat fluxes in a macroscopic sense. Other approaches require alternative approaches to accurately analyze microscale and nanoscale heat transfer.

Given the influence of the size effects and evaluating the process from a microscopic perspective, the fundamental approach to understanding and analyzing conduction heat transfer in small systems must begin with an evaluation of the molecular or atomic motion, as required by the problem and the energy transport of the thermal energy carriers. Thus, the problem of heat diffusion from regions of higher temperatures to regions of lower temperatures is directly related to the energy levels of the heat carriers, which is associated with the free electrons and vibration of the lattices. Essentially, the problem is related to the analysis of the motion of the fundamental building blocks of matter, and there is no demarcation between the heat transport problem and the dynamics problem. Not considering regimes where matter–energy transactions become important, the molecular dynamics analysis deals with the application of Newton's laws of motion to the molecules or atoms, after appropriately determining the spatial and temporal states of these, so that the equilibrium conditions are satisfied. Thus, the analysis of conduction in the microscale, and especially the nanoscale, should consider the motion of the heat carriers, which is determined by the space and timescales and the energy level related to the temperature. The limitations of this approach are the computational complexity involved, and the large number of computations required to benchmark observations from large-scale domains. At the upper end of the microscale regime, the most practical approach is to analyze the heat transfer using a specific thermal conductivity and Fourier's law, and then introduce corrections for the size effects as determined through measurements or other theoretical models.

Hyperbolic heat conduction models have been used as a substitute for the standard parabolic models, to analyze conduction problems (Gembarovic and Majernik 1988). These utilized a series solution, and a physical interpretation described in terms of traveling thermal waves. Examination of Fourier and non-Fourier approaches for the analysis of microscale conduction problems, considering the influences of the thermal length scale and relaxation time, can be found in the literature (Vedavarz et al. 1991).

Bai and Lavine (1993) used a modified hyperbolic heat conduction equation, which is consistent with the second law of thermodynamics, to consider the finite speed of thermal propagation. Because the hyperbolic heat conduction equation displays interesting and peculiar behavior in very thin domains, when the domain is of the order of the mean free path of the carrier, the continuum approach becomes questionable. The authors indicated that it may be feasible to apply the continuum equations in the interior, with jump boundary conditions at the surface to perform a transition from the kinetic boundary condition at the wall to the macroscopic description of the interior region.

Considering the heat transfer analysis of electronic components utilizing a structure of submicron dimensions, regime maps have been developed as mentioned in Chapter 1, that identify boundaries between the macroscale and microscale heat transfer regimes (Flik et al. 1991). These maps have been illustrated earlier in Figure 1.5, and have produced excellent results in the design of electronic devices. The maps establish a relationship between the smallest geometric dimension and temperature, for conduction in solids, temperature and pressure for convection in gases, and the temperature of the emitting medium for radiative transfer. The material purity and defect structure were shown to strongly influence the regime boundaries and the microstructures pertaining to a given technology could be identified on these maps to determine whether a macroscale heat transfer approach was required. This makes it possible to identify research needs in microscale heat transfer, by identifying regions on the maps where future development of microtechnologies is probable. Among the conclusions drawn from this work was that the thermal conduction in a layer is microscale if the layer thickness is approximately 7 times the mean free path for conduction across a layer or less, and 4.5 times the mean free path for conduction along the layer, where the mean free path is that of the dominant carrier of heat.

2.3   SPACE AND TIMESCALES

Experimental evidence exists that highlights the incapacity of continuum modeling of heat conduction problems in regions where microscopic length scales are applied (Tamma and Zhou 1998). This implies that the Fourier model is applicable only for the space and timescales or regimes that come under the purview of the macroscopic viewpoint. As explained earlier, modifications to the Fourier model such as the hyperbolic heat conduction equation and other alternative constitutive equations (utilizing the concept of a relaxation time to compensate for the propagation of the heat or temperature front in the domain) have been explored, but these are also valid only in cases where the continuum assumption is applicable. Other than modifications of the governing equations, other theories have been proposed to address this issue, such as the wave propagation of thermal energy transport (Tamma and Namburu 1997). In general, continuum formulation can be applied only for certain regimes or scales, and more fundamental approaches based on particle behavior in conduction heat transfer are preferable in the microscale and nanoscale domains.

The analysis of microscale heat transfer must consider the effects of the appropriate space and timescales, in relation to the physical dimensions of the domain, the speed at which the process occurs, and the temperature ranges under which the heat transfer process takes place. Considerable effort has been devoted to understanding these effects, as it is essential whether the application is related to microfabrication processes or the performance prediction occurring as part of the design of microscale conduction systems. The analysis of conduction heat transfer is not limited to the phenomenon of conduction in solids, but extends to the convective heat transfer domain, as the basis of energy transport from surfaces to fluids is pure conduction in most of the situations except for cases like slip flow. With the advent of microminiaturization, the space domains have become smaller and the associated speed of the transient processes considerably larger. This has led to the formulation and analysis in the microscales for both space and time.

Useful approaches to address the space issue at the microscale have been presented by Tien and Chen (1994) for heat conduction in thin films. Two regimes were identified: one the classical size-effect domain, which could be useful for microscale heat transfer in micron-sized environments and the other the quantum size-effect domain, which may be more relevant in nanoscale conduction analysis (Tamma and Zhou 1998). The classical microscale size-effect domain is characterized as

$$\frac{h}{\Lambda} < O(1) \quad \text{or} \quad \frac{d_{\mathrm{r}}}{\Lambda} < O(1) \tag{2.13}$$

and the quantum size-effect domain is given by the condition

$$\frac{h}{\lambda_{\mathrm{c}}} > O(1) \tag{2.14}$$

where
  h is the characteristic device dimension
  d_{r} is the penetration depth of the temperature into the domain

As shown in Equations 2.13 and 2.14, two characteristic length scales are used: one to represent the mean free path of the heat carriers (Λ) and the other to represent the characteristic wavelength of the electrons or phonons (λ_{c}). It is obvious that the ratio (h/λ_{c}) provides guidelines for determining if the analysis should pertain to the classical microscale domain or the quantum size-effect domain, and various methods of analysis have been suggested for these two. For instance, the Boltzmann transport equation or the molecular dynamics modeling could be a suitable approach for the classical microscale domain, while phonon transport theory could be used for the quantum size-effect domain (Tien and Chen 1994).

The continuum formulation and the classical heat conduction equation will deviate from the real situation when the transient time interval involved is comparable to a prespecified relaxation time, normally designated by τ. The concept of relaxation time was introduced into the process of modifying the Fourier heat conduction equation. In the physical problem, the relaxation time represents a time lag for establishing equilibrium within an elemental volume considered for analysis. It should be noted that the propagation within an elemental volume is not considered in the classical parabolic heat conduction equation, and the model is restricted to the temperature growth of the volume with respect to time, as represented by the time derivative of temperature (Tamma and Zhou 1998).

As discussed in Chapter 1, following the principles of fundamental physics, the transport of thermal energy in an ideal solid can be described in terms of two primary mechanisms: the excitation of the free electrons and the lattice vibrations or phonons. These two effects are both quantized phenomena, and the resistive heat transfer is thought to be the result of the dissipative nature of the collisions of various quanta. The relaxation constant is associated with the statistical average of the time between successive collisions. The existence of finite thermal conductivity in the medium is related to the resistive flow of energy in the medium.

Three types of interactions occur in a given medium: the phonon–electron interaction, the phonon–phonon interaction, and the electron–electron interaction. In a majority of pure metals, the electron–phonon interaction is the dominant scattering process for electrons, and the conduction of heat by phonons is negligible. The collision time associated for a phonon–electron interaction is estimated to be on the order of 10^{-11} s, and for electron–electron and phonon–phonon interaction is 10^{-13} s (Flik et al. 1992). Various expressions based on different theories have been proposed for the relaxation time for solids (Tamma and Zhou 1998). For instance, for a crystalline solid,

$$\tau = \frac{3\alpha}{v_p^2} \tag{2.15}$$

where
  α is the thermal diffusivity
  v_p is the phonon velocity

For imperfect scattering, another expression from the kinetic theory has been proposed:

$$\tau = \frac{1}{\beta\varphi\eta v} \tag{2.16}$$

where
  β is a constant (approximately equal to 1)
  φ is the scattering cross section
  η is the number of scattering sites per unit volume
  v is the velocity of sound (Tamma and Zhou 1998)

TABLE 2.1

Values of Relaxation Time for Different Materials

Nature of Material	Material	Relaxation Time (s)
Metals	Na	31×10^{-15}
	Cu	27×10^{-15}
	Ag	41×10^{-15}
	Au	29×10^{-15}
	Ni	10×10^{-15}
	Fe	10×10^{-15}
	Pt	9×10^{-15}
	Cr	0.003×10^{-12}
	V	0.002×10^{-12}
	Nb	0.004×10^{-12}
	W	0.01×10^{-12}
	Pb	0.005×10^{-12}
Nonhomogeneous materials	Ballotini (2–2.5 mm)	13.34
	Sand (0.3–0.5 mm)	3.61
	Granule $CaCO_3$ (1.3–1.6 mm)	8.59
	H acid	24.5
	$NaHCO_3$	28.7
	Sand	20.0
	Glass ballotini	10.9
	Ion exchanger	53.7
	Processed meat	15.5

Source: From Tamma, K.K. and Zhou, X., *J. Thermal Stresses*, 21, 405, 1998. With permission.

Table 2.1 shows the estimated values of relaxation time for various homogeneous and nonhomogeneous materials, as listed by Tamma and Zhou (1998).

In addition to the relaxation time, there are other time parameters that can be of value in the analysis of micro- and nanoscale systems. Significant among these is a time representation referred to as the retardation time (K), which is a function of the relaxation time, and has relevance when dealing with heat waves in viscoelastic fluids. This term has also been introduced in certain modified heat transfer models such as the Jeffrey's type heat flux model (Tamma and Zhou 1998). Interested readers are referred to the literature (Tamma and Zhou 1998) for an in-depth understanding of the application and physical interpretation of the concept of the retardation time.

2.4   FUNDAMENTAL APPROACH

Extensive discussions on the fundamental theory of conduction energy transport, studied from the perspective of the interactions between energy carriers, are available in a number of books and articles, several of which have been reviewed by

Kumar and Mitra (1999), and only a brief introduction to the important points is attempted here.

The energy transport in crystalline substances can be explained through the phonon scattering model. Analysis is based on simple spring-mass representations for phonons, and quantum mechanical consideration to obtain the phonon frequency–wavelength relationship. This explains the physical existence of phonons with wave vectors (phonon wave vector $k_{ph} = 2\pi/\lambda_{ph}$, where λ_{ph} is the phonon wavelength) in the range $0 < k_{ph} < \pi/a$ (a is the lattice constant or the size of the unit cell of the crystal; the value is between 3 and 7 Å for most crystals). This range is termed the first Brillouin zone (Kumar and Mitra 1999). Phonons of wave vectors larger than this (or corresponding smaller wavelengths) are not physically possible.

Two types of phonons (phonon branches) have been identified, namely acoustic phonons and optical phonons. The mode of production of these two branches based on the fundamental relationships will be discussed here, as presented by Kumar and Mitra (1999).

For crystalline structures with similar adjacent atoms, the following relationship between phonon angular frequency and wavelength is obtained from quantum mechanics (Kumar and Mitra 1999):

$$\omega_{ph} \propto \left(1 - \cos\frac{2\pi a}{\lambda_{ph}}\right)^{1/2}; \quad 2a < \lambda_{ph} < \infty \tag{2.17}$$

In the first Brillouin zone itself, phonons having wave vectors near the upper limit (lower wavelength limit where k_{ph} is close to π/a) produce standing waves, as the atomic vibrations of adjacent atoms will be out of phase by π radians. These will not contribute to the propagation of energy. In the large wavelength (small wave vector) range ($\lambda_{ph} \gg a$ or $k_{ph} \ll \pi/a$), the relationship between ω and $1/\lambda_{ph}$ is linear, so that the phonons propagate at a constant velocity, equal to sonic velocity in the crystal. Phonons of this nature are called acoustic phonons (Kumar and Mitra 1999).

The fundamental relationship when adjacent atoms are dissimilar is given by (Kumar and Mitra 1999)

$$(\omega_{ph})^2 \propto A + B \pm \left(A^2 + B^2 + 2AB\cos\frac{2\pi a}{\lambda_{ph}}\right)^{1/2}; \quad 2a < \lambda_{ph} < \infty \tag{2.18}$$

where A and B are constants. The two roots of Equation 2.18 explain the existence of two phonon branches with two (lower and higher) frequency levels. In the limiting condition, $\cos 2\pi a/\lambda \to 1 - 0.5\,(2\pi a)^2$, these phonon branches are (Kumar and Mitra 1999)

$$\omega_{ph} \propto 1/\lambda_{ph}; \quad 2a \leq \lambda_{ph} < \infty \tag{2.19}$$

$$\omega_{ph} \approx \text{constant} - O(2\pi a/\lambda_{ph})^2; \quad 2a \leq \lambda_{ph} < \infty \tag{2.20}$$

The phonons satisfying Equation 2.19 are acoustic phonons, and the phonons that satisfy the condition given in Equation 2.20 are referred to as optical phonons, as these are the type of high-energy phonons that are produced when photons (corresponding to electromagnetic radiation) get converted to phonons upon impingement on a surface, so that the momentum and energy are conserved (Kumar and Mitra 1999). Thus, the optical phonons are important in the study of radiative transfer in the microscale; though they do not propagate energy directly due to their low group velocities, they subsequently produce acoustic phonons that conduct heat in the medium. Optical phonons are characterized by their high frequencies (and energy) compared to acoustic phonons. The solutions given above can be extended to three-dimensional lattices. This gives rise to longitudinal and transverse wave modes for both acoustic and optical phonon branches (Kumar and Mitra 1999).

Assuming that the phonons have a linear dispersion relation over all the wavelengths (this is called the Debye model), a maximum cutoff frequency has been obtained and a corresponding Debye temperature θ_D has been defined as follows:

$$\theta_D = h\omega_D/2\pi k_B \tag{2.21}$$

where

h is the Planck's constant

k_B is the Boltzmann constant

ω_D is the cutoff frequency or Debye frequency (Kumar and Mitra 1999)

The Debye temperature can be visualized as the value such that, at temperatures above this value, all phonons get excited, that is, when $T > \theta_D$, or correspondingly, $k_B T > h\omega_D/2\pi$.

The discrete quantum of energy contained in a phonon is given by $h\omega/2\pi$. The total energy in n_{ph} phonons in any phonon mode, which considers the zero point energy also (equal to 0.5 $(h\omega/2\pi)$ from the uncertainty principle), is given by (Kumar and Mitra 1999)

$$\Xi = \left(n_{ph} + \frac{1}{2}\right)\frac{h}{2\pi}\omega_{ph} \tag{2.22}$$

The equilibrium distribution of the number of phonons (n_{ph}) in any phonon mode, at energy $\Xi = h\omega/2\pi$ is the Bose–Einstein distribution, which is also the Planck's distribution function (Kumar and Mitra 1999):

$$n_{ph} = \frac{1}{\exp\left(\frac{\Xi_{ph}}{k_B T}\right) - 1} = \frac{1}{\exp\left(\frac{h\omega_{ph}}{2\pi k_B T}\right) - 1} \tag{2.23}$$

The momentum of a phonon, based on the wave-particle duality principle, is given by h/λ for a phonon having energy $\Xi = h\omega_{ph}/2\pi$.

The interaction of phonons is responsible for the transfer of energy in the medium. Two types of interactions have been explained. In the normal (N) process, either two phonons interact to form a third phonon, or a phonon is scattered into two phonons, conserving the net momentum and energy, thus involving no resistance to heat flow (thermal conductivity is infinite). The conduction phenomenon with finite thermal conductivity in a medium is explained as due to what is termed the Umklapp (U) process, which takes place when the third phonon produced from the interaction of two phonons has a resultant wave vector in a physically impossible zone (wavelength less than the minimum possible in the first Brillouin zone explained earlier). This system does not conserve the net momentum when mathematically expressed, and represents the physical case of a finite thermal conductivity material (Kumar and Mitra 1999).

Energy transfer in metals, and the related properties can be explained using the concept of a free-electron model. In this approach, the weakly bound valance electrons (conduction electrons) are assumed to move freely through the solid medium, thus serving as the primary energy carriers. In this ideal model, interactions between the free electrons and ions are neglected, under the assumption that the electrons are not deflected by the periodically arranged ions in the medium. Another assumption is that the electrons are only infrequently scattered by other electrons, and the fundamental particles of the free-electron gases (Fermi gas) do not interact. The following discussions pertain to the ideal case (Kumar and Mitra 1999).

The Fermi–Dirac distribution which gives the number density of the noninteracting ideal electron gas particles at equilibrium is given by

$$f_o(\Xi) = \frac{1}{\exp\left(\Xi - \frac{\zeta}{k_B T}\right) + 1} \tag{2.24}$$

where the energy is given by

$$\Xi = \frac{1}{2}mv^2 \tag{2.25}$$

where
   m is the electron mass
   v velocity
   ζ is the Fermi level defined below
   T is the absolute temperature
   k_B is the Boltzmann constant

The Fermi level is the chemical potential of the electrons given by (Kumar and Mitra 1999)

$$\zeta = \Xi_F - \frac{\pi^2 k_B^2 T^2}{12 \Xi_F} - \ldots; \quad \Xi_F = \frac{h^2}{8m}\left(\frac{3\bar{n}}{\pi}\right)^{2/3} \tag{2.26}$$

where
   Ξ_F is termed the Fermi energy at absolute zero temperature, which is the energy of the highest energy level at absolute zero

n represents the total number of free electrons per unit volume
h is Plank's constant

The Fermi temperature (also called the degeneracy temperature) is the value given by Ξ_F/k_B. The significance of this value is that, if the temperature is below this, f_o is almost constant in the energy range $0 < E < E_F$. Electrons in states where the energy $(k_B T)$ is within the Fermi level (Ξ_F) only are thermally excited while heating from absolute zero to temperature T (Kumar and Mitra 1999).

The distribution of the number of electrons per unit volume per unit energy is given as a function of the energy as follows (Kumar and Mitra 1999):

$$n(\Xi) = \frac{8\pi(2m^3)^{1/2}}{h^3}\Xi^{1/2}f_o(\Xi) \tag{2.27}$$

It should be noted that the above expression gives the value at any particular absolute temperature, as $f_o(\Xi)$ contains ζ evaluated at any particular T, ζ being a function of T. The total number of electrons per unit volume is obtained by integrating the above distribution in this range between 0 and Ξ_F, as (Kumar and Mitra 1999)

$$\bar{n} = \frac{\pi}{3}\left(\frac{8m}{h^2}\right)^{3/2}\Xi^{3/2} \tag{2.28}$$

The following relations give the momentum and energy:

$$\text{Momentum, } mv = \frac{h}{2\pi k} = \frac{h}{\lambda} \tag{2.29}$$

$$\text{Energy, } \Xi = \frac{1}{2}mv^2 = \frac{h\omega}{2\pi} = h\nu = \frac{1}{2m}\left(\frac{h}{2\pi}\right)^2 k^2 \tag{2.30}$$

The above equation for energy indicates that the angular frequency ω and the wave vector k have a parabolic relationship.

The ideal free-electron model should be modified to account for interaction of the electrons with the lattice, if this exists (Kumar and Mitra 1999). If there are lattice interactions, the energy–wave vector relationship will deviate from parabolic, introducing energy bands, which, when plotting between the energy and the wave vector, gives a deviation in curvature from a perfect parabola. The concept of the energy bands explains the energy transport behavior of insulators and semiconductors. In insulators, and in semiconductors at absolute zero (which behave as insulators), all the energy bands are completely filled (valance bands) or completely empty (conduction bands). In semiconductors, electrons in the valance bands can be thermally excited to the conduction bands, and according to this view, holes are thought about as equivalent to the absence of electrons in an otherwise completely filled energy band, which produces the semiconductor behavior (Kumar and Mitra 1999).

2.5 THERMAL CONDUCTIVITY

The simplest definition of the thermal conductivity of a heat transfer medium is the macroscopic definition from Fourier's law of heat conduction. Thus, it is the proportionality constant appearing in Fourier's law, and can be considered as the reciprocal of the thermal resistivity, which represents the resistance to heat flow in the medium. The expression for the thermal resistance of a conducting medium, thus, has the thermal conductivity appearing in the denominator. This definition of the thermal conductivity and calculations using the thermal resistance formulation for heat conduction are the essence of macroscopic conduction heat transfer analysis. Readers are referred to classical heat transfer books for the practical use of thermal conductivity for macroscopic heat transfer analyses (Arpaci 1966).

In the macroscopic definition of thermal conductivity, it is considered to be a property of the material, which is independent of the bulk material or size of the domain. The effect of temperature and local variations are the only factors considered in the analysis. This macroscopic definition, however, is not completely valid in the microscopic size-affected domain, as indicated by experimental observations in which the measured thermal conductivity values deviate from the bulk values. In order to analyze problems in the size-effect domain, an estimation of the thermal conductivity based on a more fundamental perspective is required. Estimation using approaches such as molecular dynamics modeling, electron–phonon transport, thermodynamic models from the kinetic theory, and the solution of the Boltzmann transport equation, among many other methods, are recommended by investigators in such cases. Sophisticated experimental investigations to accurately determine the thermal conductivity have also been proposed, though this area still requires significant refinement in the techniques and data analysis methods applied.

The electron–phonon transport theory assumes that the transport of thermal energy in an ideal solid is due to the excitation of free electrons, and lattice vibrations or phonons. The fundamental theoretical approach considers the propagation of heat in the medium as the manifestation of three different types of interactions of different natures—the electron–electron interaction, the electron–phonon interaction, and the phonon–phonon interaction. The resistive transport of energy is due to the scattering and the dissipative effect on the heat carriers. In metals, where free electrons are present, the conduction by phonons is weak and the major effect is the electron–phonon interaction. In the case of nonmetals and fluids, all the interactions might be significant.

The fundamental approaches to the problem of heat conduction have resulted in different models developed for thermal conductivity of solids and fluids, and also of mixtures and suspensions of solids in liquids. A very extensive and philosophical treatment of the concepts and models of thermal conductivity can be seen in the book by Tye (1969). Some of the important models for thermal conductivity resulting from statistical thermodynamics as well as from molecular and electron–phonon considerations are listed in the following sections.

2.5.1 Thermal Conductivity Models

Several theoretical expressions for the calculation of thermal conductivity of substances have been derived through the fundamental physics approaches. General kinetic theory-based expression for thermal conductivity was evolved for calculating the value of the gas phase, but could be used, to a limited extent, for other media to include the size effect in the microscale domain. Thermal conductivity models have also been developed for specific states-of-matter; models appropriate for the determination of thermal conductivity of solids and liquids are available. Some of the important expressions which find common use in the calculation of thermal conductivity of substances (Tye 1969) are given in the following subsections. As extensive discussions on fundamental physics is beyond the scope of the subject matter of the book, readers are advised to refer to the literature for better understanding of the premises and definitions on which these models have been developed (Tye 1969).

2.5.1.1 Thermal Conductivity Expression from Kinetic Theory

A simple expression for thermal conductivity is obtained when the phenomenon of conduction of heat in gases is considered based on the kinetic theory (McGaughey 2004). It can, however, be used for other states of matter as well, in obtaining the size-dependent thermal conductivity values, though the results may suffer from approximation due to the assumptions involved.

The expression for thermal conductivity based on the kinetic theory is given by

$$k = \frac{1}{3} n c_v v^2 \tau = \frac{1}{3} n c_v v \Lambda \tag{2.31}$$

where
  c_v is the specific heat per particle
  v is the phonon velocity
  τ is the relaxation time
  Λ is the phonon mean free path
  n is the number density of atoms

The mean free path is the average distance that energy carriers travel between successive collisions, such as the phonon–phonon collision in a dielectric material and the electron–phonon collision in a semiconductor or conductor. The derivation of this model assumes that all particles are identical and all of them have the same average specific heat, mean free path, relaxation time, and average velocity. This is consistent with Einstein's model of a solid as a collection of identical harmonic oscillators. If there are i different modes available for particles, the above equation can be rewritten as

$$k = \sum_i \frac{1}{3} n_i c_{v,i} v_i^2 \tau_i = \sum_i \frac{1}{3} n_i c_{v,i} v_i \Lambda_i \tag{2.32}$$

The principal disadvantage in the application of these equations is that the exact values of mean free path and relaxation time cannot be determined without *a priori* knowledge of the thermal transport in a given system. Direct calculation of the mean free path is quite difficult.

2.5.1.2 Thermal Conductivity Models for Solids

The thermal vibrations of the atoms about their equilibrium position have an important role in determining the thermal properties of materials. The restoring forces acting on a displaced atom are proportional to the displacement, which indicates that Hooke's law is obeyed and the atom executes a harmonic vibration. If an atom were constrained to vibrate at a fixed site, the atomic vibrations would be independent and the solid on the whole could be treated as an ordered group of harmonic oscillators.

The interaction processes that attempt to restore the equilibrium distribution and the phonon distribution that changes due to a temperature gradient are trends with opposing effects. These two effects attain a balance between them causing a non-equilibrium distribution resulting heat current, which describes the thermal conductivity. At high temperatures, the temperature dependence of thermal conductivity indicates the temperature dependence of the phonon mean free path. At low temperatures, it also depends on the phonon frequency (Tye 1969).

As mentioned before, the following expression for thermal conductivity can be derived based on kinetic theory:

$$k = \frac{cv\Lambda}{3} \qquad (2.33)$$

where

v is the average particle velocity
Λ the mean free path
c the specific heat per unit volume

In the case of solids, we visualize the crystal structure to be static and consider only the excitations that produce energy to account for their thermal conductivity. This energy produces the thermodynamic properties and since the excitations are mobile, they produce heat conduction. Each of the different types of excitations in solids makes its own contribution to the heat conduction and hence to the thermal conductivity. So in effect, the thermal conductivity can be formulated by summing up all kinds of excitations, to yield the following expression:

$$k = \sum \frac{c_\alpha v\alpha \Lambda_\alpha}{3} \qquad (2.34)$$

where α denotes the kind of excitation.

The lattice waves and the electrons are the most significant among the excitations in solids. Some other carriers of energy are electromagnetic waves, spin waves, etc. Interestingly, if we explain the lattice and elastic waves and the electromagnetic waves, in terms of their quantum equivalent, namely the phonons and photons respectively, the thermal conductivity proposed by kinetic theory and that for solids can be found to be analogous.

Obtaining the exact value of the mean free path or the associated relaxation time is substantially difficult even though Equations 2.33 and 2.34 appear to give an exact value for the solid thermal conductivity. The difficulties in predicting the thermal conductivity of solids are also due to its temperature dependence, which in effect is due to the highly random scattering between the lattice waves and the electrons. The mean free paths of carriers are reduced during the scattering processes. At the same time, lattice waves are scattered by each other. Scattering can also be due to imperfections and defects in the crystal lattice, which affect also the conduction electrons. On the whole, the uncertainty and the diversity involved in all the scattering processes make the thermal conductivity of solids highly parameter dependent. It is a very significant effect that the thermal conductivity is a highly temperature-dependent property. This would mean that the thermal conductivity at higher temperatures cannot be deduced by an extrapolation of the values at lower temperatures, because the conduction mechanism could be completely different in the two cases.

2.5.1.2.1   Thermal Conductivity of Metals

The concept of the electron gas is used to describe the thermal conductivity of metals. The conduction is mainly due to the electron–phonon interaction, which is usually very difficult to estimate from the first principles. The intrinsic thermal resistance and the associated intrinsic thermal conductivity are primarily due to the phonon–electron interactions, or in other words, due to the phonon-scattering of conduction electrons. The scattering in electrons can also be due to static imperfections. Both of the effects add up to the relaxation time. Considering the effects, at low temperatures, the intrinsic thermal conductivity has been expressed as (Tye 1969)

$$k(T) = \frac{L_o T}{\rho_0} \tag{2.35}$$

where
ρ_0 is the residual resistivity
L_o is called the Lorentz number

This implies that k varies linearly with temperature.

A lattice component of thermal conductivity is observed in the less highly conducting substances like alloys, transition metals, and semimetals. Thus, it would be an approximation to say that the thermal conductivity of metals is the electronic thermal conductivity alone. For instance, an expression is given, relating the lattice component (k_g) and the intrinsic thermal conductivity (k) for chromium as follows (Tye 1969):

$$k(T) + k_g = \frac{L_o T}{\rho} \tag{2.36}$$

More discussions on lattice thermal conductivity are presented in a later section.

2.5.1.2.2   Thermal Conductivity of Dielectric Solids

The lattice thermal conductivity of dielectric solids can be expressed in terms of the relaxation time. The various factors having significant influence on the thermal conductivity of dielectric solids are outlined by Tye (1969). The value of thermal conductivity is given by (Tye 1969)

$$k = \int d\omega \left[C(\omega)\tau(\omega)v^2 \right] \tag{2.37}$$

where $C(\omega) \, d\omega$ is the spectral contribution to the lattice specific heat.

Taking into account the intrinsic resistance at high temperatures, the thermal conductivity has been expressed as (Tye 1969)

$$k \approx \frac{3\mu v^2}{2\gamma^2 \omega_D T} \tag{2.38}$$

where
μ is the shear modulus
v is the velocity
γ is a coefficient dependent on the nature of strain and the polarization and direction of the wave
ω_D is the Debye frequency (the theoretical maximum frequency of vibration for the atoms that constitute the lattices)

The intrinsic conductivity at low temperatures is given by (Tye 1969)

$$k(T) = A k_o T^3 \frac{e^{\frac{\theta_D}{\alpha T}}}{\theta^3} \tag{2.39}$$

where

$$k_o = \theta_D \frac{k(T')}{T'} \tag{2.40}$$

In the above, $k(T')$ is the high-temperature intrinsic conductivity (at temperature T'; α and A are coefficients).

If the mean free path and the shortest linear dimensions of the specimen represented by l are comparable, the thermal conductivity in the low temperature limit is shown to be (Tye 1969)

$$k = \frac{c(T)vl}{3} \tag{2.41}$$

where
v is the phonon velocity
$c(T)$ is the specific heat at the given temperature

Considerable effort has been taken to identify and account for the dependence of thermal conductivity on defects and imperfections, by obtaining expressions for deformed materials. If $k_o(T)$ is the thermal conductivity in the absence of point defects, then at high temperatures, $k(T)$ in the presence of point defects has been shown to be (Tye 1969)

$$k(T) = k_o(T) \frac{\omega_o}{\omega_D} \tan^{-1} \frac{\omega_D}{\omega_o} \qquad (2.42)$$

where ω_D is the Debye frequency.

With further approximations, another expression has been obtained as follows (Tye 1969):

$$k(T) = k_o(T) \frac{\omega_o}{\sqrt{2}\omega_D} \tan^{-1} \frac{\omega_D}{\sqrt{2}\omega_o} \qquad (2.43)$$

The above equation is found to predict well the thermal conductivity behavior of germanium–silicon alloys.

The phonon mean free path is highly dependent on the orderliness of crystals. In highly disordered solids and noncrystalline solids, the phonon mean free path is considerably shorter than that in perfect crystals, which explains their lower thermal conductivity. For example, in the case of glass, the phonon mean free path observed is of the order of 10 Å, which indicates that lattice waves contribute much less to the thermal conductivity (Tye 1969).

2.5.1.2 3   Lattice Thermal Conductivity of Metals and Alloys

The U processes mentioned earlier have an important effect on the lattice thermal conductivity (k_g) of metals. The lattice thermal conductivity (k_g) of alloys is most influenced by the resistance due to dislocations. It is also dependent on point defect scattering. The scattering is more prevalent at low temperatures. At low temperatures, the following expression holds good (Tye 1969):

$$k_g \propto \frac{T^2 N k_B v_e M v^2 \xi}{\theta_D^2 k_\xi C^2} \qquad (2.44)$$

where

k_ξ is the radius of the Fermi sphere

C is an electron–phonon interaction parameter

θ_D is the Debye characteristic temperature

ξ is the Fermi energy with the bottom of the conduction band as the datum

v_e is the electron velocity

v is the sound velocity

N is the number density of atoms

M is the atomic mass

k_B is the Boltzmann constant

The factors taken into consideration for expressing the thermal conductivity of semimetals and degenerate semiconductors are similar to the factors considered for metals, but in these factors, the electron density and the scattering of phonons by electrons are comparatively reduced. The lattice thermal conductivity k_g is explained by the same factors as in the case of dielectrics, except that the scattering of phonons by electrons becomes significant only at lower temperatures.

In superconductors, the electronic thermal conductivity (k_e) is reduced by a factor, which is a function of T/T_c, where T_c is the transition temperature for superconductivity. The reduction in k_e assumed to be related to the reduction in the entropy content and specific heat in superconductors (Tye 1969).

2.5.1.2.4 Atomic Thermal Conductivity Models

Models for thermal conductivity have been developed where the vibrational states are assigned to the atoms themselves and not the phonons. In Einstein's thermal conductivity model, the atoms are assumed to be on a simple cubical lattice. The expression given below is the derived statement of Einstein's thermal conductivity model (Cahill and Pohl 1989):

$$k_E = \frac{n^{-1/3} k_B^2}{\pi \hbar} \Theta_E \frac{x_E^2 e^{x_E}}{\left(e^{x_E} - 1\right)^2} \tag{2.45}$$

where

Θ_E is called the Einstein's temperature (defined as $h\omega_E/k_B$)
$\hbar$ is the Plank's constant divided by 2π
k_B is the Boltzmann constant
n is the number density of atoms
x_E is the displacement of an atom, defined as $\hbar\omega_E/k_B T$

An extension of the Einstein's model is the Cahill–Pohl (CP) model. In this, a range of frequencies is used instead of the single frequency as used in the Einstein model.

The Cahill–Pohl expression for thermal conductivity is (Cahill and Pohl 1989):

$$k_{CP} = \left(\frac{\pi}{6}\right)^{1/3} k_B n^{2/3} \sum_i v_i \left(\frac{T}{\Theta_i}\right)^2 \int_0^{\frac{\Theta_i}{T}} \frac{x^3 e^x}{\left(e^x - 1\right)^2} \, dx \tag{2.46}$$

where Θ_i is defined as

$$\Theta_i = v_i \frac{h}{k_B} \left(6\pi^2 h\right)^{1/3} \tag{2.47}$$

and

v_i is the low frequency speed of sound
k_B is the Boltzmann constant
n the number density of atoms
h the Plank's constant

2.5.1.3 Thermal Conductivity of Liquids

Various models have been proposed for theoretical estimation of the thermal conductivity of liquids (Tye 1969). Some of the most important models are listed below.

1. Horrocks and Mclaughlin model
   In this model, the liquid is assumed to have a lattice structure and the excess energy between layers is transferred down the temperature gradient by the mechanisms of the convective distribution and the vibration of molecules. The corresponding thermal conductivities can be represented by k_{conv} and k_ϕ, respectively.

$$k_{\text{conv}} = \frac{\sqrt{2}k_0 C_v}{a} \tag{2.48}$$

$$k_\phi = \frac{\sqrt{2}\nu C_v}{a}; \quad a = \left(\frac{\sqrt{2}V}{N}\right)^{\frac{1}{3}} \tag{2.49}$$

Here
   a is the nearest neighbor distance
   C_v is the specific heat capacity at constant volume
   V is the volume
   ν is the lattice frequency
   k_0 is the frequency of diffusive displacement

2. Zwanzig model
   In the Zwanzig model, the collision events in a liquid are considered to be analogous to Brownian motion. The expressions for k_{conv} and k_ϕ are obtained as follows:

$$k_{\text{conv}} = \frac{nk_{\text{B}}^2 T}{2\zeta} - \frac{k_{\text{B}}^2 T^2}{6\zeta}\left(\frac{\partial n}{\partial T}\right)_p \tag{2.50}$$

$$k_\phi = \frac{n^2\pi k_{\text{B}} T}{3\zeta}\int_0^\infty R^3\left(R\frac{\mathrm{d}\phi}{\mathrm{d}R} - \phi\right)g_0^{(2)}(R)\frac{\mathrm{d}}{\mathrm{d}R}\left(\frac{\partial}{\partial T}\ln g_0^{(2)}(R)\right)_p \mathrm{d}R$$

$$+ \frac{n^2\pi k_{\text{B}} T}{\zeta}\int_0^\infty R^2\left(\phi - \frac{R}{3}\frac{\mathrm{d}\phi}{\mathrm{d}R}\right)\left(\frac{\partial}{\partial T}g_0^{(2)}(R)\right)\mathrm{d}R \tag{2.51}$$

Here
   $g_0^{(2)}(R)$ is the equilibrium radial distribution function
   $\phi(R)$ is the intermolecular potential
   ζ is the friction coefficient
   k_{B} is the Boltzmann constant
   R is the radial coordinate
   n is the number density

3. Rice and Kirkwood model
   Rice and Kirkwood proposed a theory, incorporating a simpler approximate formula for the intermolecular force contribution.
   The expressions for k_ϕ and k_{conv} are given below:

$$k_\phi = -\frac{k_B T}{12\zeta} \frac{\partial}{\partial T}\left[\left(\frac{N}{V}\right)^2 \int_0^\infty R^2 \nabla^2 \phi(R) g_0^{(2)}(R) \mathrm{d}^3 R\right] \qquad (2.52)$$

$$k_{conv} = \frac{k_B T}{2\zeta V}\left(\frac{N k_B T_\alpha}{3} + N k_B\right) \qquad (2.53)$$

and,

$$k = k_{conv} + k_\varphi \qquad (2.54)$$

where
   α is the coefficient of thermal expansion
   $g_0^{(2)}$ is the equilibrium radial distribution function
   $\phi(R)$ is the intermolecular potential
   ζ is the friction constant
   k_B is the Boltzmann constant
   N is the total number of molecules
   V is the total volume
   R is the radial coordinate

2.5.1.4 Thermal Conductivity of Solid–Liquid Suspensions

Proposed models for the thermal conductivity of solid–liquid suspensions are generally based on effective medium theory. The thermal conductivities in this case will have to take into account the influence of the two media and the distribution of the solid phase in the liquid medium, thus giving an effective value for the thermal conductivity which will explain the observed conduction heat transfer effect. On the basis of the assumption of the particle distribution in the suspension, various expressions have been proposed. The distribution and the unit cells considered in obtaining these expressions, and the derivations of the effective thermal conductivity have been discussed extensively in the literature (Xue and Xu 2005). Some of the thermal conductivity expressions are reproduced below.

1. Maxwell model

$$\frac{k_{eff}}{k_m} = 1 + \frac{3(\alpha - 1)v}{(\alpha + 1) - (\alpha - 1)v} \qquad (2.55)$$

2. Hamilton and Crosser model

$$\frac{k_{\text{eff}}}{k_{\text{m}}} = \frac{\alpha + (n-1) - (n-1)(1-\alpha)v}{\alpha + (n-1) + (1-\alpha)v} \qquad (2.56)$$

3. Jeffery model

$$\frac{k_{\text{eff}}}{k_{\text{m}}} = 1 + 3\beta v + \left(3\beta^2 + \frac{3\beta^2}{4} + \frac{9\beta^3}{16}\frac{\alpha+2}{2\alpha+3} + \cdots\right)v^2 \qquad (2.57)$$

4. Davis model

$$\frac{k_{\text{eff}}}{k_{\text{m}}} = 1 + \frac{3(\alpha-1)}{(\alpha+2) - (\alpha-1)v}\left[v + f(\alpha)v^2 + O(v^3)\right] \qquad (2.58)$$

5. Lu and Lin model

$$\frac{k_{\text{eff}}}{k_{\text{m}}} = 1 + \alpha v + bv^2 \qquad (2.59)$$

In the above expressions, k_{eff} is the effective thermal conductivity of the solid/liquid suspension. The notations n and v represent the particle shape factor and the particle volume fraction, respectively. Where k_{m} and k_2 are the thermal conductivity of the base fluid and the suspended particle, respectively; $\alpha = k_2/k_{\text{m}}$ and $\beta = (\alpha-1)/(\alpha+2)$.

It has been noticed that all of the models given above have failed to account for the anomalous enhancement of the thermal conductivity in nanoparticle suspensions (nanofluids) compared to base fluids, which has been determined experimentally. It is believed that a perfect theory is still required to be postulated to correctly account for the observed thermal conductivity values of nanoparticle suspensions. More discussions on these lines are presented in Chapter 6.

2.5.2 THERMAL CONDUCTIVITY PREDICTION USING MOLECULAR DYNAMICS

Using the molecular dynamics method introduced in Chapter 1, it is possible to generate the velocity trajectories of all the molecules in the system under study. This is done by obtaining the force between two interacting particles as the spatial derivative of the interaction potential chosen and thus calculating the accelerations from the forces. The velocities at successive time steps are obtained through suitable integration schemes. Two methods to obtain the physical properties of the system under consideration from the molecular dynamics simulation are briefly discussed below (McGaughey 2004).

1. Green–Kubo method:
   Green–Kubo (GK) method is the most widely used technique to predict the thermal conductivity, once the velocities of all the molecules are obtained

through the molecular dynamics simulation. GK method is based on the fluctuation–dissipation theorem of statistical mechanics. This theorem, in general, relates the thermodynamic response functions to the most appropriate time autocorrelation functions, and the GK formula is given as follows:

$$k = \frac{1}{k_B V T^2} \int_0^\infty \frac{\langle S(t) \cdot S(0) \rangle}{3} \, dt \tag{2.60}$$

where S is called the heat current vector. It can be written as

$$S = \frac{d}{dt} \sum_i r_i E_i \tag{2.61}$$

Here the summation designates the consideration of all particles in the system. r_i is the position vector of a particle and E_i is the total energy of the particle (sum of the kinetic energy and the potential energy). S fluctuates about zero at equilibrium. The thermal conductivity or any physical property in general is linked to the time required for the fluctuations to dissipate. $S(t) \cdot S(0)$ is called the heat current autocorrelation function (HCACF). The decay rate of the HCACF is what influences the value of the thermal conductivity of a specific material. If the decay rate is very high, it means that the fluctuations in the heat current vector are very small or short lived, which, in turn, is due to the small mean phonon relaxation times. The thermal conductivity which is linked to integral of HCACF is therefore small in value for such materials.

If the potential used to account for the interparticle interaction forces is a simple Lennard–Jones potential (McGaughey 2004), then the heat current can be rewritten as follows:

$$S = \sum_i e_i v_i + \frac{1}{2} \sum_{i,j} (F_{ij} \cdot v_i) r_{ij} \tag{2.62}$$

where
  e_i is the energy and v_i is the velocity vector of a particle
  F_{ij} and r_{ij} are the force vector and the interparticle position vector, respectively, for two interacting particles i and j

This form of heat current vector can be easily implemented in a molecular dynamics simulation. The GK method has been successfully used to predict the thermal conductivities of argon, β-silicon carbide, diamond, silicon, amorphous silicon, germanium-based materials, and nanofluids. More discussions on the application of molecular dynamics simulations for estimating thermophysical properties and analyzing thermal phenomena are presented in Chapter 6, in relation to the computational study of nanofluids.

2. Direct method:

In the direct method, a temperature is imposed in the simulation cell in a single dimension. The Fourier's law of heat conduction is then used to determine the thermal conductivity from the resulting heat flux. In the cases of nanoscale systems, the application of large temperature gradients may result in very high heat fluxes and nonlinear temperature profiles as the outcome of the simulation and so, the direct method is not suitable for such cases. However, the direct approach has been successfully used to predict the thermal conductivity of amorphous silica, zirconia, and yttria-stabilized zirconia, and various other liquids in microscale systems.

2.6 BOLTZMANN EQUATION AND PHONON TRANSPORT

Solution of the Boltzmann transport equation, introduced in Chapter 1, has been tried out as a method for estimation of thermal conductivity. The idea of using such an equation is to approach the problem theoretically, and not from a macroscopic law, to possibly obtain results applicable in the size-effect domain.

Kumar and Vradis (1991) examined the effects of transverse thickness on the thermal conductivity of thin metallic films. A temperature gradient was imposed along the film and the transport of thermal energy was considered to be predominantly due to electron motion. The small size allowed the Boltzmann equation of electron transport to be evaluated along with appropriate electron-scattering boundary conditions. Simple expressions for the reduction of conductivity due to increased dominance of boundary scattering were presented and the results were compared with experimental data from the literature. It was noted that the scattering of electrons at the boundaries of the film became increasingly dominant as the film thickness decreased. At thicknesses on the order of the mean free path thickness or less, the boundary scattering was predicted to dominate the conduction process as compared to the bulk scattering processes. The complete Boltzmann equation with the Fermi energy distribution of electrons was solved to evaluate the reduction in thermal conductivity due to the increased dominance of interactions of electrons with boundaries and its effects on the transport process. One of the assumptions made in this analysis was that of a pure, homogeneous metal having a perfect lattice without grain boundaries, i.e., scattering from sources such as impurities and grain boundaries was neglected. Results indicated that the reduction in electrical and thermal conductivities were identical for cases of practical importance in thin film technology. It has also been suggested that a more rigorous solution of the Boltzmann equation that would consider other scattering processes would match existing experimental data such as that of Nath and Chopra (1974) in real films.

In the phonon transport theory, the transport of thermal energy in an ideal solid is postulated as the manifestations of the excitations of free electrons, and lattice vibrations (termed phonons). These two effects are quantized phenomena, and the resistive heat transfer can be expressed in terms of the energy dissipation due to the collisions of quanta. For phonon heat transport, several possibilities of the boundary conditions based on diffusion theory, the analogy to gas kinetic theory, and the time relaxation model have been examined, along with modified hyperbolic heat conduction equation (Bai and Lavine 1993). Utilizing these boundary conditions reduced the differences between the conventional and modified hyperbolic heat conduction equations. This resulted in the following conclusions:

1. When the domain thickness is on the order of a carrier mean free path, jump boundary conditions have a significant effect on the solution of the conduction problem.
2. Even with the jump boundary conditions (this assumes a discontinuity of temperature at the boundary, and are explained further in subsequent chapters), the hyperbolic and parabolic solutions are significantly different when the domain is thin.
3. Even though the hyperbolic heat conduction equation violates the second law of thermodynamics, it can still be used in engineering applications.

This conclusion was based on the observation that the conventional and modified hyperbolic heat conduction equations agreed well, provided either the domain was large compared to a mean free path or appropriate jump boundary conditions were used. However, no definite conclusion has been made from the analysis, on which of the various jump boundary conditions was best suitable.

Citing critical issues in the design of electronic devices and packages, Majumdar (1993) investigated microscale heat conduction in dielectric thin films. This paper indicated that in the microscale regime, heat transport by lattice vibrations or phonons could be analyzed as a radiative transfer problem. On the basis of the Boltzmann transport theory, an equation of phonon radiative transfer (EPRT) was developed. For transient heat conduction, the EPRT suggests that a heat pulse is transported as a wave, which becomes attenuated in the film, due to phonon scattering. The EPRT was used to evaluate heat transport in diamond thin films. A simple expression for thermal conductivity was developed. In the regime of thicknesses in which Fourier's law is inappropriate, this study concluded that heat conduction occurs by lattice vibrations or phonons and behaves like radiative transfer.

Fushinobu and Majumdar (1993), taking into account the nonequilibrium nature of electrons and phonons, developed a simple model for thermal and electrical modeling of metal semiconductor field-effect transistors (MESFETs). When applied to an n-doped GaAs MESFET with a doping concentration of 3×10^{-17} cm^{-3} and gate length of 0.2 mm, the model predicted that due to the presence of very high electric fields a hot spot would be created beneath the drain side of the gate. The model also predicted the distributions of electron temperature, lattice temperature, current density, electron concentration, and electric field within the device.

2.7 CONDUCTION IN THIN FILMS

One of the major application areas of microscale heat conduction is electronics. Because of the advent of miniaturization of electronic circuits, thermal problems associated with such circuits also have become important. This is because modern components and systems deal with larger fluxes of heat to be dissipated, both due to the high levels of generated heat and due to the smaller surface areas associated with dissipation. The classes of electronic components that use silicon and insulator films, called silicon on insulator (SOI) devices, require a good understanding of the conduction heat transfer in thin films for their thermal management. Apart from the operating conditions, a good knowledge of heat conduction in the microscale, based on measurements as well as analysis, is required to tackle the thermal problems associated with the manufacture of SOI devices and other electronic devices.

The existence of a size effect on transport phenomena in thin metallic films, when the physical dimensions are of the order of magnitude of scattering mean free path of electrons, has been recognized long ago. Tien et al. (1968) had calculated the thermal conductivity of metallic films and wires at cryogenic temperatures. A reduction in electrical and thermal conductivities was expected in thin films and wires, because of the reduction of the mean free path near the surface boundary. Size effects were expected to be more significant at cryogenic temperatures, as the mean free path increased with a reduction in temperature. The method outlined for calculating thermal conductivity consisted of seeking an appropriate expression for the electrical conductivity including the size effect, determination of the mean free path corresponding to the electrical conductivity, and the application of the electrical–thermal transport analogy for calculating the size effect on thermal conductivity. The investigation revealed that size effects may become increasingly important in the design and analysis of thin metallic films. Given that the thin film deposition techniques available at the time of these investigations were not as well developed as the present-day techniques and still demonstrated significant size effects at room temperatures, these observations were quite important. Also, because thin films are used extensively in optical components and solid-state devices and systems, this discovery stimulated additional investigations into the effect of size on the energy transport including the thermal conductivity measurements for single crystal silicon (Savvides and Goldsmid 1972) and copper films (Nath and Chopra 1974). While the methods of measurement of thermal conductivity of thin films will be dealt with in later sections, it is helpful at this point to review some of the more important results of these thin film investigations, from both theoretical and experimental perspectives.

Calculation of thermal conductivity using an electrical–thermal transport analogy, in copper films, indicated an increase in the value with an increase in film thickness, as shown in Figure 2.1 (Nath and Chopra 1974, Duncan and Peterson 1994), and this effect was partially attributed to boundary scattering of the conduction electrons, and partially to the scattering due to lattice impurities and structural defects in the films. Measurement of thermal conductivity for thin film materials including Al_2O_3 and SiO_2 deposited in high vacuum indicated that the values were significantly lower than bulk values—for instance, the value for SiO_2 thin film was found to be lower than the bulk value of silica by a factor of 5–8. In Al_2O_3, the reduction was almost half of the bulk value (Decker et al. 1984). A temperature dependence of thin film thermal conductivity has also been observed. In Bismuth films (20–400 nm, 80–400 K, thermal conductivity measured in the direction of the film plane), this was characterized by a decrease in thermal conductivity with increasing temperatures in the region below 250 K and an increase, with an increase in temperature above 300 K (Volklein and Kessler 1986).

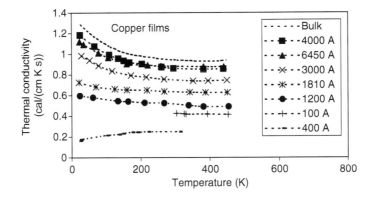

FIGURE 2.1 Variation of thermal conductivity with temperature in copper films. (Adapted from Duncan, A.B. and Peterson, G.P., *Appl. Mech. Rev.*, 47, 397, 1994. With permission.)

The experimental observations were validated in most of the cases using physical models such as that of phonon scattering. Other hypotheses were also postulated for modeling the phenomena by various investigators. Several such models were proposed to explain the reduction in thermal diffusivity and thermal conductivity of thin optical coatings—one which assumed the coatings contained microcracks, and another that consisted of a random distribution of voids in the coating (Redondo and Beery 1986).

The analysis and estimation of the thermal conductivity in thin films were based on fundamental equations such as the Boltzmann equation. This is particularly useful in films of very small physical dimensions with considerably low domain size for computation, and can be applied along with the appropriate electron-scattering boundary conditions. Such analyses have given evidence to believe that the reduction of thermal conductivity is a manifestation of the dominance of the scattering of electrons at the boundaries the film (Kumar and Vradis 1991), at thicknesses comparable to the mean free path. Microscale heat conduction in thin films also has been analyzed based on the phonon transport theory, which describes heat transport through lattice vibration or phonons, treating the phenomenon essentially as a radiative heat transfer problem (Majumdar 1993).

Measurement of thermal conductivity has also been the focus of various investigations. The thin film conductivity has been shown to be relatively insensitive to the method used in the preparation of the film. Comparative techniques used in the measurement of thin dielectric films of optical materials in high power laser applications indicated that the values are lower than bulk values (Lambropoulos et al. 1989).

The measurement of electrical conductivity and its correlation with thermal conductivity (Wiedemann–Franz law) has been of considerable value in the estimation of thermal conductivity for sputtered rare earth-transition metallic films, which have applications in thermomagnetic recording processes used for recording data on magneto-optic data storage disks. Here, the thermal conductivity values are expected to be dependent on the chemical composition of the film and its method of preparation, unlike other kinds of films (Anderson 1990).

2.7.1 SUPERCONDUCTING FILMS

Investigations have been carried out on thin films of superconducting materials to study the thermal performance and thermal conductivity behavior. Some of the common structures are $YBa_2Cu_3O_7$ films, Bi–Sr–Ca–Cu–O sheets and gallium arsenide (GaAs) quantum well structures. Studies have been conducted to understand the bolometric responses of high-temperature superconducting films (Flik et al. 1990), and to measure and computationally estimate the thermal conductivity (Crommie and Zettl 1990, Chen and Tien 1993, Goodson and Flik 1993). A difficulty encountered while studying the thermal response of thin film superconductors using laser pulses is the confusion over whether the responses are due to a thermal mechanism or caused by a nonequilibrium process created due to the laser pulse. Experimental measurements aimed at determining the normal state interactions of electrical resistivity and thermal conductivity of single crystal Bi–Sr–Ca–Cu–O sheets have indicated that the thermal conductivity in the metallic regime has substantial electron and phonon contributions (Crommie and Zettl 1990). Measurement in the polycrystalline form indicated a peak just below the critical temperature, similar to that observed in other high-temperature superconducting materials, as shown in Figure 2.2. The measured thermal conductivity and electrical resistivity variations are shown in the figure. It has been interpreted that the existence of the peak is due to an enhancement of the phonon contribution caused due to a reduction of phonon–electron scattering.

Volklein and Baltes (1992) utilized a differential method with a pair of oxide microbeam test structures in order to determine the thermal conductivity of polysilicon films sandwiched between oxide layers of a free-standing silicon oxide microbeam. Their thermal conductivity was determined by a differential method using oxide beams with and without polysilicon films. The measurements were reported to be the first of the thermal conductivity of polysilicon films produced in a standard IC process. Applying only one maskless postprocessing step, thermal test structures were fabricated that allowed accurate measurement of the thermal conductivity directly on the process-specific films. The measured values of thermal conductivity of the polysilicon films were much smaller than that of the monocrystalline silicon. This was primarily attributed to the phonon scattering at the grain boundaries.

A particular class of electronic devices, made from superconducting materials, is known as Josephson junctions. These devices cycle between resistive and nonresistive states. Heat generated in the resistive state causes a temperature rise, which may affect the electrical behavior by reducing the critical current. Lavine and Bai (1991) investigated the heat transfer phenomena associated with Josephson junction devices and found that the temperature distributions and resulting reductions in critical current could be calculated for Josephson junctions made from low- and high-temperature superconductors. In addition, an unacceptable reduction in critical current was found to occur for junctions made from high-temperature materials. The authors agreed that this problem can almost certainly be overcome, but perhaps at the expense of the size advantages associated with Josephson junctions. The intent of this study was to (1) calculate the temperature distribution in a Josephson junction

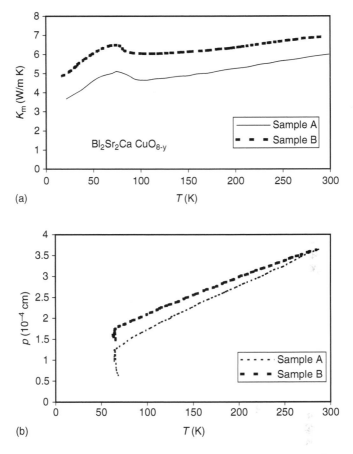

FIGURE 2.2 (a) Measured thermal conductivity as a function of temperature for two metallic crystals of $Ba_2Sr_2CaCu_2O_8$. (Adapted from Duncan, A.B. and Peterson, G.P., *Appl. Mech. Rev.*, 47, 397, 1994. With permission.) (b) Measured electrical resistivity as a function of temperature for two metallic crystals of $Ba_2Sr_2CaCu_2O_8$. (Adapted from Duncan, A.B. and Peterson, G.P., *Appl. Mech. Rev.*, 47, 397, 1994. With permission.)

as a function of the various governing parameters, (2) evaluate the effect of the temperature distribution on the critical Josephson current, and (3) assess the important differences between junctions made from low- and high-temperature superconductors. A generic geometry of an infinite square array of square Josephson junctions on a substrate was considered. Heat transfer from the bottom electrode to the substrate was modeled using an effective heat transfer coefficient, which accounted for nearly all of the resistances between the electrode and the coolant. In conclusion, the analysis indicated that Josephson junction devices made from high-temperature superconductors may generate a temperature rise sufficient to cause unacceptable reductions in critical current. Solutions that overcome these problems may be at the expense of certain benefits of Josephson junctions such as their compactness and speed.

Investigations focused at electron and phonon thermal conduction in super-conducting films and quantum well structures have introduced a strong theoretical basis for analyzing the phenomenon of heat conduction in the micro-scale, and approached the problem from the background of the kinetic theory and fundamental models such as the Boltzmann transport equation. Electron and phonon thermal conduction in superconducting films has been studied by Goodson and Flik (1993). In the absence of boundary scattering, the thermal conductivity and the mean free path of each carrier type were calculated as functions of temperature using kinetic theory, a two-fluid model of the superconducting state, and experimental data for the thermal conductivity and electrical resistivity of a single crystal. The effective thermal conductivity reduced by boundary scattering along an epitaxial $YBa_2Cu_3O_7$ film was predicted as a function of tempera-ture and film thickness. The size effect on the phonon conductivity was found to dominate over the size effect on the electron conductivity. The predicted electron mean free path was limited by scattering on defects and was in very good agreement with experimental data from infrared spectroscopy. This inves-tigation developed a new technique for calculating the a–b plane thermal conduct-ivity of superconducting films from measurements performed on a crystal in which boundary scattering was negligible. In the absence of boundary scattering, the mean free path of normal state electrons was calculated from normal state electrical resistivity data. The electron thermal conductivity in the absence of boundary scattering in the superconducting state was calculated from the electron mean free path using kinetic theory and the two-fluid model for electrons in superconductors. The phonon conductivity is the difference of the electron contribution and the conductivity measured in a large crystal. The phonon mean free path was determined using kinetic theory. The effects of boundary scattering for a thin film were analyzed separately for the electron and phonon contributions.

Interesting theoretical investigations have been reported on quantum well structures. Gallium arsenide (GaAs)-based quantum well structures represent a rapidly expanding area of research and development. Due to the small dimen-sions associated with quantum well structures, it is likely that thermal conductivity cannot be determined from bulk material behavior. Chen and Tien (1993) analyzed the size and the boundary effects of a GaAs-based quantum well structure on the thermal conductivity of the well material. Calculations showed that the order of phonon mean free path is equal to or even longer than the typical dimension of the well (~200 Å or less). The EPRT mentioned earlier, developed from the Boltzmann transport equation, was utilized to determine the heat transport in the quantum well structure. Approximate solutions were obtained for the cases of heat flow perpendicular and parallel to the well. Results indicated that the thermal conductivity of the quantum well could be one order of magnitude lower than that of its corresponding bulk form at room temperature. The size and boundary effects also caused anisotropy of the thermal conductivity, even though the unit cell of GaAs is cubic. This investigation also presented a theoretical study of the size effects on the thermal conductivity of the active region in a quantum well.

2.7.2 Laser-Induced Heating

An area where microscale heat conduction analysis becomes important is that of high power laser systems. The principal constraint in the design of high power laser systems stems from laser-induced damage considerations, due to the interaction of intense coherent light with optical components in the system. The interaction of the intense laser light with optical elements is primarily governed by optical absorption. This can lead to a distortion of optical components, reduction of optical power, and, if carried far enough, fracture as a result of a thermally induced failure process. These types of damage are primarily observed in continuous wave (CW), high energy, or high average power systems. In contrast, lasers operating in the short pulse mode may undergo nonlinear optical processes such as transient, small-scale self focusing, also inducing beam or wave front distortion. A significant amount of research has been conducted in recent times as an attempt to understand the mechanisms of laser damage and laser heating in materials.

In connection with a study of laser-induced degradation of light-emitting semiconductor material with a temperature rise, Lax (1977) reduced the spatial distribution of the temperature rise induced by a laser beam absorbed in a solid, to a one-dimensional integral, which was evaluated numerically. The solution for a general laser intensity distribution was specialized to the case of a Gaussian beam. A closed form expression in terms of tabulated functions was obtained for the maximum temperature rise. Analytical details of the solution of the heat equation include determining the maximum temperature rise in the solid and determining the spatial distribution of the temperature rise.

Burgener and Reedy (1982) derived the Green's function solution of the heat equation for a two-layer structure and described how to extend the technique to derive the Green's function for multilayered structures. The Green's function was applied to the interaction of a two-layer structure with scanning circular energy beams. The mathematics necessary to reduce the resultant solutions from a quintuple integral to a double integral were presented. Numerical integration allowed the authors to obtain predictions of the temperature profiles generated during laser irradiation of two-layer structures. These profiles were shown to reduce, in the appropriate limits, to previously reported profiles for the single material case. Normalized linear temperature profiles as well as actual temperature profiles for silicon on sapphire were presented and discussed. The problem was considered for zero and nonzero beam velocities. A need was pointed out by the authors for future efforts to concentrate on deriving the Green's function for three-layer structures and for a discontinuity in a layer (such as silicon covering an oxide or nitride step).

Ristau and Ebert (1986) reported the development of a thermographic laser calorimeter for the measurement of absorptance in optical coatings. The purpose of this work was to investigate the usefulness and the difficulties of an infrared scanning laser calorimeter (ISLC). The absorption loss in coated dielectric layers is generally higher than in the corresponding bulk material and depends on the specific coating procedure. The authors reported that it is possible to utilize a contact-free temperature measurement by recording the infrared radiation of the sample. Using this technique, it is possible to report the entire temperature profile on the surface of the film. The calorimeter described in this paper was, in principle, such a line scanning device applied to the laser-heated coating sample. The heat equation was solved for a single layer coated on a disk-shaped substrate to analyze the influence of the absorption, the thickness, and the thermal conductivity of the film on the temperature field. Temperature profiles of In_2O_3 layers were experimentally recorded and compared with the theoretical data. A calibration of the ISLC was carried out with moderately absorbing films by utilizing a conventional laser calorimeter. Samples with Al_2O_3, HfO_2, SiO_2, Ta_2O_5, and TiO_2 coatings were measured. The thermal conductivities obtained were orders of magnitude smaller than those of the bulk material.

Abraham and Halley (1987) presented calculations of heat flow in an absorbing film deposited on a nonabsorbing substrate. The energy source was a Gaussian laser beam. The effects of film thickness, thermal conductivity, spot radius, and forced cooling on the temperature profile and, in a special case, the time evolution of that profile were considered. The emphasis was on applications in optical bistability. It was stated that current interest in nonlinear optical processes, optical bistability in particular, as a possible basis for processors for digital optical computing and high-speed optical communication systems inspires the continued development of models intended to predict temperature distributions of materials subjected to laser-induced heating. An example given is a family of thermo-optic devices based on thin film interference filters. In non-thermo-optic devices, thermal effects can also significantly affect behavior. The particular aim of this work was to present some simple calculations, applicable to such devices, based on the linear heat equation. The model developed consisted of an absorbing thin film on a nonabsorbing substrate irradiated by a Gaussian beam. The film was infinite in both the x and y directions whereas the substrate was semi-infinite. The heat equation was solved separately for the film and substrate. A linear theory of the heat flow and temperature rise in the absorbing film on a nonabsorbing substrate irradiated by a Gaussian laser beam was presented. Analysis of the steady-state case indicated that when the film thickness was much smaller than the spot size and the film thermal conductivity was significantly less than that of the substrate, the temperature increased with the absorbed power and was inversely proportional to both the spot radius and the thermal conductivity of the substrate. As the film thermal conductivity was allowed to increase, transverse heat flow also increased and the substrate became less effective in controlling the temperature change. The same occurred if the film was allowed to broaden in relation to the Gaussian input. Far from the beam center, the temperature was found to decrease with the inverse of the distance when the surface heat loss was negligible. The two-layer system could be approximated by a semi-infinite heat sink with the heat dumped in the surface if the radial heat flux from the source region was small, relative to the longitudinal flux into the substrate at the source. This required that the conductivity of the substrate be greater than that of the film and the radius of the absorption region be greater than the depth of the film. The time-dependent solutions indicated how the profiles evolved in time for the special case of layers of similar diffusivity. It became apparent that the temperature change became greater and occurred more quickly as the spot radius decreased. In addition, a more insulating substrate was found to speed up the heating of the film while slowing down the cooling process. Under conditions of forced external cooling, the temperature dropped with the inverse of the cube of the distance.

Guenther and McIver (1988) analyzed the pulsed laser-induced damage of optical thin films. A heat flow analysis of this process stressed the importance that the thermal conductivity of both the thin film host and the substrate play in establishing the laser-induced damage threshold. The authors cited investigations which found that the thermal conductivity of thin films can be several orders of magnitude lower than that of a corresponding material in bulk form as a consequence of the film structure that results principally from the deposition process. It was noted that the importance of thermal conductivity would be related to parameters such as absorption mechanisms, film materials, composition, and other variables. Bulk optical materials can have absorption coefficients ranging from 10^{-4} to less than 10^{-6} cm^{-1}. Thin films, however, may commonly have absorption coefficients as great as 10^{-1} cm^{-1}. Fortunately, these films are thin and thus the total energy absorbed by the film is small. However, if this heat is not dissipated rapidly enough, unacceptable consequences usually occur as a result of the rapid localized temperature rise.

In an encounter of pulsed lasers with optical thin films, initial pitting is frequently followed by a number of structural changes, which may include delamination and increased surface roughness. These mechanisms result in a change of optical properties. The threshold of energy required to induce damage generally decreases as absorption increases, pulse duration decreases, wavelength decreases, films become thicker, areas increase, or surface roughness increases.

The following are the conclusions drawn:

1. The thermal conductivity of thin films may be from one to several orders of magnitude lower than the corresponding bulk values, affecting the laser-induced damage threshold of these films as this reduction would severely limit the dissipation of heat absorption sites.
2. Thin film thermal conductivity is very structure dependent and is thus related to the substrate finish, material selection, deposition technique, post-processing, aging, and the environment. Spectroscopic ellipsometry offers a convenient technique for studying this aspect of the problem.
3. Due to the wide variations in structural properties, which occur in specific thin film manufacturing processes, the thermal conductivity of films must be determined on virtually an individual basis.
4. On the basis of very limited data, it appeared that the thermal conductivity of fluoride thin films is similar to that of the bulk while that of oxide thin films is generally orders of magnitude lower than corresponding bulk values.
5. Any comprehensive model of thin film thermal conductivity must include and understand the anisotropies and inhomogeneities associated with the film as well as temperature-dependent behavior.

Anderson (1988) described a method to calculate the heating that occurs when a Gaussian laser beam is scanned across the surface of a flat object which may be coated with one or more layers of differing materials. The laser beam was characterized by its power, radius, and scan velocity. Each layer of the object was characterized by its thickness, density, heat capacity, and thermal conductivity. The absorption of laser light was assumed to vary linearly across any one layer, and was characterized by two dimensionless constants. Temperature rise versus time was calculated as a function of position within the multilayer stack. Use of the technique was illustrated by considering the case of an optically opaque organic dye/polymeric binder coated onto a polymeric substrate. Such a construction is typical of one which may be used in optical data storage applications. The analysis assumed constant physical properties for the layers of the structure and applied only to the case of a beam with constant total power. A detailed physical description of the modeled system was presented. This consisted of an n layered object, each layer characterized by thickness, density, specific heat, and thermal conductivity. Details of the mathematical solution were provided and an example calculation was carried out. The calculation of temperature rise anywhere in the structure was reduced to the evaluation of a two-dimensional Fourier transform. The example case considered was that of an optically opaque layer of an organic dye/polymeric binder coated onto a plastic substrate of similar thermal properties. In this case, when considering points directly under the scanning beam, the results were compared with existing experimental data and good agreement was obtained. In addition, curves displaying temperature rise versus time for points located off the scan line of the beam were presented.

Citing that picosecond and subpicosecond lasers had become important tools in the fabrication and study of microstructures, Qiu and Tien (1992) presented an analytical investigation of the size effect on nonequilibrium laser heating of metal films. The duration of laser pulse has decreased to the same order of magnitude as the characteristic time of energy exchange among microscopic carriers. Thus, the excited carriers are no longer in thermal equilibrium with the other carriers. The presence of interfaces in metals provides additional scattering processes for electrons, which in turn affects the nonequilibrium heating process. This work studied the size effects due to both surface scattering and grain-boundary scattering, on the thermal conductivity and the energy exchange between electrons and the material lattice during nonequilibrium laser heating. A simple formula was established to predict the influence of film thickness, grain size, interface scattering parameters, and the electron and lattice temperatures on the effective thermal conductivity of thin metal films. The model predictions agreed well with the available experimental data. A three-energy level model was developed to characterize the energy exchange between electrons and the lattice. This study indicated that the size effects reduced the effective thermal conductivity and increased the electron–phonon energy exchange rate. The results were considered useful for improving processing quality, interpreting diagnostic results, and preventing thermal damage of thin films in the short-time regime.

Grigoropolous et al. (1988) reported the results of an experimental study on the phenomena associated with laser annealing of thin silicon layers. In pursuit of the production of a single crystal, thin silicon layers on amorphous insulators, one of the most promising techniques is such a laser annealing process. Direct heating of thin silicon layers on substrates was shown to produce a variety of different silicon melting patterns and a systematic study of these phase change phenomena was performed. The important parameters were found to be the laser beam power, laser beam intensity distribution, and the speed of the translating silicon layer. Unstable silicon phase boundaries were observed to break up and form regions where solid and melted silicon would coexist. Complicated silicon phase boundary patterns were shown. The experimental results showed the occurrence of organized patterns of alternating solid and liquid silicon stripes for two-dimensional heating distributions. The temperature fields for the experimental operating conditions were calculated using an enthalpy model. Part of the intent of using laser recrystallization was to improve the crystalline structure of chemical vapor deposited (CVD) silicon layers. The material obtained in a CVD process has a definite crystalline structure; however, it presents imperfections in the form of grain boundaries that can act as barriers to the movement of charge carriers across junctions. The method of laser recrystallization may be effectively used to improve the structure. Results indicated that the basic instability mechanism was the abrupt silicon surface reflectivity increase during the occurrence of melting. In the neighborhood of phase boundaries, the solid silicon absorbs twice as much heat as the adjacent liquid material. Thus, there exists a strong tendency toward solid superheating and liquid supercooling.

Continuing that investigation, Grigoropoulos et al. (1991) has presented a detailed experimental account of the phase change process during the melting of a semiconductor deposited on an amorphous insulating substrate. A computational conductive heat transfer model was also presented. The numerical predictions were compared to the experimental results and good agreement was obtained. In this work, sufficient experimental data were obtained to validate a three-dimensional transient heat transfer algorithm for predicting the phase change process. The experimental uncertainty was reduced by direct measurement of the laser beam parameters. It was therefore possible to compare the heat transfer solutions with the experimental results. This was the first time that such a detailed comparison was reported.

Cole and McGahan (1992) developed a solution of the heat conduction equation for multilayer samples heated by a periodically modulated axisymmetric laser beam. The sample was allowed to contain any number of thin layers on a thick substrate. This new theory combined exact optical absorption, multilayers, contact resistance between layers, and ease of calculation. The solutions were based on a local Green's function formulation, and no approximations were made as to the form of absorption of energy in the sample from the heating laser. The method involved integral transforms (Fourier and Hankel) and the normal component of the heat fluxes across the layer boundaries could be found analytically in transform space. To calculate temperatures in real space, one numerical integration was required. Temperature results were presented in the form of magnitude and phase plots at the frequency of the modulated laser beam. One application for these results is photothermal deflection experiments for the measurement of thermal conductivity in multiple films on a thick substrate. The temperature and beam deflection calculations are very efficient and can be carried out easily on a personal computer. The methodology presented was as follows:

1. Necessary Green's functions were discussed for both layers and semi-infinite regions and Fourier and Hankel transformations were applied to these functions.
2. The absorption of energy from the heating laser beam was treated analytically in multilayer systems.
3. Temperature solutions were given for the two-body case in which the sample was assumed to be infinitely thick.
4. The two-body solution was extended to systems with any number of layers.

On the experimental front, the use of optical techniques has shown great promise as a tool to investigate the thermal transport properties of very thin liquid and solid films. Rosencwaig et al. (1985) conducted an investigation to show that thermal wave detection and analysis could be performed in a noncontact and highly sensitive manner, through the dependence of sample optical reflectance on temperature. Applications to the study of microelectronic materials were illustrated by measuring the thickness of thin metal films. Iravani and Wickramasinghe (1985) presented a technique based on Fourier optics, which was utilized to explain the propagation, as well as loss, of three-dimensional thermal waves in isotropic, homogeneous materials. Using this, the dependence of temperature distribution at any arbitrary infinite parallel plane on the aperture distribution was derived. In addition, the temperature distribution at the aperture plane, due to a given perpendicular source, was formulated by applying the boundary conditions in the spatial frequency domain. A scattering matrix theory was developed to analyze the propagation of thermal waves in multilayered structures. This directly related the heat source characteristics to the temperature distribution at any level. The contrast mechanism in the subsurface mode of operation was explained, and the dependence of the response of a typical system on depth was explored. In addition, the theoretical amplitude and phase images of a cylindrical void were presented. The results were in good agreement with the published experimental subsurface images of voids. Finally, transform techniques in both temporal and spatial frequency domains were employed to analyze the pulse response of layered structures. Among other conclusions, it was found that the decay rate of the surface temperature is strongly influenced by the presence of inclusions within a sample. It was stated, however, that as the number of layers increased in a multilayer structure, the system of algebraic equations became prohibitively involved. The authors stated that the theoretical analysis presented in this paper was sufficient to provide a sound understanding of the various photothermal phenomena. McDonald (1986) published a comprehensive theoretical and conceptual review of the physical processes involved in photoacoustic and photothermal techniques, and in related techniques employing particle beams. Thermal imaging techniques were reviewed, with attention to questions of resolution, definition, and sensitivity.

In a series of investigations intended to develop an understanding of the thermal properties of thin optical coatings, Swimm (1983, 1987, 1990) and Swimm and Hou (1986) described the photoacoustic-based technique of measuring the thermal properties of a thin film deposited on a substrate. Data obtained for nickel coatings

(Swimm and Hou 1986) indicated that thin film transport parameters were slightly smaller than corresponding bulk values, and clearly displayed the effects of a nonideal thermal bond between the coating and the substrate.

In an attempt to test and refine the measurement system (Swimm 1987), an additional investigation was carried out that consisted of the study of a simple model system consisting of a thin metal coating deposited on a substrate. A summary of the progress made in this long-term investigation was also published (Swimm 1990). Data were presented for single-layer coatings of Ti/SiO_2 and Ni and were compared with a one-dimensional bulk model. Data were also presented for a two-layer coating of nickel on silica. The developed single-layer coating model that was presented contained four adjustable parameters; these were the coating thermal conductivity, the coating thermal diffusivity, the thermal contact resistance, and the coating thickness. By adjusting the parameters of the model, the measured data could be fit, and the thermal transport properties of the coating were determined. It was indicated that the measurement system was incapable of characterizing single-layer metal coatings, but the improvements necessary to allow such coatings to be characterized were reported to be in progress.

McGahan and Cole (1992) presented exact expressions for the deflection of a laser beam passing parallel to and above the surface of a sample heated by a periodically modulated axisymmetric laser beam. The sample could consist of any number of planar films on a thick substrate. These exact expressions were derived from a local Green's function treatment of the heat conduction equation, and contained an exact analytical treatment of the absorption of energy in the multilayered system from the heating laser. The method was based on a calculation of the normal component of the heat fluxes across the layer boundaries, from which either the beam deflections or the temperature anywhere in space could be easily determined. A central part of the calculation was a tridiagonal matrix equation for the $N + 1$ normal boundary fluxes, where N equals the number of films in the sample, with the beam deflections given as simple functions of the normal heat flux through the top surface of the sample. Even though any layer or layers in the sample could be optically absorbing, the final results were considered to be remarkably simple both in form and ease of calculation, even for large numbers of layers. In the case of an infinitesimal probe beam, the beam deflections were given by an expression involving a single numerical integration, which could be eliminated for data analysis by transforming the experimental data using a Fourier technique. A general expression for the measured signals for the case of four-quadrant detection is also presented and compared with previous calculations of detector response for finite probe beams.

Anthony et al. (1990) conducted an investigation that used photothermal techniques to measure the thermal diffusivity of isotopically enriched 12C diamond. Diamond single crystals containing approximately 0.1%, 0.5%, and 1.0% 13C were synthesized and their thermal diffusivities were measured at room temperature. They exhibited the highest room temperature thermal diffusivity of any solid naturally occurring or previously synthesized. The authors also observed a laser damage threshold, which was more than an order of magnitude greater than that of natural diamond.

2.8   HEAT CONDUCTION IN ELECTRONIC DEVICES

Electronic devices such as integrated circuits, photovoltaic cells, sensors, and actuators utilize numerous structures and circuits of submicron dimensions. This makes the analysis of heat transfer in these devices essentially the analysis of heat transfer in the microscale, and it is essential in the design of these devices. As described earlier, Flik et al. (1991) considered and classified the heat transfer analysis involved in the design of electronic devices. The intent was to provide information to the designer that would determine whether macroscale heat transfer theory could be applied to a given microstructure. The regime maps to identify the boundaries between the macroscale and microscale heat transfer regimes, discussed earlier, have been the outcome of this analysis.

Brotzen et al. (1992) utilized an analytical model with experimental data to predict the thermal conductivities of silicon dioxide films. Thermal conductivities were determined for silicon dioxide films of four thicknesses. The films were deposited by plasma-enhanced chemical vapor deposition (PECVD) on monocrystalline silicon substrates. The thermal conductivities of the films were found to be significantly lower than those of bulk SiO_2. As the film thickness decreased, the thermal conductivity was lowered considerably. Three primary conclusions were drawn from the investigation:

1. The thermal conductivity of SiO_2 films deposited on silicon substrates was markedly lower than that measured in bulk SiO_2.
2. The thermal conductivity of these SiO_2 films was lowered substantially when the film thickness was decreased.
3. The thermal conductivity of SiO_2 films was found to be sensitive to the techniques employed in their deposition, because the technique of deposition strongly affects the physical and chemical nature of the films.

Noting that the performance and reliability of electronic circuits are affected by thermal conduction in the amorphous dielectric layers, Goodson et al. (1993) conducted an investigation designed to analyze the influence of boundary scattering on the effective thermal conductivity normal to amorphous silicon dioxide layers using a model for the frequency dependence of the phonon mean free path, which is in agreement with sound attenuation data. A steady-state experimental technique, which utilized aluminum resistance thermometers as illustrated in Figure 2.3, was adapted to measure the effective conductivity of amorphous silicon dioxide layers. The predictions agreed with the data and indicated a reduction in thermal conductivity of more than 5% for layers thinner than 0.094 μm at 300 K and 2.2 μm at 77 K. The authors indicated that lower than bulk conductivities of thicker layers measured elsewhere were not a result of boundary scattering, but were due to interfacial layers or differences in microstructure or stoichiometry. Conduction along layers of amorphous dielectric materials (i.e., silicon dioxide and silicon nitride) is not important due to the much higher conductivities associated with the bonding of aluminum and silicon. Conduction normal to the layers, however, is critical in the cooling of devices and interconnects. In SOI circuits, devices are separated from the substrate by a silicon

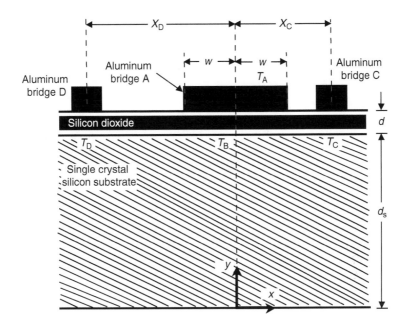

FIGURE 2.3 Experimental test structure for determination of the thermal conductivity of thin silicon dioxide films. (Adapted from Duncan, A.B. and Peterson, G.P., *Appl. Mech. Rev.*, 47, 397, 1994. With permission.)

dioxide layer having a thickness of 0.3–1.0 μm. It is important from a design standpoint to know the effective thermal conductivity normal to amorphous layers. This investigation defined the effective thermal conductivity normal to a layer as its thickness divided by its total thermal resistance, including boundary resistances. The thermal conductivity in the normal direction of an amorphous layer may be as much as an order of magnitude less than bulk values. Data indicate that the effective thermal conductivity decreased with decreasing layer thickness and is dependent upon the fabrication technique. Data analysis was complicated by the fact that different experimental techniques have been used, having unknown or large uncertainties associated with them. In a separate investigation, Goodson and Flik (1992) considered the buried silicon dioxide layer in SOI electronic circuits. This silicon dioxide layer inhibits device cooling and reduces the thermal packing limit, that is, the largest number of devices per unit substrate area for which the device-operating temperature is acceptable. Thermal analysis yielded a packing limit of SOI MOSFET devices in terms of the targeted channel-to-substrate thermal conductance. It was noted that microscale effects are generally negligible above room temperature, but may reduce the packing limit by 44% at substrate temperatures of 77 K. In the thinnest SOI devices, where the source and drain thicknesses may eventually be as small as 0.03 μm, thermal conduction from the channel may also be reduced by heat carrier boundary scattering in the source, gate, drain, and metal interconnects. Thermal conduction in a thin bridge is microscale if it is reduced significantly by the scattering of the carriers of heat on its boundaries. In silicon, phonons are the dominant transport mechanism of heat transport whereas in aluminum it is electrons. This study calculated the steady-state temperature distribution in a linear array of connected SOI MOSFET devices by modeling the source, drain, gate, and metal interconnects as cooling fins, which conduct the heat dissipated in the channel away from the device. Channel-to-substrate thermal conductance was related to the device separation and the packing limit was determined as a function of the targeted value of this conductance. The effective thermal conductivity along each fin considering boundary scattering was determined as a function of its bulk thermal conductivity, thickness, and temperature using kinetic theory and the approximate relations of an investigation by Flik and Tien (1990). The effect of microscale thermal conduction on the packing limit was investigated by comparing microscale predictions, which employ the reduced, effective thermal conductivities to macro predictions, which employ the bulk conductivities. The authors concluded that microscale thermal conduction in the aluminum interconnects and the doped silicon source, drain, and gate of SOI ultrathin devices reduced the packing limit slightly at room temperature and dramatically at low temperatures. Reduction of silicon dioxide thickness could enhance cooling and improve the packing limit. The optimal oxide layer thickness for SOI electronics should be determined by balancing the competing aims of reliability, layer thickness, and device-operating speed.

Su et al. (1992) measured and modeled self heating effects in SOI nMOSFETS. Temperature rises in excess of 100 K were observed for SOI devices under static operating conditions. This observation agreed well with the predictions of their analytical model. The temperature rise associated with the operation of the SOI nMOSFET was found to be a function of the silicon thickness, the buried oxide thickness, and the channel–metal contact separation. SOI nMOSFETS are of great interest due to potentially alleviated short channel effects, reduced parasitic capacitances, and improved radiation hardness. Channel temperature data for SOI devices of varying silicon thicknesses as well as for a bulk device were given, and the slope of the temperature–power curve for each observed case was interpreted as a thermal resistance. Channel temperature data for varying buried oxide thicknesses were also given. As oxide thickness decreased, the rate of change of temperature demonstrated a weaker dependence than for the original thickness of the silicon. Channel temperature decreased with decreasing channel–metal contact separation; however, the range of channel–metal contact separation investigated were very small. The analysis was based on the thermal model proposed by Goodson and Flik (1994). Good agreement was demonstrated between data and model predictions of the channel–substrate thermal resistance as a function of silicon film thickness. As silicon thickness was reduced, the temperature rise in the device increased, showing the same trend as the data. This is because the majority of the power dissipated in the channel travels out through the source and drain. Thermal resistance as a function of channel–metal contact separation indicated a dependence of increasing resistance with increasing separation up to about 1.5 μm. Above this value resistance was basically constant. Device thermal resistance decreased as buried oxide thickness decreased, because the cooling to the substrate was improved. The measurements and modeling indicated that the temperature rise in SOI devices is proportional to static power dissipation with the proportionality constant dependent on device dimensions. However, the device temperature predictions and measurements utilized for steady-state performance were stated to be inappropriate for dynamic performance. In a practical digital circuit, the maximum interconnect temperature is most critical because of electromigration considerations. The authors concluded that for device sizes below ~0.25 μm the transient problem must be carefully examined, the buried oxide layer thickness may need to be optimized to prevent temperatures in the metal, which would affect reliability.

Goodson et al. (1993) investigated the annealing temperature dependence of the thermal conductivity in the direction normal to low-pressure chemical vapor deposited (LPCVD)-silicon dioxide layers. The thermal conductivity increased by 23% due to annealing above 1400 K, which agreed with a model that assumed a dependence of porosity on annealing temperature. The thermal conductivity values obtained for layers of thicknesses between 0.03 and 0.7 μm were much higher than those reported previously for CVD layers, and varied between 50% and 92% of values for bulk fused silicon dioxide. The median time to failure of transistor interconnect contacts, which is limited by electromigration, and the drain current of transistors were both found to increase with decreasing temperature, motivating the use of LPCVD silicon dioxide layers having the largest possible thermal conductivity. But the dependence of the thermal conductivity on the processing parameters, such as the annealing temperature, is not known.

Lai et al. (1993) and Majumdar et al. (1993, 1996) presented a technique for thermal imaging with submicron spatial resolution using the atomic force microscope (AFM) as illustrated in Figure 2.4. The technique was used to simultaneously obtain thermal and topographical images of sample surfaces and was found to be particularly useful for imaging electronic devices and interconnects where there could normally be different materials and voltage variations on a scan area. Observation of a biased MESFET revealed hot spots and the thermal field of the device surface. When applied to a biased polycrystalline Al–Cu interconnect, the thermal images showed heating at the grain boundaries suggesting higher electron-scattering

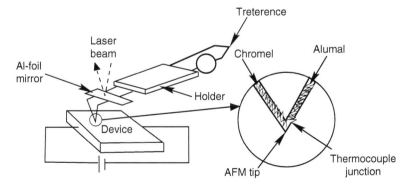

FIGURE 2.4 Schematic diagram of thermal and topographical imaging using the atomic force microscope. (Adapted from Majumdar, A., Carrejo, J.P., and Lai, J., *Appl. Phys. Lett.*, 62, 2501, 1993. With permission.)

rates at the boundaries. The ability to thermally resolve grain boundaries indicated that the spatial resolution is in the submicrometer range. Comparison with a simple analytical model indicated that the experimental technique underpredicted the temperature by approximately 5%. It was suggested that the spatial resolution of this device could be improved to the range of 50 nm.

2.9 MEASUREMENT OF HEAT CONDUCTION IN THE MICROSCALE

Determination of the size-affected thermal conductivity is very important in the design and analysis of thin films and, as a result, it is imperative that the thermal behavior be well understood. The most effective method for experimentally determining the thermal conductivity, defined in the conventional way, by Fourier's law is to measure the heat transfer rate across a known area of cross section, and the temperature difference between two known locations in the direction of heat flow, in order to calculate the thermal conductivity in that direction. In the size-effect domain, the values of thermal conductivity are expected to be highly directional, as the size effect will be different in different directions according to the physical dimensions of the domain in each direction. For instance, in the thin film, we can work with two values, one in the in-plane direction and the other in the out-of-plane direction. These values will have to be measured differently, if possible, or at least one should be estimated using a measured value for the other one, utilizing an appropriate calculation technique for estimation.

Measurement and estimation techniques for thermal conductivity of thin films, both of crystalline (such as silicon) and amorphous (such as silicon dioxide) solids, have been dealt with extensively by Ju and Goodson (1999) in their excellent treatise entitled "Microscale Heat Conduction in Integrated Circuits and Their Constituent Films." Various microscale thermometry techniques have been thoroughly described, with a special emphasis on thermoreflectance thermometry. As a result, a detailed description or analysis is not presented here, but an introduction to microscale measurement of the thermal conductivity is helpful for the sake of completeness. However, an attempt has been made to provide additional aspects wherever possible.

Microscale thermometry can be classified broadly into electrical probe techniques and optical techniques (Ju and Goodson 1999). The probe techniques normally use electrical methods, utilizing the temperature-dependent electrical properties of either an external probe or the specimen itself (which functions as a probe, in this case). Optical methods, on the other hand, are nonintrusive, and utilize electromagnetic radiations to get the information on temperature distributions. Many of the advanced methods, however, are combinations or hybrid methods, which use both electrical as well as optical signals at the same time, in order to collect the data. Some of the fundamental principles of the sophisticated methods are discussed below, while it is suggested that reference books on measurement methods may be consulted for an in-depth knowledge of these techniques and their application methodology.

2.9.1 ELECTRICAL PROBE TECHNIQUES

The selection of the microscale thermometry to be employed in a particular case is a trade-off between the measurement accuracy and the ease and cost involved in its application. Electrical techniques are relatively simple methods to set up and use, though not as sophisticated as optical methods. Moreover, though not used in their basic form, they are the fundamental measurement techniques in many of the more sophisticated measurement systems, which use modifications and combinations of these. The major electrical probe techniques for microscale measurements mentioned in the literature are the electrical resistance thermometry, scanning thermal microscopy, thermometry using electrical parameters as the thermometric properties, and noise thermometry. The basics of these methods, which are described by Ju and Goodson (1999), are briefly outlined here, with additional details in some of the cases.

1. Electrical resistance thermometry:
   This method utilizes the electrical resistance of bridges as the thermometric property. The resistance of a bridge varies with temperature, the phenomenon being due to the temperature dependence of the electron–photon scattering process in metals, and also due to the temperature dependence of the energy carrier concentration in semiconductors (Ju and Goodson 1999). Thus, measurements of the electrical resistance of semiconductors or interconnects can give the temperature levels. Sometimes, the interconnects themselves serve as the thermometer bridge and in certain cases, bridges made from other materials are used (such as polysilicon bridges in high power SOI devices).

   Electrical resistance thermometry provides accurate measurements, and has the advantage that the bridges can be calibrated to give precise values. The major limitation is that the method cannot provide local variations or distributions of temperature, as it gives only spatially averaged values. As voltage signals are used for measurement, which suffer from time limitations of response, the resolution expected in this method in the time domain is only a few hundred nanoseconds (Ju and Goodson 1999).

2. Scanning thermal microscopy:
   The high spatial resolutions provided by the scanning tunneling microscope (STM) and the AFM can be used to obtain the temperature distributions in microscale structures, and to use this information for the estimation of thermal properties. A typical arrangement uses the tip of an AFM replaced by a thermocouple, to set up a scanning probe tip (Majumdar et al. 1993). Thermocouple junctions as small as 100 nm at the tip have been developed. The method provides excellent spatial resolution. However, it also suffers from limitations in the timescale. Further, the temperature drop between the tip and the surface cannot be estimated accurately as the thermal conductance in the gap is influenced by the topological features and can vary during the scan. The capacitive effect between the probe and the surface is another problem that interferes with the test signals and makes the measurement inaccurate, especially while following fast-changing processes. The basic

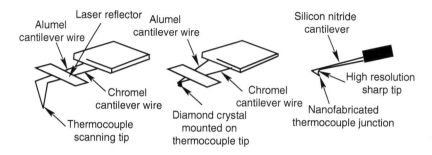

FIGURE 2.5 Evolution of the thermocouple probe design for scanning thermal microscopy. (Adapted from Majumdar, A., Luo, K., Shi, Z., and Varesi, J., *Exp. Heat Transfer*, 8, 83, 1996. With permission.)

arrangement used by Majumdar et al. (1993) is shown in Figure 2.4. The method has been refined further, using single-crystal diamond attached to the thermocouple junction, and a nanofabricated thin-film thermocouple probe in place of the wire thermocouple probe (Majumdar et al. 1996), to evolve better scanning thermal microscopy, as shown in Figure 2.5.

3. Measurement with temperature-sensitive electrical parameters:
   In many cases, the electrical parameters measured from an integrated circuit itself can indicate the temperature level it is subjected to. Utilizing these temperature-sensitive electrical parameters (TSEP) itself could be a thermometric method to extract information on the thermal performance. For instance, the thermal impedance of bipolar transistors has been obtained by measurement of such parameters (Zweidinger et al. 1996, Ju and Goodson 1999). Devices calibrated for the temperature sensitivity of these parameters, using external temperature controllers, can give information on unknown temperature levels in its actual use, using the same parameters as temperature indicators. Often, the voltages across constituent junctions of transistors are used as the appropriate parameters for measurement of temperature in a typical system. Figure 2.6 shows a typical arrangement for estimation of the thermal response, through measurement of the electrical parameters.

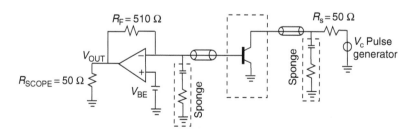

FIGURE 2.6 Setup for thermal transient response measurement of a bipolar transistor. The base current measurement is converted to a temperature response using temperature coefficients obtained on calibration. (Adapted from Zweidinger, D.T., Fox, R.M., Brodsky, J.S., Jung, T., and Lee, S., *IEEE Trans. Electron Dev.*, 43, 342, 1996. With permission.)

Temperature-sensitive electrical parameters thermometry can be applied directly to packaged semiconductor devices. As only the overall parameter value can be used in the measurement, the method gives only an average value of temperature and not the temperature distributions in the device. The typical temporal sensitivity of the method is estimated to be around 1 μs (Ju and Goodson 1999).

4. Noise thermometry:

The voltage noise induced by fluctuations of the current flow in an electrical resistor, which is dependent on the temperature, can be used as an indicator of the temperature of the resistor. The measurement of the thermal voltage noise is used for temperature measurement in noise thermometry. The method needs special devices to monitor voltage noise, and gives only the average temperature (Ju and Goodson 1999).

The thermal voltage noise is given by

$$V_{tv} = (4k_B TR\Delta f)^{0.5} \tag{2.63}$$

where
R is the resistance
k_B is the Boltzmann constant (1.38×10^{-23} J/K)
T is the temperature (K)
Δf is the frequency bandwidth associated with the measurement

The method has been demonstrated by Bunyan et al. (1992) for obtaining the steady-state temperature in a SOI MOSFET, where thermal noise in a polysilicon resistor was used as the temperature indicator. Figure 2.7 gives

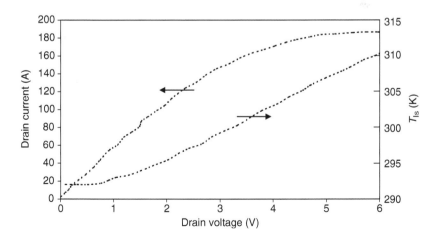

FIGURE 2.7 Typical result from noise thermometry, showing the drain current–drain voltage variation also as a function of the device island temperature. (Adapted from Bunyan, R.J.T., Uren, M.J., Alderman, J.C., and Eccleston, W., *IEEE Electron Dev. Lett.*, 13, 279, 1992. With permission.)

a typical result from noise thermometry, where the variation of the drain current with drain voltage is given also as a function of the temperature of the device. Such a result can be used to obtain the device temperature knowing the current–voltage response.

2.9.2 OPTICAL METHODS

Optical methods of measurement of temperature use optical properties or optical behavior of substances as thermometric properties. These may be the surface properties associated with the incidence of an electromagnetic radiation to the surface, such as reflection of the radiation, or may be the transmission of radiation though the solid body. Optical methods have been suggested and used as thermometric methods for semiconductor layers, dielectric layers or other films. These methods have high spatial and temporal resolutions, and are generally nonintrusive in nature. Some of the optical methods that have been used to characterize the thermal behavior of solids by measurement of temperatures are listed below.

1. Thermoreflectance thermometry:
   The optical reflectance of a surface depends on the temperature, and the thermoreflectance technique for measurement of surface temperature utilizes this. This method is highly recommended for temperature characterization of semiconductor devices, as this gives a temporal resolution as good as, or better than, 1 ns (Eesley 1983). The theoretical limit to the temporal resolution is the relaxation time associated with the scattering processes for the reflectance, which is as low as a few picoseconds for metals (Ju and Goodson 1999). By suitable selection of radiation wavelengths for which passivation layers are transparent, underlying layers can be analyzed. The method also offers good spatial resolution, limited by the diffraction of the beam by the probe, which is comparable to the wavelength of the beam.

   The thermoreflectance technique is a promising method to be applied to metal interconnects, as well as semiconductor regions in SOI devices. It is also suitable for determination of the thermal properties of thin films, including metallic films, such as the thermal conductivity. The major drawback of this method is the interaction of the incident radiation with the semiconductor device. The calibration of the thermoreflectance measurement setup is fairly complicated (Ju and Goodson 1999).

   A very extensive treatment of the thermoreflectance method, with an in-depth treatment of its principle, theoretical background, calibration, measurement procedure, and applications is presented by Ju and Goodson (1999), and therefore, not dealt with here. Figure 2.8 shows the schematic of a thermoreflectance thermometry setup using a probe laser beam. The probe beam is focused by an optical microscope and scanned over the surface using a pair of scanning mirrors.

2. Thermal expansion thermometry:
   The thermal expansion in solids is used as a measure of temperature in this method. Optical interferometry can be used to measure the expansion

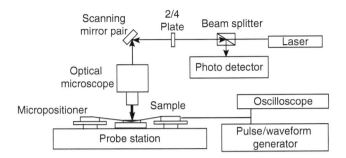

FIGURE 2.8 Schematic of the scanning laser reflectance thermometry facility. (Adapted from Ju, Y.S. and Goodson. K.E., *Microscale Heat Conduction in Integrated Circuits and Their Constituent Films*, Kluwer Academic Publishers, Boston, MA, 1999. With permission.)

accurately. The method has been used in microcircuits and field-effect transistors. Thermal expansion also has been measured using an AFM (Majumdar and Varesi 1997). Thermal expansion thermometry is a good method for overall temperature estimation, but is difficult to use for local temperature distribution measurements (Ju and Goodson 1999).

3. Micro-Raman spectroscopy:

Micro-Raman spectroscopy has been used to measure temperatures in semiconductor devices and quantum well structures effectively at steady state (Brugger and Epperlein 1990, Ostermeir et al. 1992). The method uses highly focused laser beams for measurements. The principle behind the method is the temperature-dependent photon–phonon scattering process in crystals. Inelastic scattering of incident photons by phonons in the crystal brings in a shift in energies of the scattered photons, and the creation or destruction of an optical phonon (Ju and Goodson 1999). As the process is temperature dependent, the number of optical phonons is a function of the temperature the measure of which can be used as an indication of the temperature level, knowing the nature of the incident beam.

A resolution close to 1 μm is possible by this method (Ju and Goodson 1999). Transient measurements are difficult using this method, and also there is the problem of the interference of the incident beam on the behavior of the test specimen.

4. Photothermal deflection method:

This method utilizes the effect of temperature on the refractive index of a gas layer in the vicinity of the sample surface. The deflection in an incident beam due to the gradient of refractive index, called mirage effect (Boccara et al. 1980), in the gas layer can be used to obtain data on the surface temperature. The method has been used with laser beams to measure average temperatures in semiconductor devices (Deboy et al. 1996, Ju and Goodson 1999). The method suffers from the interference of the test beam on the device behavior. Also, the sample preparation procedure is involved, and the calibration process is difficult. The experimental setup

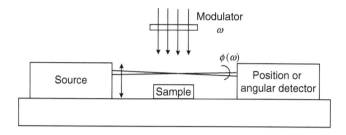

FIGURE 2.9 Experimental setup for thermo-optical thermoscopy using deflection in the incident beam. (Adapted from Boccara, A.C., Fournier, D., and Badoz, J., *Appl. Phys. Lett.*, 36, 130, 1980. With permission.)

used for thermo-optical thermoscopy using the mirage effect, presented by Boccara et al. (1980), is shown in Figure 2.9.
5. Liquid crystal thermometry:
   Liquid crystals exhibit a temperature dependence of color, and this property can be used as a means to determine the temperature itself. This method has been utilized with a spatial resolution of 25 μm and a temperature resolution of 0.1 K (Stephens and Sinnadurai 1974). The apparatus developed by them for the calibration of cholesteric liquid crystals to be used as a thermometer and temperature mapping media is given in Figure 2.10. The method can be successfully used to obtain qualitative information, but

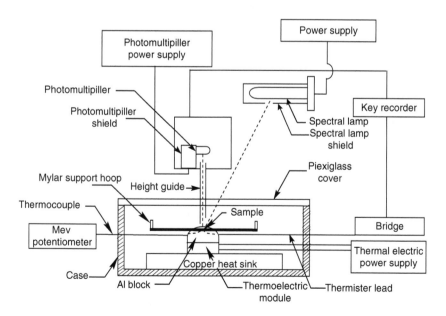

FIGURE 2.10 Calibration set up for cholesteric liquid crystals to be used in temperature mapping. (Adapted from Fergason, J.L., *Appl. Opt.*, 7, 1729, 1968. With permission.)

requires special methods to characterize the output color to obtain quantitative results. Another method of using liquid crystals in thermometry is by utilizing the polarization of the reflected light, due to the temperature-dependent transition of the crystals between the nematic and isotropic phases (Stephens and Sinnadurai 1974), where the nematic state shows polarization effect on reflected light. This method involves coating the regions on surfaces with liquid crystals, and subsequent detection of those in the isotropic and nematic states by observing the polarization effect, which gives information about the temperature levels.

6. Infrared thermometry:

It is well known that the electromagnetic radiation emitted from a solid surface is dependent on its temperature. The total emitted power, the spectrum of emission, and the peak of this spectrum are all dependent on the temperature level. Infrared thermometry utilizes monitoring of the emitted electromagnetic spectrum (thermal spectrum) to obtain information on the temperature of the solid object. An important step in this method is the calibration of the emissivity of the surface involved, as this is dependent on the surface conditions and the wavelength. The spatial resolution obtained by the infrared thermometry is limited by the wavelength of the radiation. The method is useful for qualitative thermal mapping of surfaces, and is used widely for this purpose. Infrared thermometry (or thermography) has been effectively used for semiconductor devices and interconnects (Kondo and Hinode 1995). Thermographic images of typical interconnects and their corresponding temperature distributions, as presented by Kondo and Hinode, are shown in Figure 2.11. The method has also been used for measurement of thermal diffusivity in thin films using pulsed laser heating (De Jesus and Imhof 1995). Transient measurements of thermal diffusivity of thin films also have been undertaken (Graebner 1995), as noted by Ju and Goodson (1999).

7. Photoluminescence spectroscopy:

When a system goes to a higher energy level by absorbing a photon and then comes back to a lower energy level spontaneously, photons are emitted in the process, and this leads to the phenomenon of photoluminescence (Ju and Goodson 1999). The spectrum of the emitted radiation is dependent on the wavelength of the incident radiation as well as the temperature. So, the phenomenon can be used as a measurement method for temperature fields. The method has been utilized for measurement of temperature profiles in diode lasers (Hall et al. 1992) and transistors (Landesman et al. 1998). In the first case, temperature data were inferred by monitoring the peak wavelength of the photoluminescence spectrum, and in the other case the operating temperature was obtained from the shift of this peak. The peak wavelength is found to vary almost linearly with the temperature. Figure 2.12 shows the arrangement of the test setup used by Hall et al. (1992).

The nature of the shift of the peak of the photoluminescence spectrum is further affected by the carrier concentration. So, a careful data analysis is

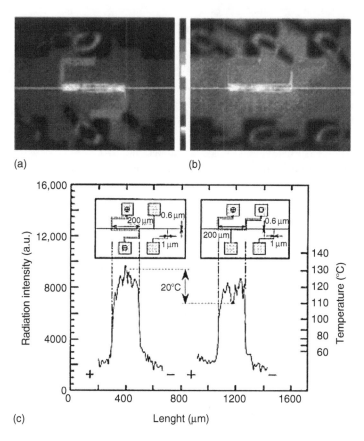

(a)   (b)

(c)

FIGURE 2.11 Infrared thermograph images of typical interconnects. The corresponding temperature profiles are also shown. (From Kondo, S.S. and Hinode, K., *Appl. Phys. Lett.*, 67, 1606, 1995. With permission.)

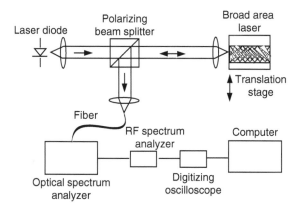

FIGURE 2.12 Setup for the photoluminescence method for lateral temperature profiling. (Adapted from Hall, D.C., Goldberg, L., and Mehyus, D., *Appl. Phys. Lett.*, 61, 384, 1992. With permission.)

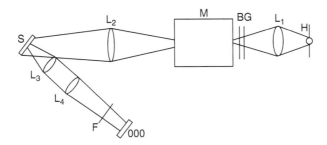

FIGURE 2.13 Temperature measurement setup using fluorescent imaging. H, arc lamp; L, lenses; M, monochromator; BG, blue filter; S, sample; F, fluorescence line filter; CCD, camera. (Adapted from Kolodner, P. and Tyson, J.A., *Appl. Phys. Lett.*, 40, 782, 1982. With permission.)

required to obtain quantitative results using photoluminescence thermometry in microelectronic devices (Ju and Goodson 1999).

8. Fluorescence thermometry:

   Fluorescence in certain materials used in coatings of thin films is found to be temperature dependent, and can be used as an indication of the temperature. One such material is europium-thenoyl-trifluoro-acetonate (Kolodner and Tyson 1982), which emits visible light at around 612 nm when ultraviolet radiation (340–380 nm) is incident on it (Ju and Goodson 1999). The intensity of the fluorescence is temperature dependent and is found to vary exponentially with the film temperature. The method of getting information about the temperature fields in a semiconductor device is comparison of the fluorescence intensity with and without electrical bias. A spatial resolution of 0.7 μm has been obtained using the method. The preparation of the sample for this method is difficult. Also, the fluorescence properties of the film deteriorate with time. The test setup for microscopic fluorescent imaging of surface temperature profiles, as presented by Kolodner and Tyson (1982), is shown in Figure 2.13.

2.10 CONDUCTION IN SEMICONDUCTOR DEVICES

SOI devices have two major films, namely the semiconductor layer and the passivation (dielectric) layer, and interconnects as their constituent parts. The material nature of the semiconductor (silicon) and the dielectric (mostly oxide of silicon) layers is different—silicon being crystalline and the oxide being amorphous (Ju and Goodson 1999). A major thermal issue in integrated chips is the difference in the thermal conductivity among the three constituents listed above, and the resulting differential heat conduction. This induces imbalance in local heat transfer, hot spots, which may affect the structure of the film or spoil it, and thermal stresses, which may disrupt the films and the interconnects. Further, in the size-effect domain, the thermal conductivity is highly directional, depending on the physical dimensions of the domain. This means that the in-plane and out-of-plane thermal conductivities in thin films have to be considered and estimated separately. Conduction heat transfer in dielectric and semiconductor layers have to be treated differently, whether the approach is theoretical or experimental.

The two thermophysical properties to be determined for conduction analysis are the thermal conductivity and the heat capacity. Often, the estimation of the thermal conductivity in the in-plane and out-of plane directions needs knowledge of the heat capacity value, and this makes the prediction of heat capacity very much essential in the task of finding out the appropriate thermal conductivity values to be used in design calculations or mathematical modeling.

Detailed discussions on the analysis and measurement aspects of thermal conductivity in the constituent films in semiconductor devices are presented by Ju and Goodson (1999). Separate and extensive chapters have been devoted to the thermal properties and heat transport in amorphous dielectric films and in silicon films. Only an overall review of the determination of film thermal conductivities, extensively discussed by them, is presented here, and readers may refer to Ju and Goodson (1999) for detailed information on these topics.

2.10.1 Conduction in Dielectric Films

A hybrid experimental–theoretical method for determination of the out-of-plane and in-plane thermal conductivities in dielectric films has been outlined in the literature (Ju and Goodson 1999). The method uses an experimental procedure for measurement of the out-of-plane component, and uses this value, along with a two-dimensional analytical formulation to determine the in-plane component by comparison with theoretical results. The method is described briefly here, referring to the material given by Ju and Goodson (1999).

First, the method to determine the out-of plane thermal conductivity is described. The physical model for this experimental setup is represented by a dielectric film (thickness d_f) on a substrate having a much higher thermal conductivity. The experimental method utilizes a small bridge introduced in the film with periodic (harmonic) heating applied to it. The resulting temperature rise will depend on the frequency, and at the same time be influenced by the thermal conductivity and heat capacity of the film. When the heating frequency is much smaller than the thermal diffusion frequency of the film ($k_f/C_f d_f^2$), the component of the steady-state temperature oscillations that is in phase with the harmonic heating will have two contributions: ΔT_f, from the film, and ΔT_{sub} from the substrate (Ju and Goodson 1999). The substrate contribution ΔT_{sub} is found to vary with the heating frequency (Lee and Cahill 1997, Ju and Goodson 1999) as follows:

$$\Delta T_{sub} = \frac{P'}{\pi k_{sub}} \left[\frac{1}{2} \ln \left(\frac{k_{sub}}{C_{sub} \left(\frac{W}{2} \right)^2} \right) + \eta - \frac{1}{2} \ln (2\omega) \right] \qquad (2.64)$$

The film contribution, ΔT_f, is found to be independent of the frequency (in the low frequency domain), and is given by (Ju and Goodson 1999)

$$\Delta T_{\mathrm{f}} = \psi \frac{P'}{W} \frac{d_{\mathrm{f}}}{k_{\mathrm{f}}} \tag{2.65}$$

where
  W is the width of the metal
  P' is the amplitude of Power dissipation/unit length of metal
  ψ is the factor describing heat spreading within the film
  η is the Joule heating frequency
  ω is the angular frequency

It should be noted that the substrate contribution (Equation 2.64) becomes negligible at high heating frequencies. The film contribution is independent of the frequency, as seen in Equation 2.65. It can also be seen that the film contribution is inversely proportional to the thermal conductivity, but independent of the heat capacity.

The variation of the amplitude of the in-phase component of the temperature oscillations, as obtained from the experimental data, with respect to the heating frequency is as shown in Figure 2.14a. The method of obtaining the thermal conductivity from the experimental data is as follows, referring to Figure 2.14a. The values of the substrate contribution, ΔT_{sub}, corresponding to the frequencies can be computed from Equation 2.64. These values when subtracted from the measured amplitudes of total in-phase temperature oscillations will give the corresponding values of the film contribution ΔT_{f}. It can be noted from Figure 2.14a that the value of ΔT_{f} is independent of the frequency, in the low frequency domain. As the film contribution ΔT_{f} is given by Equation 2.64, the value of the thermal conductivity k_{f} can be calculated readily from the value of ΔT_{f} obtained in the graph. The bulk thermal conductivity and thermal diffusivity corresponding to the substrate temperature can be used in the calculation process for ΔT_{sub} using Equation 2.64.

As mentioned before, the substrate contribution tends to be negligible at high heating frequencies. This means that at high frequencies, the amplitude of the temperature oscillations is almost fully a contribution of the film. This gives a method to determine the heat capacity of the film by comparison of the high frequency data with theoretical solutions of the heat diffusion equation in the film, once the thermal conductivity is calculated using the low frequency data. The values of heat capacities calculated by comparison with three different theoretical models are shown in Figure 2.14b, as given by Ju and Goodson (1999).

The temperature data required for use in the method can be collected using thermoreflectance thermometry.

The method described above can be effectively utilized to determine the out-of-plane thermal conductivity for dielectric films. The procedure to determine the in-plane thermal conductivity is by comparing the experimental result in a two-dimensional specimen with the analytical solution, knowing the value of the out-of-plane thermal conductivity from the experiment (Ju and Goodson 1999). The experimental part for this hybrid method employs microfabricated metal bridges, made such that the width of the bridge is negligible compared to the film width. A two-dimensional structure is approximated, by having the third dimension sufficiently large, so that variations in this dimension are negligible.

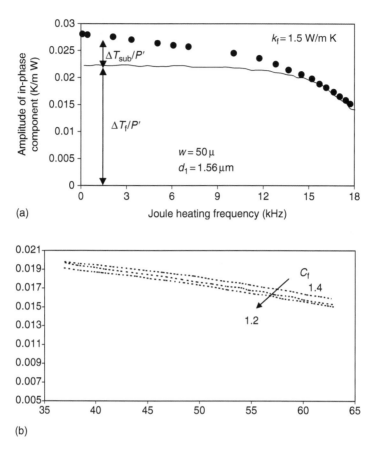

(a)

(b)

FIGURE 2.14 (a) Response of the in-phase component of the temperature oscillations to the frequency of the heating cycle. The plot shows the method for calculation of ΔT_f and subsequent computation of thermal conductivity. (Adapted from Ju, Y.S. and Goodson, K.E., *Microscale Heat Conduction in Integrated Circuits and Their Constituent Films*, Kluwer Academic Publishers, Boston, MA, 1999. With permission.) (b) Heat capacities calculated by comparison with theoretical models. (Adapted from Ju, Y.S. and Goodson, K.E., *Microscale Heat Conduction in Integrated Circuits and Their Constituent Films*, Kluwer Academic Publishers, Boston, MA, 1999. With permission.)

Adopting the continuum theory, the governing differential equation for heat conduction in two dimensions is given by

$$k_{11} \frac{\partial^2 T}{\partial x_1^2} + k_{22} \frac{\partial^2 T}{\partial x_2^2} = \rho c \frac{\partial T}{\partial t} \qquad (2.66)$$

This formulation provides for the anisotropy in the thermal conductivity. In the present case, the two thermal conductivities represent the values in the in-plane and out-of-plane directions. The right side of the equation becomes zero in the steady state, and this condition is used to obtain the experimental data to determine the

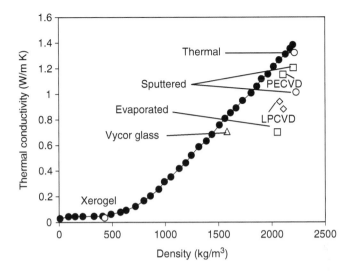

FIGURE 2.15 Comparison of the thermal conductivities of silicon dioxide, obtained from investigations on films manufactured using different processes. The theoretical prediction using effective medium theory is also given as the filled circles. (Adapted from Ju, Y.S. and Goodson, K.E., *Microscale Heat Conduction in Integrated Circuits and Their Constituent Films*, Kluwer Academic Publishers, Boston, MA, 1999. With permission.)

out-of-plane thermal conductivity, as described earlier. A typical method would be to do iterative calculations with assumed values of the in-plane conductivity, changed systematically and the equation solved using a numerical procedure, until the temperature results match measured temperature values in the domain. This procedure can give the converged result for the in-plane conductivity, which brings the theoretical formulation to match with the experimental results. Here also, thermoreflectance thermometry can be used for collecting the temperature data on the specimen.

It is noted that the thermal conductivity of amorphous silicon dioxide depends very much on the process of manufacture of the film. Figure 2.15 shows the measured values obtained by various investigators, corresponding to different processes used, as well as those predicted based on the effective medium theory, as reported in Ju and Goodson (1999).

2.10.2 CONDUCTION IN CRYSTALLINE SUBSTANCES

As seen above, in the case of amorphous dielectric layers, the heat diffusion equation could be used to deduce the in-plane thermal conductivity from the measured out-of-plane value. This was possible because amorphous materials generally exhibit short mean free paths, which permits the use of the continuum assumption. In the case of crystalline materials, the correctness of this approach needs to be verified, as such materials normally have long mean free paths, which means that if a comparison is required with the analytical solution, the test specimen should have sufficiently large dimensions.

Estimation of thermal conductivity based on a fundamental approach, such as phonon transport, may be appropriate in the case of crystalline films. The two phenomena associated with this approach are the scattering of phonons at the boundary and the dispersion of the phonons in the media. The mean free path of phonons in crystalline semiconductors is limited by scattering of phonons with crystal boundaries, impurities, vacancies, and defects. Most of the semiconductors (such as silicon, germanium, and gallium arsenide) have a covalent bond structure. As explained earlier, it has been identified that phonons in crystal structures with covalent bonding can be classified into four branches (or classes) depending on their frequency spectra. There are low frequency phonon branches named acoustic phonons, and high frequency or optical phonons, with a longitudinal and a transverse mode in each (Ju and Goodson 1999). The frequency ranges of these four branches of phonons are represented in Figure 2.16.

The kinetic theory-based model gives an expression for the thermal conductivity as a function of the mean free path. This model has been explained earlier in this chapter. The expression for the thermal conductivity is given below (Ju and Goodson 1999):

$$k = \sum \frac{(Cv_{gr})_i \Lambda_i}{3} \tag{2.67}$$

where the product Cv_{gr} is given by

$$(Cv_{gr})_i = \frac{k_B^4 T^3}{2\hbar^3 \pi^2} \int \frac{1}{v_{ph,i}^2} \frac{x^4 \exp(x)}{[\exp(x) - 1]^2} dx \tag{2.68}$$

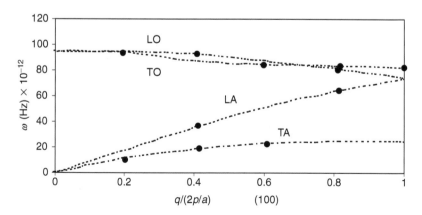

FIGURE 2.16 The four phonon branches for silicon: the transverse acoustic (TA) longitudinal acoustic (LA), transverse optical (TO), and longitudinal optical (LO) branches. (Adapted from Ju, Y.S. and Goodson, K.E., *Microscale Heat Conduction in Integrated Circuits and Their Constituent Films*, Kluwer Academic Publishers, Boston, MA, 1999. With permission.)

$$x = \frac{\hbar\omega}{k_B T} \tag{2.69}$$

where
   Λ_i is mean free path of ith phonon branch
   v_{gr} is the group velocity
   $v_{ph,i}$ is the phase velocity of phonon belonging to ith branch
   C is the heat capacity/unit volume

$$\hbar = h/2\pi$$

The value of Cv_{gr} is the temperature-dependent quantity in the expression for the thermal conductivity. The values differ greatly for the phonon branches mentioned earlier. So, the dispersion of the phonons is important in the calculation of either the thermal conductivity or the mean free path using the formula given in Equation 2.67.

Given the information from the bulk value of the thermal conductivity, Equation 2.67 can be used to find out an approximate value for the mean free path. This can be used as an indicator to decide whether measurements will have a size effect, while using an experimental method. If the physical dimensions of the specimen where measurements are done are comparable in magnitude with the mean free path, errors can be expected in the conductivity, which is computed from a hybrid model using continuum formulation. The phonon boundary scattering would be the primary effect that reduces the thermal conductivity from the bulk value in such a situation. This restricts the experimental measurements to dimensional sizes larger than the mean free path in crystalline solids.

Two other popular and useful theoretical thermophysical property models of crystalline semiconductors, especially at high temperatures, are the relaxation time model for the thermal conductivity and the two-fluid model for the heat capacity (Ju and Goodson 1999). The relaxation time model considers the scattering of phonons by other phonons and by defects in the crystal. Correspondingly, two relaxation times enter the model—τ_{pp} for the phonon–phonon scattering, and τ_I for the phonon-defect scattering. For relaxation times to be used in Equation 2.70, these have to be modeled independently using derived expressions or functions (Ju and Goodson 1999).

$$k = \frac{k_B^4 T^3}{6\hbar^3 \pi^2} \int_0^{x_2} \frac{v_{gr}}{v_{ph}} \frac{x^2 \exp(x)[\exp(x) - 1]^2}{[1/\tau_{pp}(x) + 1/\tau_I(x)]} \, dx \tag{2.70}$$

The two-fluid model for the heat capacity considers the energy to be associated with two modes. One is called the reservoir mode, which describes the energy content and which has no group velocity assigned to the phonons. The other mode is the propagation mode, which describes the transport of energy (Ju and Goodson 1999). With suitable expressions derived for the two modes, the heat capacities can be calculated for the two independent modes. The combined heat capacity can then be used in computing the thermal conductivity using the kinetic theory expression given in Equation 2.67. Detailed discussions on the computation of the heat capacity using the two-fluid model can be found in Ju and Goodson (1999).

2.11 CONCLUDING REMARKS

The developing fundamental insight into the microscale transport phenomena associated with engineering devices has emerged as a critical area requiring extensive investigation. The fundamental economic gains associated with the miniaturization of virtually all processes and devices will ensure that this remains a critically important area in transport phenomena activity. The potential uses of superconductors will also drive the technology toward lower operating temperature regimes in which microscale effects are significant in macroscale devices.

A number of both theoretical and experimental studies have concluded that the thermal conductivity values of thin film materials may be orders of magnitude less than those associated with the bulk material (Stephens and Sinnadurai 1974, Ristau and Ebert 1986, Guenther and McIver 1988, Flik et al. 1991, Brotzen et al. 1992, Goodson et al. 1993). However, in one investigation (Swimm and Hou 1986), the thin film transport parameters for nickel coatings were found to be only slightly less than the corresponding bulk values. Volklein and Kessler (1986) found that the thermal conductivity of bismuth films (20–400 nm thickness) decreased with increasing temperature in the region below 250 K and increased with increasing temperature for mean temperatures greater than 300 K. The physical and chemical nature of a particular thin film material has been found to significantly affect the degree of deviation from bulk material thermal property behavior (Guenther and McIver 1988, Brotzen et al. 1992, Goodson and Flik 1992, Lai et al. 1993). This is of particular significance in the semiconductor industry where chemically similar films may have wide variations in their microstructure, based on their method of deposition. These results indicate a need for greater understanding of the film/substrate interface microstructure (Lambropoulos et al. 1989), which critically defines the contact resistance associated with the substrate/film composite material. In optical elements, the nature of the material microstructure, voids, and cracks in the end use material significantly affect the useful life of a device by defining the parameters that control laser-induced damage. The analysis of each of these areas will require an understanding of the structural influences of transport properties at the molecular and atomic level. Macroscopically, as pointed out by Guenther and McIver (1988), comprehensive models of thin film thermal conductivity must include and account for the anisotropies, inhomogeneities, and temperature-dependent behavior associated with the film material.

The theoretical analysis of microscale conduction has produced significant information. Kumar and Vradis (1991) utilized the Boltzmann equation and found that for film thicknesses on the order of the carrier mean free path, scattering of electrons at the film boundaries was the dominant mechanism in heat transport. This study also indicated that the reduction in electrical and thermal conductivities was virtually identical for many cases of practical importance in thin film technology. It was suggested that further development of the Boltzmann equation analysis could provide beneficial insight into thin film transport property behavior. Flik et al. (1991)

developed theoretical maps defining the macro- and microscale regimes of certain materials. This investigation led to the conclusion that material purity and defect structure characteristics strongly influenced transport properties associated with microscale conduction. Thermal conduction processes were shown to fall within the microscale regime for film layers that were less than approximately 7 times the mean free path of the dominant heat carrier for conduction across a layer and 4.5 times the mean free path for conduction along the layer. Vedavarz et al. (1991) also considered these phenomena and quantified the range for which the ratio of the thermal relaxation time to the characteristic length scale of the imposed thermal conditions would induce non-Fourier conduction behavior in a material. Bai and Lavine (1993) utilized the modified hyperbolic heat conduction equation to consider propagation of heat through a solid. This study concluded that for domain thicknesses on the order of the carrier mean free path, jump boundary conditions significantly affect the solution of the conduction problem and the solution of the hyperbolic heat equation-based analysis was significantly different than that yielded by the Fourier analysis. The authors also concluded that the use of the hyperbolic heat conduction equation in engineering applications is generally justifiable.

The development of experimental methods, which may be used in microscale thermal conduction investigations, represents another critical area for future research. Optical beam deflection techniques (McDonald 1986, McGahan and Cole 1992, Cole and McGahan 1992) have been utilized with some success, though the experimental procedures are extremely sensitive to various disturbances. The use of liquid as a medium for the probe beam and the use of thin film heaters may represent mechanisms that can reduce the experimental complexity of optical beam deflection observation techniques. Use of the AFM as a thermal imaging device (Lai et al. 1993) appears to be quite promising in the area of electronic system and device temperature mapping. One mechanism of improvement in this particular technique may be that of replacing the thermocouple junction with a more accurate temperature-measuring device. A semiconductor diode, thermistor, or quartz crystal may provide more accurate temperature information.

The initial portions of the current chapter were devoted to an introduction into the size-effect domain of conduction heat transfer and a review of the state of the art in microscale conduction heat transfer. Developing from a review of the advances in the area, some key aspects in the fundamentals approaches in microscale conduction were also dealt with in this chapter. The size effect and the failure of the Fourier law-based approach of conduction in the size-effect domain were discussed. The appropriate space and timescales required for the analysis of microscale conduction were introduced. Fundamental theoretical approaches for analysis of size-effect dominated conduction were also discussed. Models suitable for the estimation of thermal conductivity in the size-effect domain have been investigated, and some of the engineering applications of the issue of microscale conduction have been addressed. Conduction analyses and thermal conductivity measurements, along with an estimation of the impact of this variation in semiconductor devices, have been discussed. In addition, the various measurement techniques and theoretical methods suitable for determining the thermal conductivities of thin films have been presented.

REFERENCES

Abraham, E. and J.M. Halley. 1987. Some calculations of temperature profiles in thin films with laser heating. *Journal of Applied Physics A* 42: 279.

Anderson, R.J. 1988. A method to calculate the laser heating of layered structures. *Journal of Applied Physics* 64: 6639–6645.

Anderson, R.J. 1990. The thermal conductivity of rare-earth-transition-metal films as determined by the Wiedemann-Franz law. *Journal of Applied Physics* 67: 6914–6916.

Anthony, T.R., W.F. Banholzer, J.F. Fleischer, L. Wei, P.K. Kuo, R.L. Thomas, and R.W. Pryor. 1990. Thermal diffusivity of isotopically enriched 12c diamond. *Journal of Physical Review B* 42: 1104.

Arpaci, V.S. 1966. *Conduction Heat Transfer*. Reading, MA: Addison Wesley Pub. Co.

Bai, C. and A.S. Lavine. 1993. Thermal boundary conditions for hyperbolic heat conduction. *American Society of Mechanical Engineers Heat Transfer Division* 37: 253.

Boccara, A.C., D. Fournier, and J. Badoz. 1980. Thermo-optical spectroscopy: detection by the mirage effect. *Applied Physics Letters* 36: 130–132

Brotzen, F.R., P.J. Loos, and D.P. Brady. 1992. Thermal conductivity of thin SiO_2 films. *Thin Solid Films* 207: 197–201.

Brugger, H. and P.W. Epperlein. 1990. Mapping of local temperatures on mirrors of gaas/algaas laser diodes. *Applied Physics Letters* 56: 1049–1051

Bunyan, R.J.T., M.J. Uren, J.C. Alderman, and W. Eccleston. 1992. Use of noise thermometry to study the effects of self-heating in submicrometer SiO MOSFETS. *IEEE Electron Device Letters* 13: 279–281.

Burgener, M.L. and R.E. Reedy. 1982. Temperature distributions produced in a two-layer structure by a scanning cw laser or electron beam. *Journal of Applied Physics* 53: 4357–4363.

Cahill, D.G. and R.O. Pohl. 1989. Heat flow and lattice vibrations in glasses. *Solid State Communications* 70: 927–930.

Chen, G. and C.L. Tien. 1993. Thermal conductivities of quantum well structures. *AIAA Journal of Thermophysics and Heat Transfer* 7: 311–318.

Cole, K.D. and W.A. McGahan. 1992. Theory of multilayers heated by laser absorption. *American Society of Mechanical Engineers Dynamic Systems and Control* 40: 267–282.

Crommie, M.F. and A. Zettl. 1990. Thermal conductivity of single-crystal Bi-Sr-Ca-Cu-O. *Physical Review B* 41: 10978–10982.

De Jesus, M.E.P. and R.E. Imhof. 1995. Thermal diffusivity measurement of thermally conducting films on insulating substrates. *Applied Physics A* 60: 613–617.

Deboy, G., G. Solkner, E. Wolfang, and W. Claeys. 1996. Absolute measurement of transient carrier concentration and temperature gradients in power semiconductor devices by internal ir-laser deflection. *Microelectronic Engineering* 31: 299–307.

Decker, D.L., L.G. Koshigoe, and E.J. Ashley. 1984. Thermal properties of optical thin film materials. National Bureau of Standards Special Publication 727. Laser induced damage in optical materials. Proceedings of Boulder Damage Symposium: 291–297.

Eesley, G.L. 1983. Observation of non equilibrium electron heating in copper. *Physical Review Letters* 51: 2140–2143.

Duncan, A.B. and G.P. Peterson. 1994. Review of microscale heat transfer. *Applied Mechanics Review* 47: 397–428.

Fergason, J.L. 1968. Liquid crystals in nondestructive testing. *Applied Optics* 7: 1729–1737.

Flik, M.I., B.I. Choi, and K.E. Goodson. 1991. Heat transfer regimes in microstructures. *American Society of Mechanical Engineers Dynamic Systems and Control* 32: 31–47.

Flik, M.I., B.I. Choi, and K.E. Goodson. 1992. Heat transfer regimes in microstructures. *Journal of Heat Transfer* 115: 666.

Flik, M.I., P.E. Phelan, and C.L. Tien. 1990. Thermal model for the bolometric response of high-tc superconducting films to optical pulses. *Cryogenics* 30: 1118–1128.

Flik, M.I. and C.L. Tien. 1990. Size effect on the thermal conductivity of high-tc thin-film superconductors. *Journal of Heat Transfer* 112: 872–880.

Fushinobu, K. and A. Majumdar. 1993. Heat generation and transport in sub micron semi-conductor devices. *American Society of Mechanical Engineers Heat Transfer Division* 253: 21–28.

Gembarovic, J. and V. Majernik. 1988. Non-Fourier propagation of heat pulses in a finite medium. *International Journal of Heat Mass Transfer* 31: 1073–1080.

Goodson, K.E. and M.I. Flik. 1992. Effect of microscale thermal conduction on the packing limit of silicon on insulator electronic devices. Proceedings of 3rd Intersociety Conference on Thermal Phenomenon, Austin, TX, Feb 5–8.

Goodson, K.E. and M.I. Flik. 1993. Electron and phonon thermal conduction in epitaxial high-T_c superconducting films. *Journal of Heat Transfer* 115: 17–25.

Goodson, K.E. and M.I. Flik. 1994. Prediction and measurement of the thermal conductivity of amorphous dielectric layers. *Journal of Heat Transfer* 116: 317–324.

Goodson, K.E., M.I. Flik, L.T. Su, and D.A. Antoniadis. 1993. Annealing-temperature dependence of the thermal conductivity of CVD silicon dioxide layers. *American Society of Mechanical Engineers Heat Transfer Division* 253: 29–36.

Graebner, J.E. 1995. Measurement of thermal diffusivity by optical excitation and infrared detection of a transient thermal grating. *Review of Scientific Instruments* 66: 3903–3906.

Grigoropoulos, C.P., R.H. Buckholz, and G.A. Domoto. 1988. An experimental study on laser annealing of thin silicon layers. *Journal of Heat Transfer* 110: 416–423.

Grigoropoulos, C.P., W.E. Dutcher, and A.F. Emery. 1991. Experimental and computational analysis of laser melting of thin silicon films. *Journal of Heat Transfer* 113: 21–29.

Guenther, A.H. and H.K. McIver. 1988. The role of thermal conductivity in the pulsed laser damage sensitivity of optical thin films. *Thin Solid Films* 163: 203–214.

Hall, D.C., L. Goldberg, and D. Mehyus. 1992. Technique for lateral temperature profiling in optoelectronic devices using a photoluminescence microscope. *Applied Physics Letters* 61: 384–386.

Holman, J.P. 1989. *Heat Transfer.* New York: Mc Graw Hill.

Iravani, M.V. and H.K. Wickramasinghe. 1985. Scattering matrix approach to thermal wave propagation in layered structures. *Journal of Applied Physics* 58: 122–131.

Ju, Y.S. and K.E. Goodson. 1999. *Microscale Heat Conduction in Integrated Circuits and Their Constituent Films.* Boston, MA: Kluwer Academic Publishers.

Kolodner, P. and J.A. Tyson. 1982. Microscopic fluorescent imaging of surface temperature profiles with 0.01°C resolution. *Applied Physics Letters* 40: 782–784.

Kondo, S.S. and K. Hinode. 1995. High resolution temperature measurements of void dynamics induced by electromigration in aluminium metallization. *Applied Physics Letters* 67: 1606–1608.

Kumar, S. and G.C. Vradis. 1991. Thermal conduction by electrons along thin films, effects of thickness according to Boltzmann transport theory. *American Society of Mechanical Engineers Dynamic Systems and Control* 32: 89–101.

Kumar, S. and K. Mitra. 1999. Microscale aspects of thermal radiation transport and laser applications. *Adv. Heat Transfer* 33: 187–294.

Lai, J., J.P. Carrejo, and A. Majumdar. 1993. Thermal imaging and analysis at sub-micrometer scales using the atomic force microscope. *American Society of Mechanical Engineers Heat Transfer Division* 253: 13–20.

Lambropoulos, J.C., M.R. Jolly, C.A. Amsden et al. 1989. Thermal conductivity of dielectric thin films. *Journal of Applied Physics* 66: 4230–4542.

Landesman, J.P., Depret, A. Fily, J. Nagle, and P. Braun. 1998. Local channel temperature measurements on pseudomorphic high electron mobility transistors by photolumines-cence spectroscopy. *Applied Physics Letters* 71: 1338–1349.

Lavine, A. and C. Bai. 1991. An analysis of heat transfer in Josephson junction devices. *Journal of Heat Transfer* 113: 535–543.

Lax, M. 1977. Temperature rise induced by a laser beam. *Journal of Applied Physics* 48: 3919–3924.

Lee, S.M. and Cahill, D.G. 1997. Heat transport in thin dielectric films. *Journal of Applied Physics* 81: 2590–2595.

Majumdar, A. 1993. Microscale heat conduction in dielectric thin films. *Journal of Heat Transfer* 115: 7–16.

Majumdar, A. and J. Varesi. 1997. Nanoscale thermometry using scanning joule expansion microscopy. Proceedings of American Society of Mechanical Engineers International Mechanical Engineering Congress and Exposition: 171–179.

Majumdar, A., J.P. Carrejo, and J. Lai. 1993. Thermal imaging using the atomic force microscope. *Applied Physics Letters* 62: 2501–2503.

Majumdar, A., K. Luo, Z. Shi, and J. Varesi. 1996. Scanning thermal microscopy at nano-meter scales: A new frontier in experimental heat transfer. *Experimental Heat Transfer* 8: 83–103.

McDonald, F.A. 1986. Photoacoustic, photothermal, and related techniques, a review. *Canadian Journal of Physics* 64: 1023–1029.

McGahan, W.A. and K.D. Cole. 1992. Solutions of the heat conduction equation in multi layers for photothermal deflection experiments. *Journal of Applied Physics* 72: 1362–1373.

McGaughey, A.J.H. 2004. Thermal conductivity decomposition and analysis using molecular dynamics simulations Part I. Lennard-Jones argon. *International Journal of Heat Mass Transfer* 47: 1783.

Nath, P. and K.L. Chopra. 1974. Thermal conductivity of copper films. *Thin Solid Films* 20: 53–63.

Ostermeir, R., K. Brunner, G. Abstreiter, and W. Weber. 1992. Temperature distribution in Si-MOSFETs studied by micro Raman spectroscopy. *IEEE Transactions of Electron Devices* 39: 858–863.

Qiu, T.Q. and C.L. Tien. 1992. Size effect on nonequilibrium laser heating of metal films. *American Society of Mechanical Engineers Dynamic Systems and Control* 40: 227–241.

Redondo, A. and J.G. Beery. 1986. Thermal conductivity of optical coatings. *Journal of Applied Physics* 60: 3882–3885.

Ristau, D. and J. Ebert. 1986. Development of a thermographic laser calorimeter. *Applied Optics* 25: 4571–4578.

Rosencwaig, A., J. Opsal, W.L. Smith, and D.L. Willenborg. 1985. Detection of thermal waves through optical reflectance. *Applied Physical Letters* 46: 1013–1015.

Savvides, N. and H.J. Goldsmid. 1972. Boundary scattering of phonons in silicon crystals at room temperature. *Physical Letters* 41A: 193–194.

Stephens, C.E. and F.N. Sinnadurai. 1974. A surface temperature limit detector using nemantic liquid crystals with an application to microcircuits. *Journal of Physics E* 7: 641–643.

Su, L.T., K.E. Goodson, D.A. Antoniadis, M.I. Flik, and J.E. Chung. 1992. Measurement and modeling of self-heating effects in SOI nMOSFETS. Proceedings of International Conference on Electron Devices, San Francisco. CA.

Swimm, R.T. 1983. Photoacoustic determination of thin-film thermal properties. *Applied Physics Letters* 42: 955–957.

Swimm, R.T. 1987. Basic studies of optical coating thermal properties. NIST Special Publication 756. Laser Induced Damage in Optical Materials: 328–337.

Swimm, R.T. 1990. Thermal properties of optical thin films. SPIE vol. 1441. Laser-induced damage in optical materials. Proceedings of 22nd Annual Boulder Damage Symposium.

Swimm, R.T. and L.J. Hou. 1986. Photothermal measurement of optical coating termal transport properties. NIST Special Publication 752. Laser Induced Damage in Optical Materials: 251–258.

Tamma, K.K. and R.R. Namburu. 1997. Computational approaches with applications to non-classical and classical thermo-mechanical problems. *Applied Mechanics Reviews* 50: 514–551.

Tamma, K.K. and X. Zhou. 1998. Macroscale and microscale thermal transport and thermo-mechanical interactions: some noteworthy perspectives. *Journal of Thermal Stresses* 21: 405–449.

Tien, C.L. and G. Chen. 1994. Challenges in microscale conductive and radiative heat transfer. *Journal of Heat Transfer* 116: 799.

Tien, C.L., B.F. Armaly, and P.S Jagannathan. 1968. Thermal conductivity of thin metallic films. Proceedings of 8th Conference on Thermal Conductivity, Oct. 7–10.

Tye, R.P. 1969. *Thermal Conductivity*, Vols 1 and 2. London: Academic Press.

Vedavarz, A., S. Kumar, and M.K. Moallemi. 1991. Significance on non-Fourier heat waves in microscale conduction. *American Society of Mechanical Engineers Dynamic Systems and Control* 32: 109–121.

Volklein, F. and H. Baltes. 1992. A microstructure for measurement of thermal conductivity of polysilicon thin films. *Journal of Microelectromechanical Systems* 1: 193–196.

Volklein, F. and E. Kessler. 1986. Analysis of the lattice thermal conductivity of thin films by means of a modified Mayadas-Shatzkes model, the case of bismuth films. *Thin Solid Films* 42: 169–181.

Xue, Q. and W.M. Xu. 2005. A model of thermal conductivity of nanofluids with interfacial shells. *Materials Chemistry and Physics* 90: 298.

Zweidinger, D.T., R.M. Fox, J.S. Brodsky, T. Jung, and S. Lee. 1996. Thermal impedance extraction for bipolar transistors. *IEEE Transactions on Electron Devices* 43: 342–346.

3 Fundamentals of Microscale Convection

3.1 INTRODUCTION

Convection is the heat transfer mode associated with the bulk movement of a fluid or medium. It occurs when a fluid is in contact with a solid surface, and when there is a temperature difference between the surface and the fluid. Whether the motion of the fluid particles is due to density differences imposed by a temperature distribution in the fluid and the resulting buoyancy device, as in the case for free or natural convection, or due to an external device such as a pump that forces the fluid past the surface, the energy transport phenomenon in convection is always closely related to the fluid motion. Thus, the analysis of convective heat transfer involves a careful analysis of the fluid flow, its behavior, and characteristics.

A number of theoretical and experimental approaches have been applied to analyze the convective heat transfer problem, in order to develop design guidelines for convective cooling systems. In conventional macroscale convective systems, the theoretical approach invariably consists of the mathematical formulation of the problem using conservation principles for the flow and heat transfer processes and solution of the governing differential equations, which are mathematical statements of these principles. The set of differential conservation equations for mass, momentum, and energy applied for a control volume, coupled with the appropriate boundary conditions, constitutes the most common theoretical description of convective heat transfer. Integral forms of the equations can be used to obtain approximate solutions for the flow and heat transfer fields. Solution of the equations, which govern and describe the fluid flow and heat transfer phenomena using analytical or numerical methods, yields the distributions of the field variables in a fluid continuum, such as the velocity components in the chosen coordinate directions, and temperature distributions and density distributions in cases where these apply. The velocity distribution in the flow field is then used to compute the frictional shear stresses, and the temperature distributions, which in turn can be used to compute the surface heat transfer rates, through a calculation of the field gradients and application of the differential form of Fourier's law. The friction coefficients and Nusselt numbers, which are useful non-dimensional parameters defined for describing the frictional and heat transfer effects, can also be computed and correlated with the flow parameters and fluid properties using such analytical techniques.

Alternatively, experimental approaches depend on the measured flow rates, pressure fields, and temperature fields in the fluid medium. Experimental observations have been used to relate the frictional effects and surface heat transfer, to the flow parameters and thermophysical properties of the fluid through empirical correlations. For complex fluid flow systems, which cannot be simplified in such a way that they can be approximated by an appropriate theoretical analysis, the experimental approach becomes a very powerful tool by which useful correlations and guidelines for design of convective heat transport systems can be obtained.

Theoretical and experimental approaches in convective heat transfer have both resulted in correlations, which are very useful for application to engineering design and analysis of external and internal flow systems. The reader is referred to classical books on heat transfer (Holman 1989, Kays and Crawford 1993) to better understand the analytical and experimental approaches and the associated results in convective heat transfer, especially internal flows, before proceeding to the discussions presented in this chapter.

Convective heat transfer in fluids flowing through channels of very small dimensions was proposed as an effective method of heat dissipation from substrates used in electronic components as early as the 1980s. While high heat transfer rates and larger pressure drops were measured, measurements using the available probes and instrumentation techniques indicated that conventional correlations for convective heat transfer were not adequate to predict the measured frictional and heat transfer data. It was not clear whether the deviations could be attributed to fundamental physical reasons, related to the fluid flow and heat transfer phenomena, or just fabrication and measurement inadequacies. However, the observed variations led to a number of investigations and hypotheses.

For very small fluid channel dimensions such as nanoscale dimensions, analysis based on the assumption of fluid continuum may not be applicable. However, the dimensions which normally designate microchannels in the literature are much larger than these. In general, channels with cross-sectional dimensions of the order of a few hundreds of microns are termed microchannels, and these physical dimensions are still far too large in comparison with the molecular mean free paths of the fluid particles to induce any noncontinuum effects in liquid flow. In gas flow, microchannel dimensions that may produce results that deviate from predictions based on purely continuum analysis can occur, and modifications such as the application of slip flow and temperature jump boundary conditions may become necessary in analyzing these cases.

One explanation suggested by investigators who noticed deviations in the behavior of frictional and heat transfer performance of fluids in microchannels from conventional predictions is the influence of the surface roughness, which becomes increasingly significant as the channel dimensions decrease. A majority of the channels analyzed in this manner have used conventional micromachining methods and, as a result, have resulted in correlations that have practical relevance when similar fabrication methods are employed.

Some of the fundamental aspects of convective heat transfer analysis are presented and discussed in the following sections, and are applicable when small channels are examined. The major findings related to convective heat transfer, with single-phase flow, phase change, and two-phase flow, and gas flow in microchannels and microtubes are reviewed and summarized in the light of the published literature. Current trends in the application of microchannels and fundamental research problems related to these applications are also described at the end of the chapter. Investigations of the heat transfer and fluid flow in microchannels related to different engineering applications are discussed in greater detail in Chapter 4.

3.2 CONVECTIVE HEAT TRANSFER IN MICROTUBES AND CHANNELS

As previously discussed, analysis of most of the practical cases of liquid flow in microchannels utilizes a continuum modeling approach. Recent measurements performed in microchannel flow have indicated that conventional correlations yield satisfactory predictions, if care is taken in the fabrication of channel arrays to avoid flow maldistribution, proper header design is utilized, and the entrance and exit losses are all properly accounted for (Rao and Webb 2000, Kendall et al. 2001). This is particularly true in the case of channels of cross-sectional dimensions of a few hundreds of microns. Typically microchannels developed for electronics cooling applications fall under this category. However, when the physical dimensions are reduced further, there are possibilities that the size effects begin to dominate and modifications are required in the governing equations. Still a clear demarcation is not obtained for the dimensions that would require an altogether different modeling methodology such as discrete computations. Examples of models suggested in analyzing microscale convection to develop appropriate theoretical approaches to predict the deviations and differences observed in experiments include slip flow models for gases, and electric double layer theory in liquids, among many others.

However, special cases would indeed exist where conventional models have to be modified to incorporate noncontinuum effects, such as gas flows in microchannels. The key fundamental concepts in analyzing channel flows, such as basic thermodynamics and fluid dynamics along with the theoretical foundation used in the analysis of convective fluid flows, are briefly discussed here. The importance of size effects in microscale convective heat transfer is also discussed, understanding which helps in suggesting modifications to conventional models to compensate for possible microscale effects.

3.2.1 THERMODYNAMIC CONSIDERATIONS

Though the most accurate analysis of convective heat transfer, like any other energy transport phenomenon, would be that which deals with the energy interactions of each discrete particle in the substance, this method is practically impossible. Application of the laws of mechanics to each particle of the fluid, describing energy transport as manifestations of the motion and potential of each discrete particle, can be applied only to cases with a limited number of particles within the limitations of computation. The continuum approach is the conventionally preferred method of analysis of macroscopic systems, which treats the fluid as a continuous expense of matter. This allows the expression of thermodynamic properties as well as thermophysical properties as mathematical functions of the coordinate system adopted to describe the flowing medium.

The system and control volume approaches can be applied to analyze a fluid medium. The system approach is the most fundamental approach in which the conservation laws are applied directly to the thermodynamic system with boundaries that separate it from the surroundings. In the control volume approach, a specific region in space with a fixed volume is considered, and the properties of the flowing fluid are attributed to spatial locations described by the coordinate system. Though the thermodynamic laws of conservation are essentially stated for a system, these can be translated to governing equations for the control volume by using the Reynolds transport theorem (Arpaci 1966). The governing equations for the control volume are expressed either in differential form or integral form. These equations that describe the conservation of mass, momentum, and energy can be derived using the Reynolds transport theorem, or by independent balances applied to elemental volumes in the fluid space. The development of the differential and integral equations of conservation is available in every classic book on convective heat transfer, to which the reader is advised to refer for details (Kays and Crawford 1993, Bejan 2004).

Fundamental analysis can be based on the motion of the individual discrete fluid particles or groups of particles following the laws of mechanics. The second approach forms the basis of the kinetic theory and statistical mechanics. Various methods such as molecular dynamics modeling, phonon transport modeling, and the Lattice–Boltzmann approach have been suggested under the category of discrete computational approaches. Though these methods become essential in treating physical dimensions comparable to the mean free path of the molecules, continuum modeling, at the most with modified governing equations, is supposed to be adequate for fluid flow and convective heat transfer in microchannels.

3.2.1.1   Continuum Approach

Conventionally, the continuum assumption is applied in solving macroscopic flows. Here, the effects of the molecular structure and arrangement are assumed to be negligibly small upon the flow and heat transfer behavior. This assumption is generally believed to be sufficiently accurate for microchannel flows as well, as for the sizes encountered in microchannel flows, which are typically larger compared to the molecular mean free path. In this way, continuum modeling can be used for analysis of flow situations where the physical dimensions are large compared to the molecular mean free path.

The most established criterion to decide whether the analytical approach to a fluid flow problem could be the continuum approach or a discrete calculation method is the magnitude of the Knudsen number (Zohar 2006). The Knudsen number is defined as the ratio between the molecular mean free path of the fluid under consideration and the characteristic length scale for the flow domain. The Knudsen number is given by

$$\text{Kn} = \frac{\Lambda}{L} \tag{3.1}$$

where the mean free path, Λ, is the average distance a molecule travels between successive collisions, and L represents an appropriate characteristic physical length.

The Knudsen number criterion has been used extensively in rarefied gas dynamics calculations to decide upon the use of governing equations; that is, to decide whether the modeling should be based on the slip flow or no-slip boundary conditions, for instance. The Knudsen number criteria for various flow regimes are given below (Zohar 2006):

0	$< \mathrm{Kn} < 0.001$	Continuum flow
0.001	$< \mathrm{Kn} < 0.1$	Slip flow
0.1	$< \mathrm{Kn} < 10$	Transition flow
Kn	> 10	Free molecular flow

Though the above criteria have been established for gas flow, these could find applications in microscale liquid flow also. It is found that most of the gas flow situations in microchannels (characteristic length scales of the order of a few microns) would produce Knudsen number values less than 0.1. Thus, a continuum assumption, or at the extreme, a slip flow model would be sufficient for the analysis of these cases (Zohar 2006). In the slip flow model, the governing equations used are still continuum equations, but the boundary conditions are modified to compensate for the incomplete interactions of the molecules with the boundaries, imposing a velocity slip-boundary condition for solving the momentum equation, and an equivalent temperature jump boundary condition for the energy equation. Slip flow modeling, which falls under the category of noncontinuum approaches in solving flow problems, will be explained further at a later stage in this chapter.

From the calculation of the Knudsen number for liquid flow in microchannels, where the channel dimensions are small (for instance the flow of water in channels with characteristic dimensions of the order of hundreds of microns), it is found that the values are smaller than 0.001, which means that the continuum assumption is valid. Thus, the Knudsen number can be used as the criterion to assess when the continuum assumption is valid for microchannel fluid flow situations.

3.2.1.2 Conservation Laws and Governing Equations

Energy transport processes are governed by both general and particular laws (Arpaci 1966). In the case of convective flow and heat transfer, the general laws that govern the process are the conservation laws of mass, momentum, and energy, expressed in forms suitable for analysis (integral or differential forms). The law of conservation of mass leads to the continuity equation. The balance of forces and rates of change of momentum yields the momentum equation (the set of momentum equations in the three dimensions constitute the Navier–Stokes equations). Similarly the law of conservation of energy is expressed as the energy equation of the fluid in convective heat transfer. The momentum equation is thus a statement of Newton's second law of motion, and the energy equation is a statement of the first law of thermodynamics. Another general law, which is not apparently visible in the formulation, is the second law of thermodynamics, which appears as the prescription of the heat flow direction from a higher temperature to a lower temperature, and remains within the energy equation written in terms of temperatures. The reader is advised to refer to textbooks in convective heat transfer to familiarize themselves with the governing equations for fluid flow and convective heat transfer (Schlichting 1955, Kays and Crawford 1993).

In the analysis of convective heat transfer, two particular laws, which describe the physical process, also become important (Zohar 2006). Fourier's law of heat conduction is essentially the law governing convective heat transfer also at the fundamental level, as this describes the heat flux associated with a temperature gradient. This is expressed in the differential form to give the heat flux in any direction, in terms of the temperature gradient and the thermal conductivity component in that direction (for an anisotropic medium). The law can generally be stated as follows:

$$q_n'' = -k_n \frac{\partial T}{\partial n} \tag{3.2}$$

A practical law, which governs the convective heat flux from a surface to a fluid, is Newton's law of cooling, which relates the temperature difference and the heat flux, through a heat transfer coefficient. Thus, for a surface with temperature greater than the bulk fluid temperature (hot surface):

$$q'' = h(T_w - T_\infty) \tag{3.3}$$

The governing equations for convective heat transfer, according to the continuum hypothesis, are the conservation of mass equation (continuity equation), the force–rate of change of momentum balance equations (momentum equations), and the energy conservation equation. These are often written in the differential form, applying the concept of a control volume. The differential equations relevant for forced convection heat transfer are given below in general form, applicable to unsteady state (Zohar 2006):

Continuity equation:

$$\frac{D\rho}{Dt} + \rho(\nabla \cdot V) = 0 \tag{3.4}$$

Momentum equations:

$$\rho \frac{DV}{Dt} = (\mu_1 + \mu_2)\nabla(\nabla \cdot V) - \nabla P + \rho B + \mu \nabla^2 V \tag{3.5}$$

Energy equation:

$$\rho C \frac{DT}{Dt} = k\nabla^2 T + \frac{DP}{Dt} + \phi + \theta \tag{3.6}$$

where the viscous dissipation term

$$\phi = 2\mu_1 \left[\left(\frac{\partial u}{\partial x} \right)^2 + \left(\frac{\partial v}{\partial y} \right)^2 + \left(\frac{\partial w}{\partial z} \right)^2 + \frac{1}{2} \left(\frac{\partial u}{\partial y} + \frac{\partial v}{\partial x} \right)^2 \right.$$
$$\left. + \frac{1}{2} \left(\frac{\partial v}{\partial z} + \frac{\partial w}{\partial y} \right)^2 + \frac{1}{2} \left(\frac{\partial u}{\partial z} + \frac{\partial w}{\partial x} \right)^2 \right]$$

Often, a one-dimensional approximation is possible in microchannels, as the cross-sectional dimensions are very small when compared to the length of the channel. In these cases, the one-dimensional simplifications of the conservation equations will be sufficient to analyze the flow and heat transfer with the accuracy required for most applications.

3.2.1.3 Solution in Size-Affected Domains

Scaling of fluid flow and convective heat transfer problems to suit small physical domains introduces effects, characteristic of the domain size. Not all physical phenomena and their respective mathematical models scale uniformly and proportionately to small length scales. Such considerations will lead to identifying whether some of the terms in a mathematical model can be neglected compared to others through an order of magnitude comparison, when microscale systems are analyzed.

Conventional continuum modeling may become inadequate while analyzing physical phenomena pertaining to size-affected domains. Modifications in continuum modeling have been used for analysis, such as the incorporation of slip flow and temperature jump boundary conditions into continuum modeling, and polar mechanics modeling, which deals with viscous fluid systems with rigid particles dispersed in them. Electric double layer theory has been employed to analyze flow problems in ionic fluids. An altogether different approach suggested in size-affected domains is discrete computation, utilizing techniques such as molecular dynamics modeling, Lattice–Boltzmann approach, or phonon transport theory, to name a few.

3.2.2 SINGLE-PHASE FORCED CONVECTION IN MICROCHANNELS

Convection in microchannels evolved as a prominent area of study in microscale heat transfer due to prospective applications in the thermal control of microelectronic devices and components. Especially, integration of microchannels into silicon substrates could provide extremely good results, as there would be no contact resistances involved between the heat sink and the electronic device. Due to miniaturization of integrated circuit devices and microelectromechanical systems (MEMS), the heat removal problem also has become increasingly demanding, so that high heat fluxes have to be handled, often at transient conditions. The two major objectives to be met in thermal management of electronic devices are the overall reduction of peak temperatures and the attainment of a more or less uniform temperature on the substrate. Microchannel heat sinks have been an attractive option to achieve these objectives, leading to optimal designs and new design methodologies supported by the analysis of fluid flow and convective heat transfer in microchannels.

A large number of studies have been undertaken on single-phase flow in microchannels, both experimental and theoretical. The experimental studies involved various geometries of cross section and various fluids—both liquids and gases. Estimation of frictional effects and convective heat transfer, and comparison of their magnitudes with conventional channels were the primary aims of these studies. Attempts were also done to examine whether conventional channel correlations can correctly predict the performance of microchannels. The theoretical studies in microchannels involved continuum analysis incorporating effects particular to small physical sizes and also exploring the possibilities and limitations in utilizing numerical methods for analysis of microchannel flow and heat transfer. Constant and varying heat transfer coefficients along the channels, and developing flow effects were also important considerations in many such investigations.

A typical configuration of a microchannel heat sink with an array of microchannels fabricated into the substrate was shown in Figure 1.7. In that example, the cooling fluid passes through the channels by forced convection, and the surface of the heat sink is attached to the heat dissipating device, producing the heat flow as shown by the thick arrows on the upper surface. Though not integrated into the substrate, such a heat exchanger can be an effective device for removal of heat from very small areas. Separate heat sinks like this have the advantage that they can be fabricated easily on metals such as copper with high thermal conductivity, with relatively simple techniques than what is needed for fabrication on silicon substrates.

Optimal design of miniature heat exchangers such as the one discussed above requires predictive techniques that are fundamentally sound and therefore based on theoretical analyses of microchannel arrays, since any optimization must be based on the geometric factors, such as channel dimensions and spacing, as well as operating parameters such as flow velocities and temperature levels. For this reason, the incorporation of size effects and noncontinuum effects is highly desired in the theoretical analysis, if the situation demands this, such as in the case of high Knudsen number flow of gases in microducts.

Some of the characteristics of single-phase flow of liquids and gases with heat transfer in microchannels will be discussed in the following sections.

3.2.2.1 Flow Regimes and Flow Transition

Conventionally, flow in closed conduits is characterized by laminar and turbulent regimes, with a transition zone in between. The transition from laminar to turbulent flow is typically determined by the flow velocity, the fluid properties (viscosity and density), and the characteristic dimensions of the pipe or channel, such as the hydraulic diameter. Thus, the Reynolds number (fundamentally defined as the ratio of the inertia force to the viscous force) is the conventional nondimensional group that decides whether the flow is laminar or turbulent, or is in the transition regime. Physically, laminar and turbulent flows show differences in the pattern of motion of the fluid particles and the transfer of momentum within the fluid. In the conventional analysis of laminar flow, the differential formulation of the governing equations based on the continuum assumption can be applied. However, to analyze turbulent flow and turbulent convective heat transfer, appropriate models should be used, so as to incorporate the effects of the fluctuating components of the flow velocities and temperatures where applicable (Kays and Crawford 1993).

In conventional internal flows, the Reynolds number, defined based on the hydraulic diameter, is normally used as the criterion for flow transition. This Reynolds number is given by

$$\text{Re} = \frac{VD_h\rho}{\mu} \tag{3.7}$$

In conventional pipe flow, it is well known that the flow ceases to be laminar around a Reynolds number of 2300 (White 2003). Thus, a value 2300 is commonly used as the critical Reynolds number. This implies that once the Reynolds number as described has been calculated, the flow regime can be assumed to be completely understood based on the resulting value. The flow transition is taken as independent of any parameter other than the critical Reynolds number.

Analysis of fully developed internal flow in the laminar regime provides a simple relationship between the friction factor and the Reynolds number, in the form

$$f\,\text{Re} = C \tag{3.8}$$

where C is a constant. The value of C for fully developed flow in the circular pipe is 64 and that for other cross-sectional shapes is different (Kays and Crawford 1993). For developing flow the product of the local friction coefficient and Reynolds number varies with the distance from the entrance of the pipe, and the value can be used to determine the hydrodynamic entrance length (the distance at which the flow becomes fully developed) when it becomes locally invariant (Kays and Crawford 1993).

It should be noted that this relationship is derived based on various simplifications applied to the Navier–Stokes equations, including the boundary layer simplifications which will not be elaborated here. The hyperbolic relationship between the friction coefficient and the Reynolds number given above is valid only for laminar flow, as the frictional characteristics will be drastically different in turbulent flow, where only a thin laminar sublayer accounts for nearly all the viscous effect in the velocity distribution (Schlichting 1955). A plot of the friction factor (f) versus the Reynolds number (Re) is very commonly used to indicate the flow transition, as the trend of the graph will be different in the laminar and turbulent regimes. This is also used to compare the experimental data with theoretical results to study whether there are deviations observed. This will be clear from Figure 3.1, the friction coefficient calculated (from a pressure drop measurement, $\Delta P = \frac{fLV^2}{2gD}$) has been plotted with respect to the Reynolds number in a typical case with a rectangular channel (Peng and Peterson 1996). In the plot given, the appearance of flow transition at a Reynolds number value much lower than what is expected from theory and the experimental observation that the values deviate from conventional prediction are obvious.

Some of the experimental observations for microchannels have indicated that the critical Reynolds number is not a constant in the laminar flow regime, as expected from conventional flow analysis. Instead, it has been observed experimentally that the critical Reynolds number depends on the inlet conditions and the duct geometry, apart from the

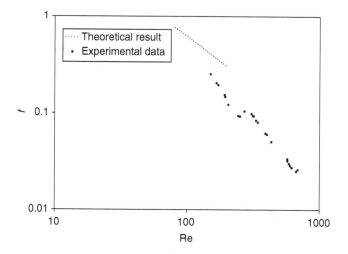

FIGURE 3.1 Experimental plot of f versus Re in microchannel flow indicating the laminar and turbulent regimes. (Adapted from Peng, X.F. and Peterson, G.P., *Int. J. Heat Mass Transfer*, 39, 2599, 1996. With permission.)

surface roughness. There have been observations, which indicated that the critical Reynolds number is further dependent on the hydraulic diameter of the channel, though the definition of the Reynolds number itself has the hydraulic diameter in it (Peng and Peterson 1996). Early studies on microchannel flow have indicated that flow transition occurs in microchannels at much lower Reynolds numbers than expected from conventional theory. However, this cannot be treated as a conclusive result, as many recent observations using sophisticated instrumentation have demonstrated that in microchannels, the results often agree with predictions for conventional channels.

Flow in microchannels is normally laminar, due to the small hydraulic diameters of the channels, which lead to low values of Reynolds numbers. Analysis based on the governing equations for laminar flow will be sufficient where the continuum approach can be applied, and this is the situation encountered in most of the liquid flow problems in microchannels. However, the relatively significant influence of the surface roughness in passages with small cross-sectional dimensions, compared to large channels, and the domination of viscous terms are problems to be addressed in microchannel analysis for design and optimization.

3.2.2.2 Hydrodynamic and Thermal Entry Lengths

The hydrodynamic entry length is the distance from the entrance to a channel to the location where the flow becomes fully developed. Beyond this, the flow is completely in the axial direction (no radial component for the velocity), and the profile of the velocity stays invariant with the axial distance. On the other hand, the thermal entry length in a channel where convective heat transfer takes place is based on the concept of a fully developed nondimensional temperature profile, that is, a nondimensional temperature profile that stays invariant with respect to the axial location (Kays and Crawford 1993) beyond a certain distance from the entrance. The nondimensional temperature mentioned above is given by (Kays and Crawford 1993):

$$\theta = \frac{t_o - t}{t_o - t_m} \tag{3.9}$$

where
  t_o is the surface temperature
  t_m is the integral bulk-mean temperature of the fluid

The concept of the thermal entry length beyond which this nondimensional temperature has an invariant profile is based on an assumption that the heat transfer coefficient is constant over the length. The assumption of the fully developed thermal boundary layer provides a means of utilizing an analogy between the velocity and temperature profiles in solving the heat transfer problem in the channel. Further reading is suggested on fully developed and thermal entry length solutions of convective heat transfer in internal flows for better understanding of the analytical approach (Kays and Crawford 1993).

The magnitudes of the hydrodynamic and thermal entry lengths with respect to the channel hydraulic equivalent diameters have been obtained theoretically, and validated against experimental results. The expressions, which are well established in classical fluid dynamics and heat transfer theory are given below (Zohar 2006):

Hydrodynamic entry length:

$$L_h = 0.056 \, \mathrm{Re} \, D_h \tag{3.10}$$

Thermal entry length:

$$L_t = 0.056 \, \mathrm{Re} \, \mathrm{Pr} \, D_h \tag{3.11}$$

It is important to note that the two entry lengths and the hydrodynamic and thermal boundary layers begin to coincide as the Prandtl number approaches unity. Thus, this provides a fairly good approximation for the flow of air in channels with convective heat transfer.

In most of the practical flow rates, with water and similar liquids flowing in microchannels of usual dimensions (for instance, rectangular channels with sides of the order of a few hundreds of microns), it can be found that the Reynolds number will be normally less than or around 100. The Prandtl number range encountered will be of the order of 1–5. These will often translate into entrance lengths which are less than 5% of the channel length, and could be comfortably ignored, assuming the flow to be fully developed both hydrodynamically and thermally throughout the channel (Zohar 2006).

3.2.3 Nonconventional Analysis Methods

It has previously been established that the continuum approach and the associated methods of solution for the governing equations can be used for analyzing flow and heat transfer of liquids in the more commonly used microchannels. Still there have been situations identified as those, which need special treatment using nonconventional methods of formulation and analysis. The interaction of the physical dimensions, thermophysical properties, and surface conditions of the channel is not well understood, and so is the effect of these on the performance of microchannels fabricated using regular methods and various cooling fluids. Whether the noticed differences from conventional predictions are a result of fundamental reasons or nonstandard fabrication, they have to be quantified and used in optimal design of microchannel heat exchangers. The analysis required for this need not always be generic in nature; it could be system specific, suitable for a particular design and application. In either case, it is required to understand the analysis methods, both conventional and nonconventional, for the problem of microchannel flow and heat transfer.

The conventional methods of analyzing internal flow and heat transfer problems using analytical and numerical methods are well established. Essentially, this involves solution of the governing conservation equations—the momentum and energy equations with the conservation of mass to obtain the velocity and temperature distributions in the flow field. Such an analysis is not discussed here, as it can be found in various fundamental books on the analysis of fluid dynamics and convective heat transfer problems (Schlichting 1955, Kays and Crawford 1993). Some of the approaches applicable to situations, which could deviate from the conventional internal flow problem and requires special treatment as suggested by investigators, are discussed here.

3.2.3.1 Electric Double Layer Theory for Ionic Fluids

Deviations in the flow and convective heat transfer characteristics have been observed for ionic fluids in microchannels, the reason for which is attributed by investigators, to the presence of the electrostatic charge difference between the surface and the fluid (Zohar 2006). A distinct electric double layer is found to exist in the fluid domain, which is of the order of a few nanometers. This is because the electrical charge of the surface attracts the ions of the opposite polarity and forms a static fluid layer near the surface. This effect will be negligible in the macroscale, but would be important in the microscale. The variation of the ionic concentration in the fluid expense will be as shown in Figure 3.2.

The presence of the electrostatic force field and the double layer is found to influence the velocity field and this influence is to be incorporated in the analysis of the flow of ionic fluids in microchannels. This can be done by introducing an additional term which represents the electric body force due to the electric double layer, in the momentum equation. The discussion below is based on the analysis presented by Mala et al. (1997), utilizing this approach.

The modified equation for one-dimensional fully developed laminar flow between parallel plates (Figure 3.3) is given by

$$\mu \frac{\mathrm{d}^2 V_z}{\mathrm{d}X^2} - \frac{\mathrm{d}P}{\mathrm{d}Z} + E_z \rho(x) = 0 \qquad (3.12)$$

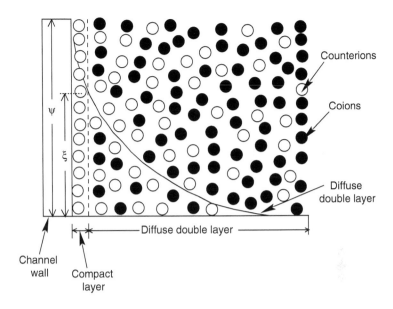

FIGURE 3.2 The electric double layer and the variation of the ionic concentration near the channel wall. (Adapted from Mala, M., Li, D.G., and Dale, J.D., *Int. J. Heat Mass Transfer*, 40, 3079, 1997. With permission.)

where
  X is the direction between the plates
  Z is the flow direction
  E_z is the electrical field strength
  $E_z\rho(x)$ is the electrical body force

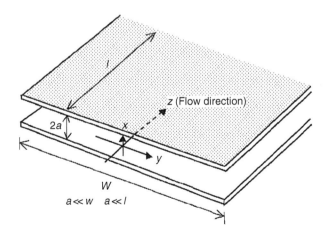

FIGURE 3.3 Microchannel formed by two parallel plates, considering the electric double layer effect. (Adapted from Mala, M., Li, D.G., and Dale, J.D., *Int. J. Heat Mass Transfer*, 40, 3079, 1997. With permission.)

ρ, the charge density, can be written in terms of the second derivative of the electrostatic potential field (ψ) from the following equation for a flat surface (Poisson equation):

$$\frac{d^2\psi}{dX^2} = -\frac{\rho}{\varepsilon_o\varepsilon} \tag{3.13}$$

Here, ε is the dielectric constant of the medium and ε_o is the permittivity of vacuum (8.854×10^{-2} C V^{-1} m^{-1}).

A nondimensional form of ψ is often used, given by

$$\overline{\psi} = \frac{ze\psi}{k_B T}$$

where
  z is the valency of the ions
  e is the electron charge (1.6021×10^{-19} C)
  k_B is the Boltzmann constant (1.3805×10^{-23} J mol^{-1} K^{-1})
  T is the absolute temperature

The distribution of the electric potential in the nondimensional form is given by the Debye–Huckel linear approximation (of an actually exponential distribution) as follows, which is applicable near the wall:*

$$\frac{d^2\overline{\psi}}{d\overline{X}^2} = -\kappa^2\overline{\psi} \tag{3.14}$$

In the above equation, $\overline{X}$ is the nondimensional distance given by $\frac{X}{a}$ where a is the half-channel spacing, and $\kappa = ak$, where "k" is called the Debye–Huckel parameter, given by

$$k = \sqrt{\frac{2n_o z^2 e^2}{\varepsilon\varepsilon_o k_b T}} \tag{3.15}$$

The characteristic thickness of the electric double layer is taken as $\frac{1}{k}$. Here, n_o is the average number of positive or negative ions per unit volume (ionic number concentration).

The zeta potential, ζ, is the electric potential at the interface between the compact layer and the diffuse layer of the EDL, as shown in Figure 3.2. This means that ζ can be used as a boundary condition to find the velocity solution for the mobile fluid domain, as the interface with the compact layer is the actual boundary of the fluid in motion. A dimensionless ζ potential is also defined as $\overline{\zeta} = \frac{ze\zeta}{k_b T}$. A final expression for the z velocity distribution (in the x direction) can be obtained by solving a non-dimensional form of Equation 3.12 as given below:

* Equation 3.14 is a linear approximation of the Poisson–Boltzmann equation $\frac{d^2\psi}{dx^2} = \frac{2n_o ze}{\varepsilon_o\varepsilon}\sinh\left(\frac{ze\psi}{k_b T}\right)$ which in the dimensionless form can be written as $\frac{d^2\overline{\psi}}{d\overline{X}^2} = -\kappa^2\sinh(\overline{\psi})$.

$$\frac{d^2\overline{V}_z}{d\overline{X}^2} + G_1 - \frac{2G_2\overline{E}_s}{\kappa^2}\frac{d^2\psi}{d\overline{X}^2} = 0 \tag{3.16}$$

In this,

$$G_1 = \frac{a^2 P_z}{\mu V_o} \quad \text{and} \quad G_2 = \frac{\zeta n_o \, zea^2}{l\mu V_o}$$

where $P_z = -\frac{dP}{dz}$; $E_s = E_z l$ where E_s is called the streaming potential, and $\overline{E}_s = \frac{E_s}{\zeta}$

$$\overline{V}_z = \frac{V_z}{V_o}$$

where V_o is a reference velocity.

Equation 3.16 is obtained by replacing the term $\rho(x)$ in Equation 3.12 using the equation

$$\frac{d^2\overline{\psi}}{d\overline{X}^2} = -\frac{\kappa^2}{2}\overline{\rho}(\overline{X}) \tag{3.17}$$

which is the nondimensional equivalent of Equation 3.13, where $\overline{\rho}(\overline{X}) = \frac{\rho(X)}{n_o ze}$.

In the process of solution, the boundary conditions are applied at the interface of the two layers (at $\overline{X} = \pm 1$, $\overline{V}_z = 0$, and $\overline{\psi} = \overline{\zeta}$). $\overline{\psi}$ is obtained as the solution of Equation 3.17 with the boundary conditions at the interface (at $\overline{X} = \pm 1$, $\overline{\psi} = \overline{\zeta}$):

$$\overline{\psi} = \frac{\overline{\zeta}}{\sinh(\kappa)}\left|\sinh(\kappa\overline{X})\right| \tag{3.18}$$

Solution of Equation 3.12 yields the dimensionless velocity distribution in terms of the streaming potential and the zeta potential in the dimensionless form as follows:

$$\overline{V}_z = \frac{G_1}{2}\left(1 - \overline{X}^2\right) - \frac{2G_2\overline{E}_s\overline{\zeta}}{\kappa^2}\left\{1 - \left|\frac{\sinh(\kappa\overline{X})}{\sinh\kappa}\right|\right\} \tag{3.19}$$

The streaming potential is the potential of the electric field generated when a liquid is forced through a channel under hydrostatic pressure (Mala et al. 1997). The zeta potential, which is the electrostatic potential at the interface of the electric double layer, can be calculated using the channel height and the volume flow rate of the fluid. The procedures for determining the streaming potential and the zeta potential in a practical flow situation are elaborated in Mala et al. (1997) and will not be discussed here. The velocity distribution has been used in calculating the friction coefficients, as well as in the solution of the energy equation to obtain the Nusselt number by Mala et al. (1997), in their excellent work. The readers are referred to these for a complete analysis of the problem and the procedure discussed above. Variation of the local Nusselt number along the channel length and the average Nusselt number as a function of the Reynolds number obtained from the analysis are shown in Figures 3.4 and 3.5.

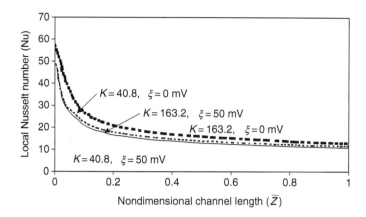

FIGURE 3.4 Variation of the local Nusselt number along the channel. (Adapted from Mala, M., Li, D.G., and Dale, J.D., *Int. J. Heat Mass Transfer*, 40, 3079, 1997. With permission.)

3.2.3.2   Augmented Equations for Micropolar Fluids

The term "micropolar fluids" describes a medium containing randomly dispersed rigid particles suspended in a viscous liquid. The dynamics of micropolar fluids should be analyzed differently from that of a simple fluid, which is governed by the momentum equation, which considers the motion of the fluid due to force imbalance. This is because of the moment distribution, which also determines the dynamics, apart from the force distribution as the rigid particles can rotate, and produce a microrotation vector, which also affects the dynamics, apart from the stress tensor used in classical fluid mechanics. Thus, the governing equations of continuum mechanics get modified, to represent the conservation of moments due to the rotation

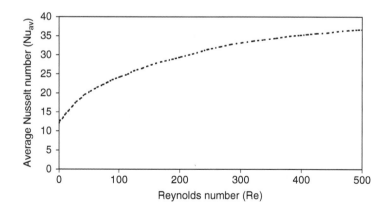

FIGURE 3.5 Variation of the average Nusselt number as a function of the Reynolds number. (Adapted from Mala, M., Li, D.G., and Dale, J.D., *Int. J. Heat Mass Transfer*, 40, 3079, 1997. With permission.)

of the rigid particles, in analyzing micropolar fluids. The modified equations are given below (Zohar 2006):

Momentum conservation:

$$\rho \frac{DU}{Dt} = \nabla \cdot \tau + \rho f \tag{3.20}$$

Angular momentum conservation:

$$\rho I \frac{D\Omega}{Dt} = \nabla \cdot \sigma + \rho g + \tau_x \tag{3.21}$$

Energy conservation:

$$\rho c_p \frac{DT}{Dt} = k \nabla^2 T + \tau : (\nabla U) + \sigma : (\nabla \Omega) - \tau_x \cdot \Omega \tag{3.22}$$

In the above equations, Ω is the microrotation vector, I is the micro-inertia coefficient, f is the body force vector and g is the couple vector per unit mass, τ and σ are the stress and couple-stress tensors. The notations, $\tau : (\nabla U)$ and $\sigma : (\nabla \Omega)$, represent scalar products. It can be shown that the equations become the classical governing equations when the stress tensor, τ, is the classical symmetric stress tensor (Zohar 2006).

3.2.3.3 Slip Flow Models for Gas Flow

As mentioned earlier, the Knudsen number, defined as the ratio of the molecular mean free path and a characteristic length dimension of the flow passage such as the diameter or equivalent diameter, can decide whether the flow deviates from the continuum assumption. A relatively small value of the Knudsen number will indicate that the continuum assumption might be valid, whereas larger values lead to cases where noncontinuum approaches become necessary. In the case of rarefied gas flow, the large values of Knudsen number may result from the rarefaction effect itself, which increases the molecular mean free path, and might lead to noncontinuum mechanics. In the case of microchannels, the same effect could be brought about by very small values of channel sizes and characteristic lengths, such as microscale cross-sectional dimensions of the channel. In the normal microchannel flow of liquids, a calculation of the Knudsen number will point to regimes where the use of continuum assumption would be acceptable, whereas in gas flows in microchannels, it is very likely that the Knudsen number is sufficiently large, leading to possible use of slip flow boundary conditions. In much smaller dimensions, the flow situation may lead to transition or free molecular flow, where altogether different discrete models will have to be used to analyze the flow and heat transfer problems.

In the slip flow regime (normally when $0.001 < \mathrm{Kn} < 0.1$), the modification in the continuum formulation is through the boundary conditions. Modified boundary conditions accounting for a velocity slip (Maxwell) and a temperature jump (Smoluchowsky) are applied (Zohar 2006), which are given below, while the governing equations pertain to the continuum formulation itself.

Velocity slip at the wall:

$$U_s - U_w = \frac{2 - \sigma_u}{\sigma_u} \Lambda \frac{\partial U}{\partial n}\bigg|_w \qquad (3.23)$$

Temperature jump at the wall:

$$T_j - T_w = \frac{2 - \sigma_t}{\sigma_t} \frac{2\gamma}{\gamma + 1} \frac{k}{\mu C_p} \Lambda \frac{\partial T}{\partial n}\bigg|_w \qquad (3.24)$$

In the above expressions, subscript w represents the wall and the gradients are in the direction normal to the wall. Here, γ is the ratio of the specific heats, μ is the dynamic viscosity, and k is the thermal conductivity of the fluid. σ_u, σ_t are called accommodation coefficients (approximately 1 from experiments), which account for the momentum and energy transfer, respectively, between the gas molecules and the solid wall.

It should be noted that both the conditions are written in terms of the mean free path and the gradients of the field variables (velocity and temperature) themselves at the wall.

A number of studies on slip flow modeling of microchannel flows have been reported in the literature. Typical results on the pressure drop along the channel, mass flow rate as a function of pressure drop, and heat transfer results in terms of local Nusselt numbers along the channel in fully developed forced convective flow of gases in microchannels are shown in Figures 3.6 through 3.8a and b. An application of slip flow modeling for the analysis of microfin arrays dissipating heat by natural convection to surrounding air medium is presented in Chapter 4.

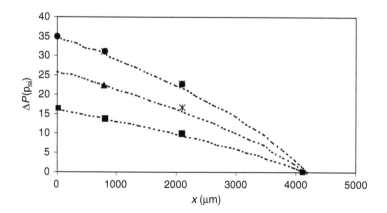

FIGURE 3.6 Comparison between calculated (dashed) and measured (symbols) streamwise pressure distributions. (Adapted from Zohar, Y., *The MEMS Handbook. 2nd ed. Applications*, Mohammed Gad-el Hak (Ed.), Taylor & Francis, Boca Raton, 2006. With permission.)

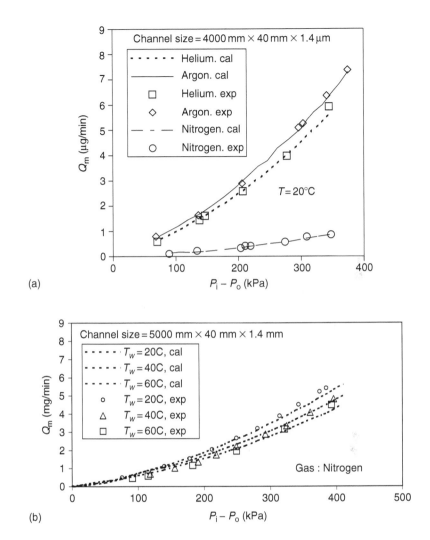

FIGURE 3.7 (a) Slip flow effect on microchannel mass flow rate as a function of the total pressure drop for various working gases, at 20°C. (Adapted from Zohar, Y., *The MEMS Handbook. 2nd ed. Applications*, Mohammed Gad-el Hak (Ed.), Taylor & Francis, Boca Raton, 2006. With permission.) (b) Slip flow effect on mass flow rate as a function of the total pressure drop, for nitrogen gas at various wall temperatures. (Adapted from Zohar, Y., *The MEMS Handbook. 2nd ed. Applications*, Mohammed Gad-el Hak (Ed.), Taylor & Francis, Boca Raton, 2006. With permission.)

3.2.4 BOILING AND TWO-PHASE FLOW

Liquid to vapor phase change and associated heat transfer in flow boiling offer an attractive option, especially in the thermal management of microelectronic devices. Among the desired qualities of optimal cooling systems for microelectronics, one of the most important is the ability to maintain an almost uniform surface temperature for the substrate, avoiding temperature gradients and local heat transfer in the

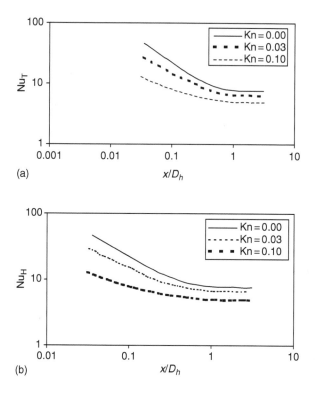

FIGURE 3.8 (a) Numerical simulations of the effect of the inlet Knudsen number on the Nusselt number, along a microchannel for uniform wall temperature (Nu_T) boundary condition. (Adapted from Zohar, Y., *The MEMS Handbook. 2nd ed. Applications*, Mohammed Gad-el Hak (Ed.), Taylor & Francis, Boca Raton, 2006. With permission.) (b) Numerical simulations of the effect of the inlet Knudsen number on the Nusselt number, along a microchannel for uniform heat flux (Nu_H) boundary condition. (Adapted from Zohar, Y., *The MEMS Handbook. 2nd ed. Applications*, Mohammed Gad-el Hak (Ed.), Taylor & Francis, Boca Raton, 2006. With permission.)

solid, which would produce failure due to thermal stresses and other undesirable phenomena. This is as important as minimizing the peak temperatures in the substrate. Utilizing sensible heat transfer in convective cooling has the constraint imposed on the temperature due to the maximum temperature to which the exchange fluid can be heated in the process. Further, single-phase convective heat transfer will heat the fluid progressively while it passes through the channels, and the resulting temperature distribution in the substrate will not be uniform. In the case of flow boiling, however, uniformity in temperature can be obtained because the process is isothermal. The heat transfer rates associated with phase change are high, and thus utilizing phase change and two-phase flow will simultaneously offer two desirable effects namely effective reduction in substrate temperature and achieving temperature uniformity. These make flow boiling an interesting option in micro heat exchangers for electronics cooling. Some of the important features of phase change flow and heat transfer in small channels will now be discussed, with reference to investigations reported in the literature.

3.2.4.1 Two-Phase Flow Patterns

Not much of investigations, either experimental or theoretical, have been reported on basic two-phase flow in microchannels, though some modeling has been done for applications such as micro heat pipes, which involve phase change among other complex heat transfer phenomena (modeling of micro heat pipes is dealt with extensively in Chapter 4). However, some work has been reported on microducts and minichannels, which is discussed here. Phase change and two-phase flow in the microscale is a topic of research which still offers quite a lot of fascination, whether the approach is theoretical modeling or experimental studies using microscale probing or visualization and optical measurements using high-speed photography to capture the effects of bubble nucleation and growth. Some of the modern trends in these topics are also discussed later in Chapter 4.

The two-phase flow patterns in various orientations of conventional conduits have been studied extensively in literature. For instance, the flow patterns for horizontal and vertical orientations of circular pipes are given in Figures 3.9 and 3.10, respectively (Collier 1980).

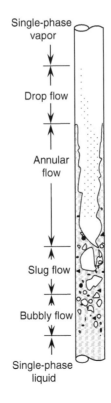

FIGURE 3.9 Two-phase flow pattern for vertical pipe. (Adapted from Collier, G.J., *Convective Boiling and Condensation*, McGraw Hill, New York, 1980. With permission.)

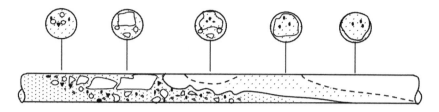

FIGURE 3.10 Two-phase flow pattern for horizontal pipe. (Adapted from Collier, G.J., *Convective Boiling and Condensation*, McGraw Hill, New York, 1980. With permission.)

Two-phase flow patterns have also been identified experimentally in microducts. Flow pattern maps also have been obtained from visualizations and measurements. For instance, flow pattern maps and void fractions have been obtained based on visual observation of two-phase air–water flow in circular and triangular microchannels of hydraulic diameters 1.1 and 1.45 mm by Triplett et al. (1999a,b). The flow pattern consisted of bubbly, churn, slug, slug-annular, and annular flows. The gravity affected influence of the duct orientation on the flow patterns has also been noticed (Stephen 1992).

Flow visualization studies on two-phase flow of air–water mixtures in tubes of circular and rectangular cross sections with small hydraulic diameters have also been reported (Coleman and Garimella 1999). Flow transition observed in the experiments was contrasted to that predicted by correlations for flow in large tubes. It was observed that the tube diameter influences the superficial gas and liquid velocities at which flow transitions take place, due to combined effects of surface tension, the tube hydraulic diameter, and the aspect ratio.

3.2.4.2 Boiling Curve and Critical Heat Flux

The common method of studying the boiling performance in conventional pool boiling or flow boiling situations is through boiling curves. The standard boiling curve is a plot of the heat flux from the heater surface to the fluid, or a boiling heat transfer coefficient, as a function of the temperature excess of the heater above the saturation temperature of the fluid. The boiling curve can be obtained through experimentation with the heater temperature excess or the heat input as the independent variable, and these are explained in books on heat transfer (Holman 1989, Incropera and DeWitt 1990).

However, in the case of phase change in mini and microchannels, investigators have used a plot of the substrate temperature (for instance the average temperature of the substrate, such as an electronic device) with respect to the heat flux dissipated as the boiling curve, as shown in Figure 3.11 (Bowers and Mudawar 1994a). The major reason for this is the difficulty in making local measurements inside channels, as it will intrude with the fluid flow. Local heat transfer coefficients and temperature excesses cannot be easily obtained through measurements in microscale flow systems.

The critical heat flux is the parameter, which decides the maximum heat transfer possible in a phase change heat exchanger. In flow boiling, the application of heat

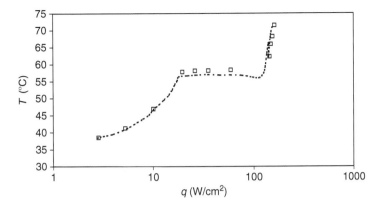

FIGURE 3.11 Boiling curve for R-113 in a microchannel of 510 μm hydraulic diameter. (Adapted from Bowers, M.B. and Mudawar, I., *Int. J. Heat Mass Transfer*, 37, 321, 1994a. With permission.)

fluxes above this value produces dry-out of the flow stream. Thus, at the critical heat flux condition, the phase change phenomenon converts all of the liquid flowing through the channel to vapor. This represents a condition, beyond which the heat transfer coefficient falls drastically and the surface temperature shoots up. The critical heat flux in a microchannel heat sink will have a strong dependence on the temperature of the substrate, represented by its mean temperature, which has been quantified experimentally for various devices (Jiang et al. 1999).

Investigations have been reported in the literature, on flow boiling heat transfer in mini and microchannels, for applications related to thermal management of electronic packages (Bowers and Mudawar 1994a,b,c). Efforts aimed at obtaining and correlating critical heat flux values in flow boiling in minichannels have also been undertaken (Celata et al. 1993, Bowers and Mudawar 1994a,b,c, Roach et al. 1999, Sturgis and Mudawar 1999a,b). The results of some of these investigations are discussed in Chapter 4.

3.2.4.3 Boiling Nucleation and Two-Phase Flow in Microchannels

Phase change heat transfer is recommended as an optimal mechanism for the removal of excess heat from electronic devices and microelectronic systems. The most important advantage of using phase change systems is that flow boiling produces only a modest increase in the device temperature, even at extremely high heat fluxes. While determination of the two-phase liquid/vapor flow patterns is critical for the design of microchannel two-phase flow heat sinks, there exists significant differences in the observations and results as measured by various investigators. Only a very few studies are found in which systematic investigations of the two-phase flow patterns in microchannel flow have been performed.

Because the shape and size of the inlet and outlet plenums of the micro-channels have been shown to influence the stability of the two-phase flow, it is entirely plausible that the observed flow patterns in many of the previously reported investigations are the result of the design of the test facility, in which the vapor phase or the vapor/liquid two-phase is significantly influenced by the entrance or exit region rather than the inherent nature of the microchannel two-phase flow. In an effort to resolve some of the questions surrounding the two-phase flow regimes in microchannels and the influence of the various parameters on these flow regimes, a unique experimental approach was designed to explore the phase change mechanisms and how they affect the flow regimes in microchannels. Using MEMS-based technology, a microscale platinum heater on a Pyrex glass wafer and a shallow, but nearly trapezoidal, microchannel fabricated on another wafer with a hydraulic diameter of 56 μm were fabricated and bonded together (Li and Peterson 2005a).

The two-phase flow patterns as a function of the local vapor generation were explored and recorded with a microscope and a high-speed CCD digital camera. The impact of the size of the microchannel and the flow rate on the incipience of the vapor generation and the two-phase flow patterns were then analyzed. Through a series of experiments, it was found that the instability associated with the phase change is not obvious and the different two-phase flow patterns are distinguished from each other and stable under certain power input and mass flow rates. By comparing the experimental observations with the results of a three-dimensional numerical heat transfer simulation, the boiling incipience temperature was found to be quite close to the theoretically predicted value obtained from the classical kinetics of boiling. A mechanism based on the stability analysis of liquid film was proposed and used to reveal the shape and resulting flow regimes of the vapor column under different flow conditions (Li and Peterson 2005b).

The test section included a Plexiglas support, into which a pressure transmitter and thermocouple are attached, along with the experimental test article, constructed from two Pyrex wafers onto which a microchannel and a small microheater have been fabricated separately and bonded together using epoxy glue after alignment. In this design of the test section, a very long single microchannel with a well-designed smooth entrance region and an outlet that is open to the environment to avoid any influence from the plenum or chamber connected to the outlet of the microchannel as used previously are adapted. The microheater used in the investigation was fabricated using MEMS-based technology and a typical lift-off metallization application. A 200 Å TiW film (for adhesion) and 1500 Å platinum film were sputtered onto a Pyrex glass wafer coated with a photoresist, after the photolithography process. After liftoff in acetone, a similar process was repeated for a gold pad with a thickness of 3000 and 200 Å for TiW. The width and length of the platinum microheater were both 100 μm. The length and width of the gold pad were 10 and 2 mm, respectively, on each side. The test section and the microheater configuration are illustrated in Figure 3.12. The microchannel and the flow guide channel were fabricated using a wet etching technique simultaneously. To accomplish this, the sides of the glass wafer were covered with a 200 Å thick layer of TiW and a 2000 Å thick layer of gold using a metal sputtering technique. After applying the pattern with a photoresist coating and the photolithography process, the wafer was immersed into a $KI:I_2:H_2O$ solution for gold etching and a hydrogen peroxide solution for TiW etching. Then using the metal pattern as the mask, the channels were fabricated by dipping the wafer into a diluted HF solution (concentrated HF

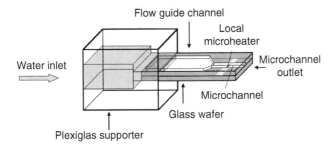

FIGURE 3.12 Test section and microheater configuration for study on boiling nucleation in microchannels. (Adapted from Li, J. and Peterson, G.P., *Int. J. Heat Mass Transfer*, 48, 4797, 2005. With permission.)

(49%): water = 1:1). Finally, the metal coating was removed to expose the glass wafer (Li and Peterson 2005b).

The experimental test facility includes a water circulating unit, a heating unit, a data acquisition unit, and the test section. Deionized, distilled water was pumped from a tank by a micropump through a valve, a 2 μm filter, a flow rate sensor, and a Plexiglas support in which a pressure transducer and a thermocouple have been inserted and finally, through the test specimen. The heating unit utilized a DC power supply and an electronic control circuit, with which the total voltage input supplied to the microheater can be adjusted and controlled, and through which the trigger signal can be produced. This trigger signal activated the data acquisition unit, which comprised of a digital oscilloscope to monitor and record the inlet pressure from the pressure transmitter, a flow sensor to determine the flow rate, and a voltage input transducer to control the microheater. A microscope with the maximum magnification of 500× and a high-speed digital CCD camera with a maximum recording rate of 2000 fps were used to observe and record the flow patterns occurring in the microchannels. The test section included a Plexiglas support to hold the chip and monitor the pressure and temperature, a Petri dish catch vessel to measure the mass flow rate, and the fabricated test specimen. The uncertainties of the flow sensor, the pressure transmitter, and the balance used to determine the flow rate have been determined to be ±1%, ±0.2%, and ±0.01%, respectively. The temperature of the microheater was obtained from the calibration relationship between the temperature and the resistance, applying a linear fit to the calibration data (Li and Peterson 2005b).

The nucleation temperature was found to be reasonably close to the theoretical values as predicted by a three-dimensional numerical heat transfer simulation with the measured bulk temperature of the microheater. The stability of the developed flow indicated three clearly distinguishable two-phase flow regimes: bubbly, wavy, and annular. The observed variations in the two-phase flow patterns were compared with the results of a model developed using a stability analysis of the liquid film.

The principal difference between the microchannel flow patterns observed in this study and the well-accepted patterns observed in macrotubes was that no stratified flow appeared to exist in the microchannels. This was believed to be the result of the

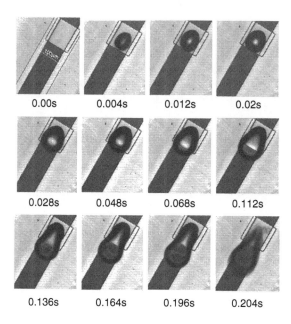

0.00s	0.004s	0.012s	0.02s
0.028s	0.048s	0.068s	0.112s
0.136s	0.164s	0.196s	0.204s

FIGURE 3.13 High-speed visualization photographs showing bubble nucleation, growth, and detachment from the microheater. (Adapted from Li, J. and Peterson, G.P., *Int. J. Heat Mass Transfer*, 48, 4797, 2005. With permission.)

combined effects of the gravitational body force, which is of relatively smaller significance and the surface tension, which plays a much more important role in the determination of the flow regime in microchannels. These two factors coupled with the influence of the shape of the interface make the flow patterns and flow regimes different for these two different situations (Li and Peterson 2005b).

Extensive discussions on the observed flow regimes, and the analysis and inferences drawn on the nucleation temperature and the two-phase flow mechanism from the experimental observations, have been presented by Li and Peterson (2005b). A typical high-speed photograph of the visualization study on nucleation and detachment of the bubble is shown in Figure 3.13.

3.2.5 CONDENSATION IN MICROCHANNELS

Condensation process can effectively be utilized as a final heat dissipation method in a liquid loop, where heat is ultimately transferred to the surroundings when a vapor condenses inside a channel. Microchannels, as in sensible heat transfer and in boiling, can be expected to provide effective heat dissipation in the condensation mode also. Due to the difficulties in measurement and flow visualization, only limited work has been reported in this area. Some of the work reported in microscale condensation phenomenon is in connection with specific applications such as refrigerating systems. In general, condensation in microchannels is a challenging and unexplored area in research and implementation at present.

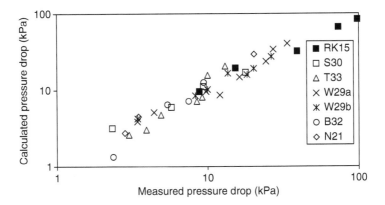

FIGURE 3.14 Comparison of pressure drop predicted by the model with experimental values for noncircular channels. (Adapted from Garimella, S., *Heat Transfer Eng.*, 25, 104, 2004. With permission.)

Garimella (2004) has presented an overview of experimental investigations of condensation flow patterns, pressure drop, and heat transfer in mini- and microchannels for various cross-sectional shapes and hydraulic diameters. Flow visualization results were presented and discussed to provide an understanding of the flow mechanisms. These studies indicated that transitions between the various flow regimes in microchannels occur at conditions different than those expected in conventional geometries. Experimentally validated models for the condensation pressure drop were also presented in this paper. Garimella et al. (2005) combined experimental results on pressure drop measurements during the condensation of refrigerant R134a in horizontal microchannels with existing pressure drop models, and proposed a comprehensive model that addresses the entire progression of the condensation process from the vapor phase to the liquid phase. Overlap and transition between the flow regimes were also addressed using an appropriate interpolation technique. A comparison of the calculated and measured pressure drops for the noncircular channels is shown in Figure 3.14.

Du and Zhao (2003) performed a theoretical analysis of the film condensation process in a vertical microtube with a thin metal wire attached to its inner surface. A two-dimensional analysis was performed to obtain the radial and axial distributions of the condensate liquid along the tube wall and in the meniscus zone. The governing equations were developed based on the principle of minimum energy in the two-phase flow system. The influence of the contact angle and the wire diameter on the condensate distribution and the heat transfer characteristics were studied. The system under analysis was found to provide better condensation heat transfer performance compared to a simple round tube. Analytical investigations have also been reported on condensation in small channels with porous boundaries. Riehl et al. (2002), in their study, analyzed the meniscus formation in condensation, and observed high values of the Nusselt number in channels of size 1.5 mm.

This analysis pertained to microchannel condensers where the pumping is provided by capillary action.

Wu and Cheng (2005) reported visualization and measurements of condensing flow of steam through silicon microchannels. Flow patterns such as fully droplet flow, annular, injection, and slug flow were all observed. It was noted that due to the bubble growth and detachment, termed as vapor injection flow, the two-phase and single-phase flows appeared alternately in the microchannels causing large fluctuations in the temperature and flow measurements. Visualization results for vapor injection flow and the associated instabilities were reported for the first time by them.

3.3 CONCLUDING REMARKS

Convection is the mode of heat transfer that finds the largest number of applications in the microscale. In most of the semiconductor-related applications, convection is sought after as a method of thermal management. Various engineering designs have been developed and proposed, which incorporate microchannels into substrates to form effective heat sinks for microelectronics. For the design of such systems, a knowledge of convective heat transfer mechanisms associated with single-phase sensible cooling, phase change, and multiphase flow is required. As the possibility of differences in flow and heat transfer in microchannels compared to conventional large channels is still under debate, and the physical conditions which would lead to such differences not clearly known, results of in-situ experimentation on similar systems become extremely helpful in designing new systems.

The objective of this chapter was to introduce the fundamental aspects of convective heat transport, which are significant while considering microscale heat transfer, along with a few modified methods of analysis proposed in the literature for microscale phenomena. Some of the significant observations from investigations were also discussed. Extensive review of research and development in microscale convective heat transfer with specific engineering applications are dealt with in Chapter 4.

REFERENCES

Arpaci, V.S. 1966. *Conduction Heat Transfer*. Reading, MA: Addison Wesley Pub. Co.
Bejan, A. 2004. *Convective Heat Transfer*. 3rd ed. New York: John Wiley & Sons.
Bowers, M.B. and I. Mudawar. 1994a. High flux boiling in low flow rate, low pressure drop mini-channel and micro-channel heat sinks. *International Journal of Heat and Mass Transfer* 37: 321–332.
Bowers, M.B. and I. Mudawar. 1994b. Two-phase electronic cooling using mini-channel and micro-channel heat sinks. Part 1: Design criteria and heat diffusion constraints. *ASME Journal of Electronic Packaging* 116: 290–297.
Bowers, M.B. and I. Mudawar. 1994c. Two-phase electronic cooling using mini-channel and micro-channel heat sinks. Part 2: Flow rate and pressure drop constraints. *ASME Journal of Electronic Packaging* 116: 298–305.

Celata, G.P., M. Cumo, and A. Mariani. 1993. Burnout in highly subcooled water flow boiling in small diameter tubes. *International Journal of Heat and Mass Transfer* 36: 1269–1285.

Coleman, J.W. and S. Garimella. 1999. Characterization of two-phase flow patterns in small diameter round and rectangular tubes. *International Journal of Heat and Mass Transfer* 42: 2869–2881.

Collier, G.J. 1980. *Convective Boiling and Condensation.* 2nd ed. New York: McGraw Hill.

Du, X.Z. and T.S. Zhao. 2003. Analysis of film condensation heat transfer inside a vertical micro tube with consideration of the meniscus draining effect. *International Journal of Heat and Mass Transfer* 46: 4669–4679.

Garimella, S. 2004. Condensation flow mechanisms in microchannels: Basis for pressure drop and heat transfer models. *Heat Transfer Engineering* 25: 104–116.

Garimella, S., A. Agarwal, and J.D. Killion. 2005. Condensation pressure drop in circular microchannels. *Heat Transfer Engineering* 26: 28–35.

Holman, J.P. 1989. *Heat Transfer. SI Metric Edition.* New York: McGraw Hill.

Incropera, F.P. and D.P. DeWitt. 1990. *Fundamentals of Heat and Mass Transfer.* New York: John Wiley & Sons.

Jiang, L., M. Wong, and Y. Zohar. 1999. Phase change in microchannel heat sinks with integrated temperature sensors. *Journal of Microelectromechanical Systems* 8: 358–365.

Kays, W.M. and M.E. Crawford. 1993. *Convective Heat and Mass Transfer.* New York: McGraw Hill.

Kendall, G.E., P. Griffith, A.E. Bergles, and J.H. Lienhard. 2001. Small diameter effects on internal flow boiling. *Proceedings of 2001 ASME International Mechanical Engineering Congress and Exposition.* New York, NY.

Li, J. and G.P. Peterson. 2005a. Geometric optimization of a micro heat sink with liquid flow. *IEEE Transactions on Components and Packaging Technologies* 29: 145–154.

Li, J. and G.P. Peterson. 2005b. Boiling nucleation and two-phase flow patterns in forced liquid flow in microchannels. *International Journal of Heat and Mass Transfer* 48: 4797–4810.

Mala, M., D.G. Li, and J.D. Dale. 1997. Heat transfer and fluid flow in microchannels. *International Journal of Heat and Mass Transfer* 40: 3079–3088.

Peng, X.F. and G.P. Peterson. 1996. Convective heat transfer and flow friction for water flow in microchannel structures. *International Journal of Heat and Mass Transfer* 39: 2599–2608.

Rao, P. and R.L. Webb. 2000. Effect of flow mal-distribution in parallel micro-channels. *Proceedings of 34th National Heat Transfer Conference ASME*: 1–9.

Riehl, R.R., J.M. Ochterbeck, and P. Seleghim, Jr. 2002. Effects of condensation in micro-channels with a porous boundary: Analytical investigation on heat transfer and meniscus shape. *Journal of the Brazilian Society of Mechanical Sciences* 24: 186–193.

Roach, G.M. Jr., S.I. Abdel-Khalik, S.M. Ghiaasiaan, M.F. Dowling, and S.M. Jeter. 1999. Low-flow critical heat flux in heated microchannels. *Nuclear Science and Engineering* 131: 411–425.

Schlichting, H. 1955. *Boundary Layer Theory.* New York: McGraw Hill Edition.

Stephen, K. 1992. *Heat Transfer in Condensation and Boiling.* New York: Springer Verlag.

Sturgis, J.C. and I. Mudawar. 1999a. Critical heat flux in a long, rectangular channel subjected to one-sided heating—I. Flow visualization. *International Journal of Heat and Mass Transfer* 42: 1835–1847.

Sturgis, J.C. and I. Mudawar. 1999b. Critical heat flux in a long, rectangular channel subjected to one-sided heating—II. Analysis of critical heat flux data. *International Journal of Heat and Mass Transfer* 42: 1849–1862.

Triplett, K.A., S.M. Ghiaasiaan, S.I. Abdel-Khalik, and D.L. Sadowski. 1999a. Gas–liquid two-phase flow in microchannels. Part I: Two-phase flow patterns. *International Journal of Multiphase Flow* 25: 377–394.

Triplett, K.A., S.M. Ghiaasiaan, S.I. Abdel-Khalik, A. LeMouel, and B.N. McCord. 1999b. Gas–liquid two-phase flow in microchannels. Part II: Void fraction and pressure drop. *International Journal of Multiphase Flow* 25: 395–410.

White, F.M. 2003. *Fluid Mechanics. 5th Revised Edition.* New York: McGraw-Hill.

Wu, H.Y. and P. Cheng. 2005. Condensation flow patterns in silicon microchannels. *International Journal of Heat and Mass Transfer* 48: 2186–2197.

Zohar, Y. 2006. Microchannel heat sinks. In Mohammed Gad-el Hak (Ed.), *The MEMS Handbook. 2nd ed. Applications.* 12:1–12:29. Boca Raton: Taylor & Francis.

4 Engineering Applications of Microscale Convective Heat Transfer

4.1 INTRODUCTION

The application of fluid flow in microchannels for thermal management of electronic components and devices has seen steady growth since the initial introduction of the idea nearly 25 years ago. The quest to understand the behavior of the flow and heat transfer in very small channels and tubes and to utilize this understanding to optimize the design of microscale heat sinks and heat exchangers has inspired a large number of investigations, the results of which are extremely useful for the development of this relatively new technology. Theoretical and experimental investigations, utilizing powerful computational techniques and sophisticated instrumentation, have been reported over the past few years, and have helped investigators make a number of interesting observations. While helpful and informative, a great deal of more information is necessary before definite and conclusive quantification of microchannel flow and heat transfer can be developed.

4.2 RESEARCH AND DEVELOPMENT

The microchannel literature published prior to the end of the last century has been reviewed by a number of authors, several of which summarized the investigations over the last decade of the century, pertaining to the various types of microchannel flow and heat transfer applications for differing geometries and substrate-working fluid combinations (Tuckerman and Pease 1981, 1982, Peng et al. 1994a,b, Peng and Peterson 1996b). As the state of the art has progressed considerably in the last few years, it is also important to review and contrast the many important publications and findings of the most recent work in this area. The discussions presented in this chapter are clearly demarked into research activities and results published in the early days of microchannel research and those reported in the recent times.

4.2.1 EARLY INVESTIGATIONS ON MICROSCALE CONVECTIVE HEAT TRANSFER

A review summarizing the published literature in microscale convective heat transfer, since the inception of the concept of using microchannels for heat dissipation from electronics and other miniature systems, until the end of the last century, is presented in detail in the sections to follow. Investigations on single-phase, two-phase, and gas flows in microchannels are described here in this chapter.

4.2.1.1 Single-Phase Liquid Flow

Since the introduction of the use of fluid flow in microchannels as an effective means of heat dissipation from electronic components in the early 1980s, by Tuckerman and Pease (1981, 1982), a number of investigations, both experimental and theoretical, have been undertaken on microchannel flow and heat transfer. The focus of these was the attempt to find out whether there are differences in the heat transfer characteristics of coolants in microchannels, compared to conventional channels. It was identified as a requirement, to quantify experimental findings and to propose predictive correlations, if differences existed in microscale heat transfer. Though the inadequacies in instrumentation limited the scope of measurements in the microscale, serious attempts were made by investigators, utilizing the available probes and techniques. Theoretical analyses were also performed to interpret the experimental findings on the basis of the fundamental physical phenomena. Later investigations with more sophisticated instrumentation including those using MEMS probes, as well as modeling based on discrete computational methods, helped in refining the early experimental findings, sometimes producing contradictory results rendering invalid many of the hypotheses; however, the contributions from early investigations in the later advances are definitely noteworthy. Some of the key investigations are discussed in this section.

Experimental investigations have been reported using various fluids in microchannels, but a large number of the systems studied focused on the use of water as the cooling fluid. Observed deviations in the heat transfer performance compared to conventional channels have been discussed, and correlations proposed based on experimental investigations. Through a series of experimental investigations, Peng et al. (1994a,b) attempted to bring out such deviations in water flowing through rectangular microchannels. A range of hydraulic diameters (0.133–0.367 mm) and aspect ratios (0.33:1) were used on stainless steel substrates. The friction factor—Reynolds number graphs were found to deviate from the conventional shapes. In obtaining correlations, the friction factor was found to be proportional to $\mathrm{Re}^{-1.98}$ in laminar flow and $\mathrm{Re}^{-1.72}$ in turbulent flow, compared to Re^{-1} and $\mathrm{Re}^{-0.25}$ for conventional channels. The hydraulic diameter and the aspect ratio were found to greatly influence the flow. An important observation was that flow transition from laminar to turbulent flow occurred in microchannels at much lower Reynolds number values (in the range 200–700) than those expected in conventional channel flows (around 2300 based on the hydraulic diameter). The critical Reynolds number was further found to be an increasing function of the channel hydraulic diameter. The trend is shown in Figure 4.1. The transition Reynolds number was identified from the friction factor—Reynolds number plots. From the corresponding heat transfer characteristics, the Nusselt number values were correlated as follows (Peng and Peterson 1996b):

$$\mathrm{Nu} = 0.1165 \left(\frac{D_{\mathrm{h}}}{W_{\mathrm{c}}}\right)^{0.81} \left(\frac{H}{W}\right)^{-0.79} \mathrm{Re}^{0.62}\mathrm{Pr}^{1/3} \text{ for laminar flow} \qquad (4.1)$$

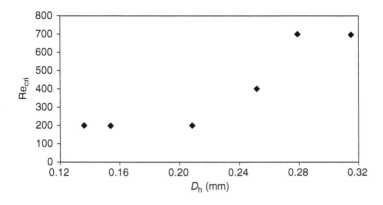

FIGURE 4.1 Variation of the critical Reynolds number with the hydraulic diameter in a rectangular channel. The critical Reynolds number is found to increase with the hydraulic diameter. (Adapted from Peng, X.F., Peterson, G.P., and Wang, B.X., *Exp. Heat Transfer, 7*, 249, 1994a. With permission.)

$$Nu = 0.072 \left(\frac{D_h}{W_c} \right)^{1.15}$$

$$\times \left[1 - 2.421(Z - 0.5)^2 \right] Re^{0.8} Pr^{1/3} \text{ for fully developed turbulent flow}$$

$$(4.2)$$

In the above equations, D_h is the hydraulic mean diameter, W is the microchannel width, and W_c is the pitch of the microchannels. Z is a dimensionless variable which brings in the effect of the aspect ratio, for which an optimum value was obtained as 0.5, giving maximum heat transfer rates.

Among various other observations, an anomalous reduction in the Nusselt number with an increase of Reynolds number had been observed by investigators in certain microchannel experiments (Wang and Peng 1994, Peng and Peterson 1995, 1996b). Tso and Mahulikar (1998, 1999) put forward the explanation for this behavior by relating the convective heat transfer to the Brinkman number, thus bringing in the influence of the temperature dependence of viscosity also into the picture. Experiments on turbulent flow of water and methanol in stainless steel channels with various widths and pitches indicated that the turbulence onset was strongly influenced by the liquid temperature and the microchannel size, apart from the liquid velocity (Wang and Peng 1994). This experimental investigation was used to modify the constant in the Dittus–Boelter correlation from 0.023 to 0.00805 for the microchannel configurations. Figure 4.2 shows a typical experimental result with methanol as the coolant. Further discussions on the effect of thermophysical properties on convective heat transfer in microchannels have been reported by Peng and Peterson (1995) which brings out the influence of the liquid temperature on the flow transition. Lower transition Reynolds number values compared to conventional large channels have also been noticed by various other investigators. In their study of

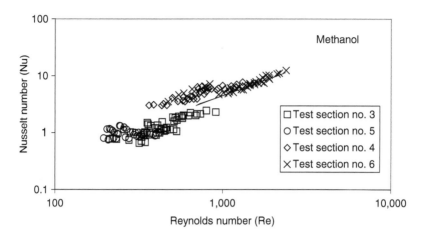

FIGURE 4.2 Variation of the average Nusselt number with Reynolds number, for single-phase forced convection flow of methanol in a rectangular microchannel. The reduction in Nusselt number with an increase in the Reynolds number may be noticed in the low Reynolds number regime. (Adapted from Wang, B.X. and Peng, X.F., *Int. J. Heat Mass Transfer,* 37, 73, 1994. With permission.)

water flow in deep rectangular microchannels (with a width of 251 μm and a depth of 1030 μm), a critical Reynolds number of 1500 was identified by Harms et al. (1997) for turbulence onset. However, they attributed this to the inlet manifold conditions and channel roughness, rather than any fundamental difference in the flow structure.

Higher values of turbulent Nusselt numbers than those predicted by conventional correlations were observed by Adams et al. (1998) in experiments with water in circular microchannels of 0.76 and 1.09 mm diameter fabricated in a copper substrate. The enhancement was found to increase with an increase of the channel diameter, and a decrease in the Reynolds number. A correlation was proposed with a scatter for the experimental data within $\pm18.6\%$. This correlation included additional data from small diameter tubes from the literature.

$$Nu = Nu_{Gn}(1 + F) \qquad (4.3)$$

In the above expression, F is an enhancement factor (which represents the increase in the experimental values compared to predicted values), and is given in graphical form with respect to the Reynolds number in Figure 4.3 for three channel diameters. Nu_{Gn} is the Nusselt number calculated as proposed by Gnielinski (1976) as follows:

$$Nu_{Gn} = \frac{(f/8)(Re - 1000)Pr}{1 + 12.7(f/8)^{1/2}(Pr^{2/3} - 1)} \qquad (4.4)$$

The friction factor "f" in this expression is given by Filonenko (1954).

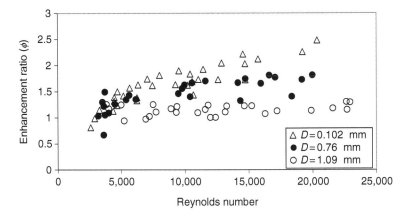

FIGURE 4.3 Variation of the "enhancement ratio" for the experimental Nusselt number over the Gnielinski correlation with Reynolds number. (Adapted from Adams, T.M., Abdel-Khalik, S.I., Jeter, S.M., and Qureshi, Z.H., *Int. J. Heat Mass Transfer*, 41, 851, 1998. With permission.)

$$f = [1.82 \log{(\text{Re})} - 1.64]^{-2} \qquad (4.5)$$

The correlation covers a Reynolds number range from 2,600 to 23,000 and a Prandtl number range of 1.53–6.43.

The experiments mentioned above were extended to a noncircular microchannel of hydraulic diameter 1.13 mm (Adams et al. 1999). Here, the Gnielinski correlation was found to predict the Nusselt number well. On the basis of this observation, a hydraulic diameter of approximately 1.2 mm was suggested as a lower limit for the applicability of conventional correlations for turbulent heat transfer.

Mixtures of water and methanol have been investigated, as cooling fluids in microchannels to examine the influence of the volume fraction. Experiments have been reported by Peng and Peterson (1996a) on channels with hydraulic diameter ranging from 0.133 to 0.367 mm. The general trend of lower transition Reynolds number in microchannels was observed in these cases also, giving a range of 200–700 where the flow ceased to be laminar. Influence of the hydraulic diameter (the transition Reynolds number decreasing with the hydraulic diameter) was also observed. The influence of the channel hydraulic diameter and aspect ratio was found to be significant in deciding the heat transfer performance. The results indicated that the influence of the aspect ratio on heat transfer varied as a function of the mole fraction of the mixture, smaller mole fractions of the more volatile component giving larger heat transfer coefficients. The observation is shown in Figure 4.4.

Theoretical and numerical investigations have also been reported at the early stages of the development of microscale heat transfer, with a view to explain the deviations observed experimentally, and to suggest optimal designs of microchannels. A model based on electric double layer (EDL) at the interface between the solid surface and liquid was presented by Mala et al. (1997a), for microchannels formed

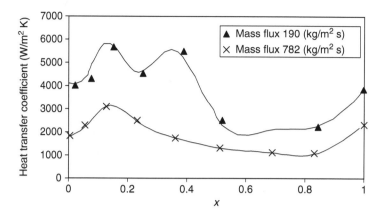

FIGURE 4.4 The effect of the concentration of the more volatile component, and the mass flux of a binary mixture of water and methanol, on the heat transfer coefficient, in single-phase forced convection flow through a rectangular microchannel. (Adapted from Peng, X.F. and Peterson, G.P., *Exp. Thermal Fluid Sci.*, 12, 98, 1996a. With permission.)

between two parallel plates, as explained in Chapter 3. The interaction of the EDL field on the flow and heat transfer was studied treating the channel size also as a parameter. The EDL field was obtained by a linear approximate solution of the Poisson–Boltzmann equation, together with a numerical solution of the modified energy equation to incorporate the effect of the EDL. The results compared well with the results from experimental investigations on the flow of aqueous solutions of potassium chloride through the microchannels (Mala et al. 1997b). This investigation was extended to the EDL field in rectangular microchannels by Yang et al. (1998). For aqueous solutions of low ionic concentration and a solid surface of high electrical potential, the liquid flow and heat transfer in the microchannels were found to be significantly influenced by the electric field and the induced electro kinetic flow, as shown in Figure 4.5.

A thermal resistance model was used by Kleiner et al. (1995) to conduct an optimization study of heat exchangers with cooling fins and microchannels formed between them. Measurements were also performed and the thermal resistances were found to agree with the predictions. A new concept of a "two-fluid model" was proposed by Takamatsu et al. (1997) for the flow of superfluid helium in capillary channels of small diameters. This model assumed that the dynamics of helium flow can be represented by two components—a superfluid component without entropy and viscosity, and a normal fluid component possessing the entropy and viscosity, and that the two-fluid components flow through each other without friction. The model indicated an optimum channel diameter for the maximum mass flow rate.

An optimization study of a flat plate micro heat exchanger has been presented by Bau (1998), where the axial variation of the cross-sectional area has been optimized for minimum thermal resistance. The assumption here was a fully developed velocity profile for the incompressible fluid flow. Lee and Vafai (1999) have presented a comparison of the relative performance of jet impingement and microchannel

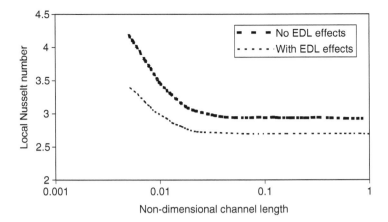

FIGURE 4.5 Comparison of the local Nusselt number variation along the channel length, from computational solutions with and without EDL effect incorporated in the formulation. (Adapted from Yang, C., Li, D., and Masliyah, J.H., *Int. J. Heat Mass Transfer,* 41, 4229, 1998. With permission.)

cooling for high heat flux applications. A procedure for optimizing the channel geometry was discussed by them. It was observed that microchannel cooling was a preferable option for targets smaller than 7×7 cm while jet impingement was comparable or better than this method for larger dimensions of the target surface.

Due to the special emphasis on the thermal control of electronic equipment and the possibility of integrating flow channels to substrates, a large number of the early investigations dealt with single-phase flow of common coolants such as water flowing in rectangular microchannels. Experimental investigations, though with the limitations imposed by the available instrumentation, always pointed toward the deviations of the flow and heat transfer characteristics from conventional predictions. However, sophisticated instrumentation techniques that followed later have now proved that in most of the microchannel flow situations, conventional predictions are applicable. Deviations and variations observed in early experiments have been attributed to errors in measurements and handicaps in fabrication techniques. Modern trends in measurements and more interesting results in microscale flows will be discussed extensively in later sections dealing with recent investigations.

4.2.1.2   Gas Flow in Microchannels

The major criterion in determining whether gas flow in a channel requires special consideration in its analysis is the Knudsen number. The Knudsen number compares the mean free path of the gas molecules with a characteristic dimension, such as the channel diameter, and decides whether conventional treatment is applicable. When the channel dimensions approach the mean free path dimensions, slip flow and temperature jump boundary conditions are recommended to be used in modeling the system mathematically. Rarefied conditions may lead to such modeling even in conventional channels, and in the case of microchannels, even normal working pressures can produce large values of the Knudsen number (as the physical dimensions of the channel are small), which will require a similar treatment. Theoretical investigations using noncontinuum modifications of the governing momentum and energy equations, introducing boundary conditions such as slip flow, have been tried to analyze gas flow through microchannels. Besides, various experimental investigations to characterize the effects of the channel dimensions and operational parameters on friction and heat transfer in gas flows through microchannels have also been performed.

Experimental results on the rate of helium flow in rectangular microchannels have been compared with theoretical predictions based on the solution of the Navier–Stokes equations with no-slip and slip-boundary conditions in an interesting work by Arkilic et al. (1994, 1997). The inference was that the no-slip boundary condition underestimated the mass flow while the model with the slip-boundary condition predicted it well, as shown in Figure 4.6. This observation indicates that microscale gas flow requires an approach different from conventional modeling. The details of the theoretical model and the experimental set up are given by Arkilic et al. (1997).

Experimentation leading to correlations has been performed by Yu et al. (1995) on nitrogen flow in microtubes of diameters 19, 52, and 102 μm, in Reynolds number ranges of 250–20,000. Comparisons were also performed with theoretical predictions for conventional channels. In the laminar as well as the turbulent regime, the flow friction was overpredicted by the conventional model. Correlations were obtained based on the experimental data, following the same forms as in conventional channels:

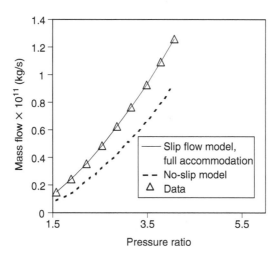

FIGURE 4.6 Variation of the mass flow rate of helium with the pressure ratio for a microchannel of depth 1.33 μm, compared with prediction from models with and without the assumption of slip-boundary conditions. (Adapted from Arkilic, E.B., Schmidt, M.A., and Breuer, K.S., *IEEE J. Microelectromechanical Systems,* 6, 167, 1997. With permission.)

$$f = 50.13/\text{Re} \quad (\text{Re} < 2000) \qquad (4.6)$$

$$f = 0.302/\text{Re}^{0.25} \quad (6{,}000 < \text{Re} < 20{,}000) \qquad (4.7)$$

An interesting observation from this experimental study was that heat transfer in microchannel flow was larger, and the Nusselt number much higher than conventional prediction, in the turbulent flow regime. The following correlation was proposed for the Nusselt number:

$$\text{Nu} = 0.007\text{Re}^{1.2}\text{Pr}^{0.2} \quad (6{,}000 < \text{Re} < 20{,}000) \qquad (4.8)$$

Slip flow analysis of gas flow in microchannels has been presented by Kavehpour et al. (1997). In this work, compressible momentum and energy equations were solved using finite differences with velocity slip and temperature jump boundary conditions expressed in terms of the Knudsen number. The results were found to compare well with experimental results (Pfahler et al. 1991, Arkilic et al. 1997) for nitrogen and helium. The Nusselt number and the friction coefficient were found to be substantially reduced for slip flows compared to continuum flows. The effect of compressibility was important at high Reynolds numbers, and the effect of rarefaction was significant for lower Reynolds numbers in supersonic flow, leading to a reduced wall heat flux. Another slip flow analysis has been presented by Niu (1999), for two-dimensional microchannels and three-dimensional straight and spiral grooves, which find applications in turbo molecular pumps. The nonlinear nature of the pressure variation obtained was attributed to the density variations in the flow. Slip-boundary conditions along with compressibility effects were considered in the numerical investigation of nitrogen and helium flow in microchannels with large width to height ratios, assuming a two-dimensional flow, by Chen et al. (1998). The numerical results compared well with experimental data. Small velocities and high-pressure gradients due to large wall shear stresses were noticed as characteristics of microscale gas flow. The two-dimensional axial and radial momentum equations and the energy equation pertaining to gas flow in a smooth microtube were solved by Guo and Wu (1997), using a finite difference forward marching technique. The formulation involved the Mach number as a parameter, and an ideal gas assumption. The friction coefficients and Nusselt numbers were calculated. The friction factor–Reynolds number product was not a constant, but depended further on the Reynolds number. This as well as the dependence of the local Nusselt number, which increased along the length of the tube, were understood to be the effects of the compressibility, and the absence of fully developed flow due to this.

For the analysis of two-dimensional microchannel gas flows, a different approach was adopted by Mavriplis et al. (1997). They applied the direct simulation Monte Carlo (DSMC) technique to analyze rarefied flows in a range of continuum to transitional regime of the Knudsen number. The method was used to simulate supersonic, subsonic, and pressure driven low speed flows. The channel wall heat flux was found to be inversely dependent on the Knudsen number. Further, the heat flux was found to depend on the channel length. The effects were explained as results of shock-boundary interactions, and the broadening of the shock when the Knudsen number and the channel length increased.

The foregoing discussions indicate that most of the research on gas flows in microchannels focused on the deviations due to velocity slip and temperature jump boundary conditions through modified continuum modeling and associated experimental verification. The major geometric parameters investigated have been the tube diameters and channel aspect ratios. Experimental investigations have attempted to investigate the phenomenon in terms of the flow rates, Reynolds numbers, pressure ratios, and channel dimensions as the major parameters.

4.2.1.3 Phase Change and Two-Phase Flow

The idea of utilizing the high heat transfer capability of phase change phenomena in the cooling of substrates has been of great importance ever since microscale heat transport became the focus of attention of investigators. The high latent heat values of common cooling fluids such as water, added to the possibility of integrating channels on to substrates, made phase transition in microchannels an attractive option. Investigations on these lines were reported to understand the flow characteristics and to quantify and correlate the findings. Theoretical investigations also were performed. Some of the early investigations are discussed below.

Experimental investigations to understand the influence of liquid velocity, subcooling, property variations, and microchannel configuration (width, pitch, and number of channels) on the heat transfer behavior in flow boiling in rectangular microchannels have been performed by Peng et al. (1995). Experiments on nucleate flow boiling indicated that the liquid velocity and subcooling did not affect fully developed nucleate boiling, but greater subcooling increased the velocity and suppressed the initiation of flow boiling. The boiling characteristics of subcooled water flowing in rectangular microchannels of cross section 0.6×0.7 mm, machined into stainless steel, were experimentally studied by Peng and Wang (1993). The wall superheat required for flow boiling in microchannels was found to be much smaller, compared to conventional channels for an identical wall heat flux. The transition between single-phase convection and nucleate boiling was observed to be characterized by the absence of partial nucleate boiling. However, some of these observations were found to be the results of the incapability of instrumentation at the time when the experimental research was undertaken. Peng et al. (1996) also reported experimental results on the flow boiling of binary mixtures of water and methanol in microchannels studying eight rectangular microchannel configurations, each of length 50 mm, and the hydraulic diameter varying from 0.15 to 0.267 mm. Nine different mixture mole fractions were used in the experiments. The liquid subcooling ranged from 38°C to 82°C, and the liquid velocity varied from 0.1 to 4.0 m/s. Boiling curves were plotted and boiling heat transfer coefficients were also presented. It was observed that the heat transfer coefficients at the onset of flow boiling and in partial nucleate boiling were greatly influenced by the liquid concentration, microchannel and plate configuration, flow velocity and subcooling, while these parameters had little effect in the fully nucleate boiling regime. Mixtures with small concentrations of methanol gave larger heat transfer compared to those with large methanol concentrations. The experimental investigation was further extended to flow boiling in V-shaped microchannels (Peng et al. 1998). The channels analyzed had hydraulic diameters in the range 0.2–0.6 mm and groove angles from 30° to 60°. An optimum combination of these parameters was found to exist, resulting in a maximum possible flow boiling heat transfer.

Bowers and Mudawar (1994a,b,c) reported extensive work in microscale flow boiling and associated two-phase flow for application in the cooling of electronic equipment. Mini channels (diameter 2.54 mm) and microchannels (diameter 0.51 mm) were used in heat sinks, with R 113 as the coolant. Critical heat flux (CHF) values of the order of 200 W/cm^2 were obtained for a range of inlet subcooling from 10°C to 32°C, for flow rates less than 95 mL/min. A heat sink thickness- to-channel diameter ratio of 1.2 was found to provide a good compromise between minimizing the overall thermal resistance and obtaining structural integrity; a value for this ratio of less than two provided negligible surface temperature gradients even at heat fluxes of 200 W/cm^2 (Bowers and Mudawar 1994a). A pressure drop model, which accounted for the pressure drop in the single-phase inlet region, the single- and two-phase heated region, and the two-phase unheated outlet region, was found to predict the experimental results with good accuracy. Figure 4.7 shows a comparison of the theoretical predictions of pressure drop with the experimental data presented by Bowers and Mudawar (1994c). The effect of the acceleration of the fluid due to evaporation, the compressibility effect in high heat flux applications when the Mach number exceeded 0.2, and the effect of the channel erosion due to flow boiling in mini and microchannels were all discussed, based on the investigations. Channel erosion in the mini channel geometry was found to be at acceptable levels but the erosion effects in the microchannel at high heat fluxes exceeded allowable limits.

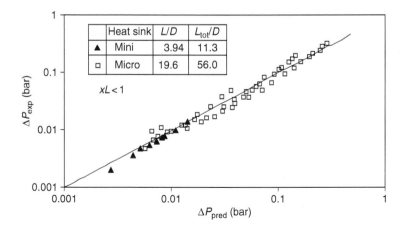

FIGURE 4.7 Comparison of theoretical predictions of pressure drop in mini and microchannels with experimental data. (Adapted from Bowers, M.B. and Mudawar, I., *Int. J. Heat Mass Transfer*, 37, 321, 1994c. With permission.)

A single correlation for the CHF was developed from experimental results for both the mini and microchannels (Bowers and Mudawar 1994c) as follows:

$$\frac{q_{m,p}}{Gh_{fg}} = 0.16 \text{ We}^{-0.19} \left(\frac{L}{D}\right)^{-0.54}$$

(4.9)

where

$q_{m,p}$ is the CHF
G is the mass velocity
h_{fg} is the latent heat of vaporization
We is the Weber number
L and D are the length and hydraulic mean diameter of the channel

In an attempt to analyze the conditions related to CHF in mini channels, Sturgis and Mudawar (1999a,b) studied flow boiling of FC-72 in long rectangular channels of cross section 5.0×2.5 mm. Flow visualization investigations indicated the coalescence of vapor bubbles at the CHF. A series of wavy vapor patches, which propagated along the heated wall, was observed, which allowed the liquid to contact the wall only at discrete locations. Further, the length and height of the vapor patch increased along the flow direction, but decreased with increasing subcooling and velocity (Sturgis and Mudawar 1999a). An analysis to predict CHF was also presented (Sturgis and Mudawar 1999b), modeling the periodic distribution of the vapor patches along the surface by a sinusoidal interface with amplitude and wavelength increasing in the flow direction. The model predicted the near-saturated CHF data from the experiments in long channels within a mean absolute error of 10%. Experimental investigations on the CHF associated with the flow of subcooled water in heated microchannels have been presented by Roach et al. (1999). Experimental investigations on subcooled flow boiling of water in stainless steel tubes of diameter 2.5 mm and wall thickness 0.25 mm, under the application of very high heat fluxes of the order of 60 MW/m^2, have been performed by Celata et al. (1993). Fluid pressures ranging from 0.1 to 2.5 MPa and velocities from 10 to 40 m/s were used. The experimental data on CHF were compared with existing correlations and theoretical models, and the possibility of modifying some of the correlations to give acceptable predictions of the high heat flux data was discussed.

Two-phase flow of refrigerant 124 in a microchannel heat exchanger was experimentally investigated by Cuta et al. (1996). The heat exchanger consisted of 54 microchannels of rectangular cross section (1 mm deep by 0.27 mm wide) with a hydraulic diameter of 425 μm. The experiments were carried out in the Reynolds number range of 100–750 at a uniform heat flux of up to 40 W/cm^2 with wall superheat ranging from 0°C to 65°C. Experimental investigations were also performed for single-phase flow. The Nusselt number (ranging from 5 to 12) increased with an increasing Reynolds number in single-phase flow, and appeared to be approximately constant (around 20) in two-phase flow. Ravigururajan (1998) compared the two-phase heat transfer performance of two designs of microchannel geometry in a copper substrate, using refrigerant 124 as the cooling fluid. The microchannels were of two patterns (parallel and diamond), and had a hydraulic diameter of 0.425 mm. The influences of the wall superheat and the exit vapor quality on the heat transfer coefficient were investigated. For the diamond pattern, the heat transfer coefficient was found to decrease from 12,000 to 9,000 W/m^2 K corresponding to an increase in wall superheat from 10°C to 80°C. An increase in exit vapor quality from 0.01 to 0.65 produced a decrease of 30% in the heat transfer coefficient. The heat exchanger with the diamond pattern yielded heat transfer coefficients that were 20% lower at corresponding conditions of inlet subcooling and wall superheat. Mertz et al. (1996) have reported the results of experimental investigations on flow boiling in narrow rectangular channels of 1–3 mm width and aspect ratios of up to 3, fabricated in copper, used as heat exchanger elements. Saturated flow boiling in a vertical orientation was studied, with water and R-141b as the working fluids. Boiling curves and the variations of the heat transfer coefficient with local and average heat fluxes were obtained.

Coleman and Garimella (1999) performed flow visualization investigations on two-phase flow of air–water mixtures in tubes of circular and rectangular cross sections with small hydraulic diameters. Flow patterns were observed and flow regime maps obtained for superficial velocities in the range 0.1–100 m/s for the gas and 0.01–10 m/s for the liquid. Flow transition observed in the experiments was contrasted to that predicted by correlations for flow in large tubes. It was concluded that the tube diameter influences the superficial gas and liquid velocities at which flow transitions take place, due to combined effects of surface tension, the tube hydraulic diameter, and the aspect ratio.

An experimental investigation of two-phase air–water flow in circular and triangular microchannels of hydraulic diameters 1.1 and 1.45 mm was conducted by Triplett et al. (1999a,b). Flow pattern maps and void fractions were obtained based on visual observation which indicated bubbly, churn, slug, slug-annular, and annular flows. Comparison of the data with flow regime transition models and correlations did not show good agreement, indicating deviations in microchannels (Triplett et al. 1999a). The models and correlations overpredicted the measured channel void fraction and pressure drop in annular flow, suggesting that in this regime the liquid–gas interfacial momentum transfer and wall friction in microchannels may be significantly different from those in larger channels (Triplett et al. 1999b).

Ma et al. (1994) presented a mathematical model for the liquid pressure drop in liquid–vapor flow in triangular microgrooves. The model considered the interfacial shear stresses due to liquid–vapor frictional interaction. The channel angles ranging from 20° to 60° and contact angles (angle between the liquid–vapor meniscus and the groove wall) from 0° to 60° were used. A dimensionless vapor–liquid interface flow number was introduced to incorporate interfacial shear stress, and used as a parameter. The predicted friction factor–Reynolds number products were found to be strongly dependent on the channel angle, the contact angle, and the dimensionless interface flow number. The predictions were found to compare well with experimental data (Sparrow et al. 1964, Ayyaswamy et al. 1974). Figure 4.8 shows the friction factor–Reynolds number product as a function of the contact angle and the groove angle, as obtained from the investigation by Ma et al. (1994).

Analytical investigations on phase change in microgrooves have been performed by Ha and Peterson (1996, 1998a). An investigation of the heat transfer in

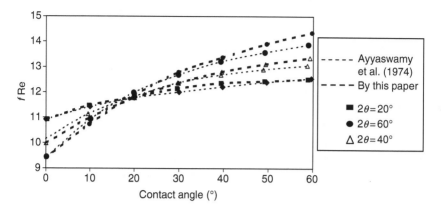

FIGURE 4.8 Variation of the friction factor–Reynolds number product with the interfacial contact angle for flow in a triangular microgroove. (Adapted from Ma, H.B., Peterson, G.P., and Lu, X.J., *Int. J. Heat Mass Transfer*, 37, 2211, 1994. With permission.)

evaporating thin liquid films in V-shaped microgrooves with nonuniform input heat flux has been reported (Ha and Peterson 1996). Liquid conduction and interfacial vaporization were both considered in determining the local interfacial mass flux, and hence in defining a local heat transfer coefficient. An expression was developed for the axial variation of the thickness of the evaporating film. The axial flow of the evaporating film was further investigated by Ha and Peterson (1998a) taking into account a gravity term to incorporate the effect of small tilt angles.

In most of the early investigations on phase change and two-phase flow discussed above, attention was focused on determining (and minimizing) the overall thermal resistance, as well as on obtaining and correlating the heat transfer and CHF data. Friction models also have been developed to be used in the analytical solutions. A few investigations have also reported flow rate measurements and visualization of flow regimes in microchannel flows (Richter et al. 1997, Meinhart et al. 1999).

As has been indicated before, the review of the literature given in the above sections mostly scans the investigations on microchannel flow and heat transfer reported in the last decade of the twentieth century. The investigations, especially intrusive experimental studies reported during that period, though carefully done with available probes, had their limitations in measurement accuracies. The beginning of the twenty-first century witnessed tremendous progress in instrumentation and measurement techniques, due to the advent of MEMS technology in a large scale. Experimental investigations on microchannel flow and heat transfer have seen a dramatic improvement, as more sophisticated probing and visualization techniques have been employed for obtaining data. Many of the hypotheses and interpretations underwent modifications and reconsiderations, due to the more extensive and sometimes contradictory data and results obtained from accurate and reliable measurements. Some of the newer investigations have ascertained the use of conventional approaches in microchannels, and have reestablished the physical dimensions where such conventional approaches could prove wrong. The following section deals with the results of some of these recent investigations, which would help to understand

and interpret experimental and theoretical treatment of microscale convective heat transfer in an altogether new dimension.

A summary of the early investigations on microscale convective heat transfer has been presented in a compact form in Table 4.1, as reproduced from Sobhan and Garimella (2001) and Garimella and Sobhan (2003).

4.2.2 RECENT ADVANCES IN MICROSCALE CONVECTIVE HEAT TRANSFER

Although a majority of the recent theoretical investigations have utilized conventional modeling methodologies, a few have focused on the introduction of new modeling strategies as part of the analysis. The primary aim of many of these theoretical investigations was to compare the results of experiments and continuum analysis, to determine the applicability of conventional theoretical approaches in the analysis of microchannel flows. Sophisticated experimental investigations utilizing conventional temperature and pressure measurements with intrusive sensors, when coupled with modern micro fabrication techniques, have revealed a number of interesting phenomena, not heretofore observed. In addition, a number of investigations, which have utilized flow visualization techniques, have also been reported. Although a few experimental investigations have reported deviations from conventional behavior, the majority of the recent investigations, particularly those that employ sophisticated measurement techniques, support the conclusion that conventional theories can typically be used to accurately predict and explain microscale flow and heat transfer, at least down to some yet to be determined limit of the channel characteristic cross-sectional dimensions. In the following sections, a review is presented on recent advances in microscale heat transfer, involving sophisticated instrumentation and powerful computational methods. In addition, some of the more recent and ongoing research involving both theoretical and experimental investigations of microchannel flow and heat transfer are also discussed.

4.2.2.1 Single-Phase Liquid Flow

A number of significant publications have appeared in recent years on computational and experimental investigations of single-phase fluid flow and heat transfer in microchannels. The following section reviews the important findings in the area of single-phase flow in microchannels.

Many of the recent theoretical models are based on the solution of Navier–Stokes governing equations in the microscale domain and a few of these are three-dimensional computational investigations that provide insight to the temperature distributions and heat transfer patterns in the fluid as well as the substrate. As one of the key areas of application is electronics cooling, various researchers have studied water–silicon systems most frequently.

Fedorov and Viskanta (2000) presented a three-dimensional conjugate heat transfer analysis of microchannel heat sinks for electronics cooling applications, by numerically solving the incompressible laminar Navier–Stokes equations within the framework of conjugate flow solutions. The model was validated against experimental data for thermal resistance and friction factors for a wide range of Reynolds numbers. Practical recommendations with respect to the cooling efficiency and the effect of thermal stresses in these heat exchangers were also provided.

TABLE 4.1
Summary of Early Investigations on Microchannel Heat Transfer and Fluid Flow

Configuration/Parameters	Nature of Work	Observations/Conclusions	Reference
		Microchannel Concepts and Early Work	
Rectangular cross section; water in silicon, $W = 50$ μm, $H = 300$ μm, $Q = 4.7, 6.5, 8.6$ cm³/s	Experiments on integral heat sink for silicon integrated circuits	Demonstrated use of microchannels for very high convective heat transfer in cooling integrated circuits (790 W/cm² at a substrate to coolant temperature difference of 71°C)	Tuckerman and Pease (1981)
Microchannels in cooling of integrated circuits	Microchannel fabrication and implementation details discussed	Coolant selection, packing/headering, microstructure selection, fabrication, and bonding discussed. Etching and precision sawing compared; fabrication and advantages of "micro pillars" using precision sawing discussed. Expressions for coolant figure of merit provided: $\text{CFOM} = (k_s \rho C/\mu)^{0.25}$ for given coolant pressure and $(k_s \rho^2 C^2/\mu)^{0.25}$ for given pumping power	Tuckerman and Pease (1982)
Trapezoidal; nitrogen in silicon and glass, $W = 130-300$ μm, $H = 30-60$ μm, $D_h = 55-76$ μm	Friction factors measured and compared with Moody's chart values for commercial channels	Friction factor for glass channels 3–5 times larger than smooth-pipe prediction. Flow transition occurred at $\text{Re} \approx 400$. Correlation for friction factor f: $(110 \pm 8)/\text{Re}$ for $\text{Re} \leq 900$; $0.165(3.48 - \log \text{Re})^{0.24} + (0.081 \pm 0.007)$ for $900 < \text{Re} < 3{,}000$; $(0.195 \pm 0.017)/\text{Re}^{0.11}$ $3{,}000 < \text{Re} < 15{,}000$	Wu and Little (1983)
As in Wu and Little (1983)	Heat transfer experiments	Correlation for Nusselt number in the turbulent regime: $\text{Nu} = 0.0022\text{Pr}^{0.4} \, \text{Re}^{1.09}$ for $\text{Re} > 3000$	Wu and Little (1984)

Rectangular; air in silicon, $W = 0.13$–0.25 mm, $H/W = 10$, $A_s = 47$–63 cm²/cm³	Comparison of performance with conventional heat sinks, based on correlations	Microstructured compact heat sinks attractive compared to conventional air circulation heat sinks	Mahalingam and Andrews (1987)
Rectangular; water in silicon, $W = 50$–600 μm	Theoretical model for fully developed, developing flows	Turbulent flow designs showed equivalent or better performance compared to laminar flow designs	Phillips et al. (1987)
Rectangular; N-propanol in silicon, $A_c = 80$–7200 μm²	Experiments	Channel with larger cross-sectional areas showed better agreement with theoretical predictions for the friction factor. Proposed $f = C/\mathrm{Re}$ with C given as C versus Re graphs (laminar)	Pfahler et al. (1990)
Microchannel structures for cooling applications	Microchannel applications discussed	Applications of microchannel to electronics cooling, compact heat exchangers, heat shields, and fluid distribution systems discussed	Hoopman (1990)
Microtubes; nitrogen in silica, $D = 3, 7, 10, 53, 81$ μm, $L = 24$–52 mm	Experiments on friction and heat transfer	Correlations for friction factor and Nusselt number: for laminar (Re $< 2,000$) $f = 64/\mathrm{Re}[1 + 30 (v/Dc_a)]^{-1}$ Nu $= 0.000972\mathrm{Re}^{1.17}\mathrm{Pr}^{1/3}$; for turbulent ($2,500 < \mathrm{Re} < 20,000$) $f = 0.140\,\mathrm{Re}^{-0.182}$ Nu $= 3.82 \times 10^{-6}\,\mathrm{Re}^{1.96}\mathrm{Pr}^{1/3}$	Choi et al. (1991)
Rectangular; water in etched silicon, $W = 1$ mm, $H = 176$–325 μm, $L = 46$ mm, $P = 2$ mm	Experiments	Nusselt numbers higher than those predicted from analytical solutions for developing laminar flow	Rahman and Gui (1993)

Single-Phase (Liquid) Experiments

Rectangular; deionized water in stainless steel, $W = 0.6$ mm, $H = 0.7$ mm, $T_i = 30°C$–$60°C$, $v = 0.2$–2.1 m/s	Experiments on single-phase forced convection	In single-phase convection, a steep increase in wall heat flux with the wall temperature; Heat flux for microchannels higher than for normal-size tube	Peng and Wang (1993)

(continued)

TABLE 4.1 (continued)
Summary of Early Investigations on Microchannel Heat Transfer and Fluid Flow

Configuration/Parameters	Nature of Work	Observations/Conclusions	Reference
Rectangular; water, methanol in stainless steel, $W = 0.2$, 0.4, 0.6, 0.8 mm, $H = 0.7$ mm, $T_i = 10°C$–$35°C$ (water), $14°C$–$19°C$ (methanol), $v = 0.2$–2.1 m/s	Experiments on forced convection flow and heat transfer	Heat transfer augmented as liquid temperature was reduced and as liquid velocity was increased. Fully developed turbulent convection regime starts at $Re = 1000$–1500. Correlation for turbulent heat transfer $Nu = 0.00805\ Re^{4/5}\ Pr^{1/3}$	Wang and Peng (1994)
Rectangular; water in stainless steel, $D_h = 0.133$–0.367 mm, $L = 50$ mm, $H/W = 0.333$–1, $T_i = 22°C$–$44°C$, $v = 0.25$–12 m/s, $Re = 50$–4000	Experiments on frictional behavior in laminar and turbulent flow	Flow transition occurred for $Re = 200$–700. Correlations proposed $f = C_{f,l}/Re^{1.98}$ laminar flow, $f = C_{f,t}/Re^{1.72}$ turbulent flow	Peng et al. (1994a)
As in Peng et al. (1994a)	Experiments on forced convection heat transfer characteristics	Fully turbulent convective conditions reached at $Re = 400$–1500. Transition Re diminished with a reduction in microchannel dimension. $Nu = C_{h,l}\ Re^{0.62}Pr^{1/3}$ laminar. $Nu = C_{h,t}\ Re^{0.8}\ Pr^{1/3}$ turbulent (Values for $C_{h,l}$, $C_{h,t}$) available in Peng et al. (1994)	Peng et al. (1994b)
As in Wang and Peng (1994) except $T_i = 11°C$–$28°C$ (water), $12°C$–$20°C$ (methanol), $v = 0.2$–2.1 m/s (water), $v = 0.2$–1.5 m/s (methanol)	Experiments on effect of thermofluid properties and geometry on convective heat transfer	Changes in flow regimes and heat transfer modes initiated at lower Re in microchannels compared to conventional channels. Transition zone and heat transfer characteristics in laminar and transition flow influenced by liquid temperature, velocity, Re, and microchannel size	Peng and Peterson (1995)

Geometry / parameters	Study	Results	Reference
As in Peng et al. (1994a)	Experiments on single-phase flow and heat transfer	Ratio of experimental to theoretical friction factor at critical Re plotted as a function of $Z = \min[H,W]/\max[H,W]$. Correlations proposed $\mathrm{Nu} = 0.1165(D_h/P)^{0.81} (H/W)^{-0.79} \mathrm{Re}^{0.62}\mathrm{Pr}^{0.33}$ for laminar; $\mathrm{Nu} = 0.072(D_h/P)^{1.15}[1 - 2.421(Z - 0.5)^2]\mathrm{Re}^{0.8}\mathrm{Pr}^{0.33}$ for turbulent	Peng and Peterson (1996b)
Rectangular; water–methanol mixture in stainless steel, $D_h = 0.133$–0.367 mm, $L = 50$ mm, $W = 0.1, 0.2, 0.3, 0.4$ mm, $H = 0.2, 0.3$ mm, $T_i = 14°C$–$36°C$, $v = 0.04$–3.8 m/s, Re = 6–3500	Experiments	Laminar heat transfer ceased for Re $\approx$ 70–400 depending on flow conditions; fully developed turbulent heat transfer achieved at Re = 200–700, depending on D_h. Transition Re reduced with reduction in microchannel size. D_h, H/W and mixture mole. Fraction influenced heat transfer. Heat transfer increased for smaller mole fractions of the more volatile component	Peng and Peterson (1996a)
Rectangular; deionized water in silicon, $W = 251$ μm, $H = 1030$ μm, $D_h = 404$ μm, $L = 2.5$ cm, $Q = 5.47$–118 cm^3/s	Experimental and theoretical study	Critical Re of 1500 identified for onset of turbulence. Analysis showed that flow and heat transfer performance could be improved by increasing H, and that for the same pressure drop and pumping power, thermal resistance was smaller for deeper channels	Harms et al. (1997)
Rectangular; FC-72 and transformer oil in stainless steel $H = 0.10$–0.58 mm; nozzle dimensions (mm): Length = 35, $B = 0.146, 0.210, 0.234$, height = 12, $v = 0.54$–8.45 m/s, Re = 70–170 (oil), 911–4807 (FC72)	Experiments in impingement on 2D microchannels	Empirical correlation for Nusselt number for the two liquids $\mathrm{Nu}_x = 0.429\mathrm{Re}^{0.583}\mathrm{Pr}^{1/3}(x/2H)^{0.349}(B/2H)^{-0.494}$	Zhuang et al. (1997)

(continued)

TABLE 4.1 (continued)

Summary of Early Investigations on Microchannel Heat Transfer and Fluid Flow

Configuration/Parameters	Nature of Work	Observations/Conclusions	Reference
Circular; distilled water in copper, $D = 0.102$–1.09 mm, $v < 18.9$ m/s, $Re = 2.6.10^3$–$2.3.10^4$. $Pr = 1.53$–6.43, $q'' < 3.0$ MW/m^2	Experiments on turbulent single-phase flow	Nusselt numbers higher than those predicted by large-channel correlations Gnielinski correlation modified for Nusselt number for turbulent flow in circular microchannels (f from Filonenko, 1954): $Nu = Nu_{Gn} (1+F)$ where $F = CRe[1 - (D/D_o)^2]$ $Nu_{Gn} = (f/8)(Re - 1000)Pr/[1 + 12.7(f/8)^{1/2}$ $(Pr^{2/3} - 1)]$ $C = 7.6 \times 10^{-5}$ $D_o = 1.164$ mm $f = [1.82\log (Re) - 1.64]^{-2}$	Adams et al. (1998)
Noncircular; water in copper, $D_h = 1.13$ mm, $Re = 3.9.10^3$–$2.14.10^4$, $Pr = 1.22$–3.02	Experiments on turbulent convection	Experimental Nusselt number well predicted by Nu_{Gn} $D_h \approx 1.2$ mm proposed as reasonable lower limit for applicability of standard Nusselt-type correlations to noncircular channels	Adams et al. (1999)
Rectangular; laminar and transition flow	Dimensional analysis based on experimental data in the literature	Attempted to explain the observation that Nu may decrease with increasing Re in laminar regime and may remain unaffected in transition regime. Proposed that Brinkman number may better correlate convective heat transfer	Tso and Mahulikar (1998, 1999)
Almost circular; water in aluminum $D_h = 0.73$ mm	Experiments	Laminar flow data found to correlate well using Brinkman number	Tso and Mahulikar (2000b)

Single-Phase (Liquid) Models and Optimization Studies

Geometry/Conditions	Type of Study	Findings	Reference
Triangular microgrooves, channel angle $20°–60°$	Analytical/numerical analysis	Friction factor–Reynolds number product strongly dependent on channel angle, contact angle, and dimensionless vapor–liquid interface flow number	Ma et al. (1994)
Microchannel, plate-fin heat sink; air in copper, aluminum, $W = 400, 500$ μm, $H = 2.5$ cm, $Q = 1–6$ L/s	Thermal resistance model, experiments, optimization	Thermal resistance of microchanneled heat sink lower than for heat sinks employing direct air cooling, by a factor of more than 3	Kleiner et al. (1995)
Circular capillary channels $D = 8.1–96$ μm, 0.76–4.7 μm	Numerical study on the flow of super fluid helium using a two-fluid model	Existence of an optimum channel diameter for maximum mass flow rate indicated	Takamatsu et al. (1997)
Parallel plates at 25 μm separation, dilute aqueous electrolyte $L = 10$ mm	Theoretical analysis incorporating effects of EDL field	EDL resulted in a reduced flow velocity than in conventional theory, thus affecting temperature distribution and reducing Re Higher heat transfer predicted without the double layer	Mala et al. (1997a)
Parallel plates (10×20 mm) of P-type silicon and glass at 10–280 μm separation; $\Delta P = 0–350$ mbar	Experimental study and comparison with predicted volume flow rates	For solutions of high ionic concentration as well as for $D_h >$ few hundred μm, EDL effect is negligible, EDL effect becomes significant for dilute solutions	Mala et al. (1997b)
Rectangular; dilute aqueous electrolyte in silicon, $H = 20$ μm, $W = 30$ μm, $L = 10$ mm, $\Delta P = 2$ atm, $T_i = 298$ K, $q'' = 1.0 \times 10^5$ W/m²	Numerical analysis with effects of EDL and flow-induced electrokinetic field	The EDL field and electrokinetic potential act against the liquid flow, resulting in higher friction coefficient, reduced flow rate, and a reduced Nusselt number, for dilute solutions	Yang et al. (1998)
Rectangular (flat plate micro heat exchangers)	Optimization study on microchannel shape	Width of heat exchanger conduits may be optimized to reduce maximum temperature of the uniformly heated surface	Bau (1998)

(continued)

TABLE 4.1 (continued)
Summary of Early Investigations on Microchannel Heat Transfer and Fluid Flow

Configuration/Parameters	Nature of Work	Observations/Conclusions	Reference
Microchannel cooling and jet impingement	Comparative analysis of jet impingement and microchannel cooling	Thermal performance of jet impingement without any treatment of spent flow substantially lower than microchannel cooling, regardless of target dimension. Microchannel cooling preferable for target dimension smaller than 7×7 cm	Lee and Vafai (1999)
Gas Flow			
Rectangular; helium in silicon $H = 1.33$ μm, $W = 52.25$ μm, $L = 7\,500$ μm, inlet to outlet pressure ratio $= 1.2$–2.5, Re $= (0.5$–$4) \times 10^{-3}$	Flow rates measured and compared with theoretical model	Mass flow–pressure relationship accurately modeled by including a slip flow boundary condition at the wall	Arkilic et al. (1994)
As in Arkilic et al. (1994) with pressure ratio $= (1.6$–$4.2)$, Re $= (1.4$–$12) \times 10^{-3}$	Experiments and comparison of mass flow with results from 2D analysis with slip-boundary condition	Discussions on nondimensional formulation and perturbation solution	Arkilic et al. (1997)
Microtubes; nitrogen and water in silica, $D = 19, 52, 102$ μm, Pr $= 0.7$–5, Re $= 250$–$20{,}000$	Experiments; theoretical scaling analysis	Turbulent momentum and energy transport in the radial direction significant in the near-wall zone of a microtube. Correlations proposed: $f = 50.13/$Re (laminar, Re < 2000 $f = 0.302/$Re$^{0.25}$ (turbulent, $6000 <$ Re < 20000) Nu $= 0.007$ Re$^{1.2}$ Pr$^{0.2}$ (turbulent, $6000 <$ Re < 20000)	Yu et al. (1995)

Rectangular $H = 0.5$, 5 μm, $H/W = 2.5$, 5, 10, 20 (subsonic); 5, 10, 20 (supersonic)	Numerical study using DSMC technique	Heat flux on the channel surface decreases with increase in Knudsen number and channel length in supersonic flow	Mavriplis et al. (1997)
Rectangular; helium, as in Arkilic et al. (1994), helium and nitrogen, $D_h = 1.01$ μm, $L = 10.9$ mm	2D numerical model, comparison with experiments in literature	Nusselt number and friction coefficient substantially reduced for slip flows compared to continuum flows. Effect of compressibility significant at high Re	Kavehpour et al. (1997)
Smooth microtubes gas flow	Numerical solution of gas flow in microtubes	Local Nusselt number increased with dimensionless length, due to compressibility. fRe product not constant; dependent on Re	Guo and Wu (1997)
Rectangular; nitrogen, helium in silicon, $W = 40$ μm, $H = 1.2$ μm, $L = 3$ mm(N_2), $W = 52$ μm, $H = 1.33$ μm, $L = 7.5$ mm(He)	Numerical solution with slip-boundary condition	Small velocities and high-pressure gradients due to large wall shear stresses. Comparisons with experiments of Arkilik et al. (1994)	Chen et al. (1998)
3D straight and spiral grooves	Numerical study on slip flow in long microchannels	Nonlinear pressure gradient along the microchannels due to density variations	Niu (1999)

Boiling in Microchannels

Circular; R-113 in copper $D = 2.45$ mm (mini), 510 μm (micro), $Q = 19$–95 ml/min, $\Delta T = 10$°C–32°C	Experiments on boiling and 2-φ flow; boiling curves and CHF values obtained	Microchannel yielded higher CHF (28% greater at $Q = 64$ ml/min) than mini channel, with a larger ΔP (0.3 bar for micro, 0.03 bar for mini)	Bowers and Mudawar (1994a)
As in Bowers and Mudawar (1994a)	Pressure drop model developed; predictions compared to experiments	Major contributor to pressure drop identified as the acceleration resulting from evaporation. Compressibility effect important for microchannel when Mach number >0.22. Channel erosion effects more predominant in microchannels than in mini channels	Bowers and Mudawar (1994b)

(continued)

TABLE 4.1 (continued)
Summary of Early Investigations on Microchannel Heat Transfer and Fluid Flow

Configuration/Parameters	Nature of Work	Observations/Conclusions	Reference
As in Bowers and Mudawar (1994a)	Experiments on boiling and two-phase flow	Single CHF correlation for mini and microchannels developed: $q_{m,p}/(Gh_{fg}) = 0.16We^{-0.19}(L/D)^{0.54}$	Bowers and Mudawar (1994c)
Rectangular; water in stainless steel, $W = 0.6$ mm, $H = 0.7$ mm, $T_i = 30°–60°C$, $v = 1.5–4.0$ m/s	Experiments on subcooled boiling of water	Nucleate boiling intensified and wall superheat for flow boiling smaller in microchannels than in normal-sized channels for the same wall heat flux. No partial nucleate boiling observed in microchannels	Peng and Wang (1993)
Rectangular; methanol in stainless steel, $W = 0.2, 0.4, 0.6$ mm, $H = 0.7$ mm, $L = 45$ mm, $P = 2.4–4$ mm: $T_i = 14°C–19°C$ (subcooling: 45°C–50°C), $v = 0.2–1.5$ m/s	Experiments on boiling	Liquid velocity and subcooling do not affect fully developed nucleate boiling. Greater subcooling increased velocity and suppressed initiation of flow boiling	Peng et al. (1995)
Rectangular; methanol–water mixture in stainless steel, $W = 0.1, 0.2, 0.3, 0.4$ mm, $H = 0.2, 0.3$ mm, $L = 45$ mm, $D_h = 0.133–0.343$ mm, $v = 0.1–4.0$ m/s, $T_i = 18°C–27.5°C$ (subcooling: 38°C–82°C)	Experiments on flow boiling in binary mixtures	Heat transfer coefficient at onset of flow boiling and in partial nucleate boiling greatly influenced by concentration, microchannel/substrate dimensions, flow velocity, and subcooling. These parameters had no significant effect on heat transfer coefficient in the fully nucleate boiling regime. Mixtures with small concentrations of methanol augmented flow boiling heat transfer	Peng et al. (1996)

V-shaped; water and methanol in stainless steel, Groove angle 30°–60°; D_h = 0.2–0.6 mm, v (water) = 0.31–1.03 m/s, v (methanol) = 0.12–2.14 m/s	Experiments on flow boiling	Heat transfer and pressure drop were affected by flow velocity, subcooling, D_h, and groove angle. No bubbles observed in microchannels during flow boiling, unlike in conventional channels. Experiments indicated an optimum D_h and groove angle	Peng et al. (1998)
V-shaped	Analysis of microgrooves with nonuniform heat input	Analytical expression developed for the evaporating film profile	Ha and Peterson (1996)
V-shaped	Analysis of axial flow of evaporating thin film	Used perturbation method to solve the axial flow of an evaporating thin film through a V-shaped microchannel with tilt	Ha and Peterson (1998a)
Circular and rod bundle; water in copper D = 1.17, 1.45 mm, D_h = 1.131 mm, m = 250–1000 kg/m²s, exit pressure = 344–1043 kPa, inlet pressure = 407–1207 kPa, T_i = 49°C–72.5°C	Experiments on CHF in flow of subcooled water	CHF found to increase monotonically with increasing mass flux or pressure. CHF depends on the channel cross-section geometry, and increases with increasing D	Roach et al. (1999)

Boiling in Small Diameter Tubes and Channels

Circular; water in stainless steel, D = 2.5 mm, t = 0.25 mm, v = 10–40 m/s	Experiments on subcooled flow boiling of water under high heat fluxes	Experimental data did not match predictions from CHF correlations in the literature	Celata et al. (1993)
Rectangular; water and R141b in copper, W = 1, 2, 3 mm, H/W < 3, m = 50, 200, 300 kg/m²s	Experiments on flow boiling in narrow channels of planar heat exchanger elements	Boiling curves and variations of heat transfer coefficient with local and average heat fluxes obtained	Mertz et al. (1996)

(continued)

TABLE 4.1 (continued)

Summary of Early Investigations on Microchannel Heat Transfer and Fluid Flow

Configuration/Parameters	Nature of Work	Observations/Conclusions	Reference
Rectangular; FC-72 in fiberglass $W = 5$ mm, $H = 2.5$ mm, heated length $= 101.6$ mm, $v = 0.25$–10 m/s, Re $= 2,000$–130,000 (subcooling at outlet $= 3°C$, $16°C$, $29°C$)	CHF experiments on long channels; flow visualization	Propagation of vapor patches resembling a wavy vapor layer along the heated wall at the CHF Length and height of the vapor patch found to increase along flow direction, and decreased with increasing subcooling and velocity	Sturgis and Mudawar (1999a)
As in Sturgis and Mudawar (1999a)	Theoretical model for CHF; data analysis	Effect of periodic distribution of vapor patches idealized as a sinusoidal interface with amplitude and wave length increasing in flow direction	Sturgis and Mudawar (1999b)
Two-Phase Flow			
Rectangular; R124 in copper, $W = 0.27$ mm, $H = 1.0$ mm, $D_h = 425$ mm, Re$D_h = 100$–750; $q'' < 40$ W/cm^2	Experiments on microchannel heat exchanger	Nusselt number (-5 to 12) showed an increase with Reynolds number in 1-φ flow, but was approximately constant in 2-φ flow	Cuta et al. (1996)
Rectangular; R124 in copper, $W = 270$ μm, $H = 1000$ μm, $L = 2.052$ cm, $D_h = 425$ μm, inlet subcooling: $5°C$–$15°C$; $Q = 35$–300 ml/min	Experiments with two microchannel patterns (parallel and diamond)	Heat transfer coefficient and pressure drop found to be the function flow quality and mass flux, in addition to the heat flux and surface superheat Heat transfer coefficient decreased by 20%–30% for an increase in vapor quality from 0.01 to 0.65	Ravigururajan (1998)
Circular and semitriangular; air–water mixture in glass, $D = 1.1.1.45$ mm, $D_h = 1.09$, 1.49 mm, v(air): 0.02–80 m/s, v (water): 0.02–8 m/s (superficial velocity)	Visual observation of flow pattern and pattern maps	Bubbly, churn, slug, slug-annular, and annular flow patterns observed	Triplett et al. (1999a)

Parameters	Method	Findings	Reference
As in Triplett et al. (1999a)	Frictional pressure drop measured and compared with various 2-φ friction models	Models and correlation over predicted channel void fraction and pressure drop in annular flow pattern. Annular flow interface momentum transfer and wall friction in microchannels significantly different from those in larger channels	Triplett et al. (1999b)
Circular and rectangular; air–water mixture in glass, $D_h = 1.3$–5.5 mm, v (gas) = 0.1–100 m/s, v (liquid) = 0.01–10 m/s	Experiments, flow visualization	Tube diameter influences the superficial gas and liquid velocities at which flow transitions take place, due to combined effect of surface tension, hydraulic diameter, and aspect ratio	Coleman and Garimella (1999)

Design and Testing

Parameters	Method	Findings	Reference
Rectangular; water in silicon	Numerical solution for temperature field; comparison with experiments	Design algorithm developed for selection of heat exchanger dimensions. Expression for maximum pumping power obtained as function of channel geometry	Weisberg et al. (1992)
Almost rectangular; water in copper 0.5×12 mm, 0.125×12 mm, $Q = 0.47$–5 gpm	Design and testing, microchannel heat exchanger for laser diode arrays	Thermal resistance due to solder bond estimated	Roy and Avanic (1996)
Rectangular and almost triangular; air in copper, aluminum	Parametric studies and experiments of air impingement in microchannels	Thermal resistance model developed. Parametric studies to determine influence of static pressure, pumping power, and geometric parameters on thermal resistance	Aranyosi et al. (1997)
Rectangular; diamond-shaped and hexagonal; water in silicon	3D numerical model; optimization for reducing thermal resistance	Rectangular geometry had the lowest thermal resistance	Perret et al. (1998)

(continued)

TABLE 4.1 (continued)
Summary of Early Investigations on Microchannel Heat Transfer and Fluid Flow

Configuration/Parameters	Nature of Work	Observations/Conclusions	Reference
Rectangular; water, FC72 in copper	Experiments on micro heat sink for power multichip module (MCM); 3D and 1D thermal resistance models	Power densities of 230–350 W/cm^2 dissipated with a temperature rise of 35°C, and a pumping power of about 1W per chip. Parameter "heat spread effect" defined $S = (R_{th1D} - R_{th3D})/R_{th1D}$	Gillot et al. (1998)
Rectangular; water, FC72 in copper, $W = 230$, 311 μm, $H = 730$, 3040 μm, Q (ml/min) = 1350 (water, 1-φ) and 30(2-φ); 2000(FC72 1-φ) and 300 (2-φ)	Experiments on single- and two-phase micro heat exchangers for cooling transistors	Two-phase heat exchanger provided lower thermal resistance and pressure drop compared to single-phase heat exchangers	Gillot et al. (1999)
Rectangular; air in copper, $W = 800$ μm, $H = 50$ μm, $Q = 140$ m^3/h	Experiment and thermal resistance model	Pressure drop found to have large deviation from predicted values at high air flow rates. Cooling capacity ≈1700 W at heat flux ≈15 W/cm^2	Yu et al. (1999)
Measurement Techniques			
Triangular; water in silicon, $W = 28$–182 μm, $Q = 0.01$–1000 μl/min	Optical flow measurements using microscope	Measured flow rates in good agreement with theoretical values for laminar flow through triangular channels	Richter et al. (1997)
Rectangular; water in glass, $W = 300$, $H = 30$, $L = 25$ mm	Particle image velocimetry (PIV)	(PIV) Results agreed well with analytical solutions for Newtonian flow in rectangular channels	Meinhart et al. (1999)

Qu et al. (2000a) performed experimental measurements of flow rate and pressure drop in water flowing through trapezoidal silicon microchannels with a range of hydraulic diameters from 51 to 169 μm. In this investigation, it was observed that a significant difference between the experimental results and the theoretical predictions occurred, indicating higher values of pressure drop and friction than predicted by conventional laminar flow theory. To explain these effects, the variation in the surface roughness of the channels was highlighted. Qu et al. (2000b), in another publication, reported on experimental investigations of the flow of water in trapezoidal silicon microchannels with hydraulic diameters ranging from 62 to 169 μm. A numerical solution was also obtained for the conjugate problem involving simultaneous determination of the temperature distribution in the solid and liquid domains. A significant difference was found between the theoretical predictions and the experimental results, in which the experimental Nusselt numbers were found to be much lower than the values as predicted using conventional techniques. On the basis of the roughness–viscosity model, a modified relationship, designed to account for the roughness–viscosity effects, was proposed to interpret the experimental results. Typical results showing the comparison between experimental results and theoretical predictions using this modified relationship are illustrated in Figure 4.9.

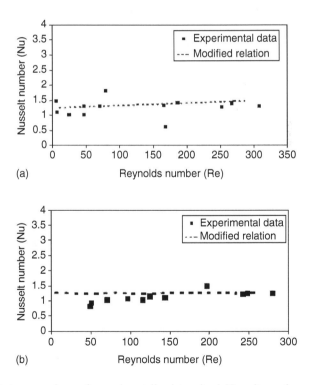

FIGURE 4.9 A comparison of experimentally determined Nusselt numbers as a function of Reynolds numbers with the predictions of the modified relationship (a) $D_h = 62.3$ μm; (b) $D_h = 63.1$ μm;

(continued)

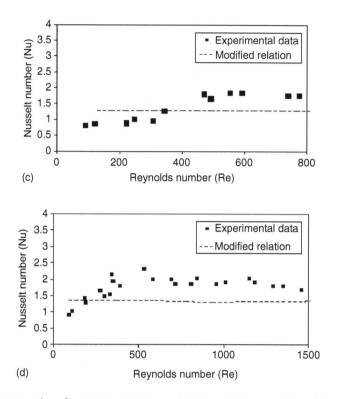

FIGURE 4.9 (continued) (c) $D_h = 114.5$ μm; (d) $D_h = 168.9$ μm. (Adapted from Qu, W., Mala, G.M., and Li, D., *Int. J. Heat Mass Transfer*, 43, 3925, 2000b. With permission.)

Kim et al. (2000) presented analytical solutions for temperature distributions in microchannel heat sinks utilizing both one- and two-equation models for the heat transfer. The effects of the variations in the Darcy number and the effective thermal conductivity, on the temperature distributions, were studied. Richardson et al. (2000) conducted an analytical investigation to investigate the existence of an optimum laminar flow regime in singly connected microchannels with complex free flow area cross sections. The objective function was represented by the entropy generation rate per unit heat capacity rate of the fluid stream to perform a thermodynamic optimization. The results indicated that the existence of the objective function minimum can only be found for very small ducts with irregular cross sections, and these minima are relatively weak, indicating that the possibility of the occurrence of an optimum flow is rare.

Ryu et al. (2002, 2003) reported numerical computations for the three-dimensional analysis of a microchannel heat sink. The analysis presented was based on numerical solution of the Navier–Stokes and energy equations. An optimization investigation was performed to determine the optimal design parameters of the microchannel heat sink that minimizes the thermal resistance for a given pumping power. The optimal dimensions and thermal resistance were found to have a power-law dependence on the pumping power. The channel width and depth were found to be the most crucial parameters affecting the heat sink performance. By performing a parametric study, the optimal dimensions and thermal resistance were also shown to have a power-law dependence on the pumping power. A finite volume method was used in the analysis, and the effect of the pumping power on the optimal dimensions was highlighted as shown in Figure 4.10.

Qu and Mudawar (2002) conducted a three-dimensional numerical analysis of microchannel heat sinks consisting of rectangular channels with a width of 57 mm, a depth of 180 mm, and a separation distance of 43 mm. A numerical code to solve the three-dimensional governing Navier–Stokes and energy equations based on the SIMPLE algorithm (Patankar 1980) was used, and the results were compared with other analytical solutions and available experimental data. The results indicated that the temperature rise in the solid and the fluid is approximately linear. The heat flux and the Nusselt number were also shown to have a peripheral variation, and exhibited the maximum values at the channel inlet. In addition, it was noted that for relatively high Reynolds numbers, i.e., in excess of 1400, the whole of the flow in the channel is under developing flow conditions.

Zhao and Lu (2002) performed analytical and numerical investigations of forced convection in microchannel heat sinks. Two approaches, namely the fin approach and the porous medium model, were utilized in the analysis. In the porous medium approach, the Darcy model for pressure and two-equation models for heat transfer between solid and fluid phases were used. Predictions from both the approaches showed that the overall Nusselt number increased with an increase in the aspect ratio and a decrease in the effective thermal conductivity ratio (between the fluid and the porous medium). It was found that there is a large difference between the predictions of the temperature distributions and the Nusselt numbers, using the two approaches.

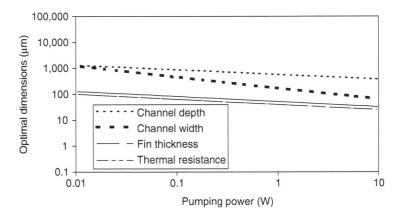

FIGURE 4.10 Effect of the pumping power on the thermal performance of a microchannel heat sink. (Adapted from Ryu, J.H., Choi, D.H., and Kim, S.J., *Int. J. Heat Mass Transfer*, 45, 2823, 2002. With permission.)

The reasons for this were explained in the paper based on the assumptions made in the models.

Toh et al. (2002) used the finite volume method to analyze three-dimensional fluid flow and heat transfer in water flowing through silicon microchannels of four different dimensions. The numerical code was validated by comparing with experimental data for local thermal resistances. The results indicate that the frictional resistances are reduced at high heat inputs, particularly at low Reynolds numbers. The reduction of frictional losses at lower Reynolds numbers, at a given heat input, was interpreted to be due to the reduction in the temperature dependent fluid viscosity.

Kim (2004) discussed and summarized three optimization methods applicable to microchannel heat sinks, namely the application of a fin model, a porous medium model, and an optimization method based on the three-dimensional numerical simulation of the microchannel heat sink to evaluate the thermal resistance. The porous medium model was found to predict the thermal performance of the microchannel heat sink more accurately than the fin model with a one-dimensional heat conduction assumption, when compared with the numerical simulation results.

Nakayama et al. (2004) presented discussions on the geometric uncertainties associated with microchannels incorporated into electronic devices. A summary of the sources of uncertainties was presented, along with a CFD simulation based methodology to assess situations involving geometric uncertainties. A model channel was defined and was used to illustrate the methodology. The results of these simulations indicate the stringency measure required in the assembly of these microchannel-cooling devices. A hypothetical experiment was reported for comparing the performance of a channel of optimal height, with that of a nonoptimal channel under a given pumping power. Van Male et al. (2004) used CFD simulations to determine the heat and mass transfer characteristics of laminar and plug flow of methane in a silicon square microchannel with asymmetrical heating. The simulation results were validated against experiments and used to propose correlations for different Nusselt and Sherwood numbers.

Tchikanda et al. (2004) developed a model of the flow in open rectangular microchannels to obtain the mean velocity of the liquid flow. The solutions for the mean velocity were based on asymptotic expressions for the governing simplified Navier–Stokes equations as applied to pressure driven and shear driven flows, and blending these with an approach that decomposes the flow into additive components, driven by the pressure gradient and the shear stress. Numerical computations of velocity were performed for various contact angles, channel aspect ratios, and wetting regimes, and were then used to develop the appropriate analytical expressions. Min et al. (2004) conducted a numerical investigation to investigate the influence of the tip clearance in a microchannel heat sink (the gap between the tip of the microstructures forming the microchannels and the top plate of a heat sink). Using the thermal resistance of the heat sink as an indication of its performance, an optimal value of the tip clearance was obtained for a given pumping power. It was observed that the presence of a small tip clearance can improve the performance of the heat sink, provided that its value is less than the channel width.

Koo and Kleinstreuer (2004) investigated the effects of viscous dissipation on the temperature and friction factor in microconduits of different geometries using dimensional analysis and experimentally validated computer simulations. Various working fluids (water, methanol, and isopropanol) were considered in the investigation. The results indicated that viscous dissipation is a strong function of the geometric parameters (aspect ratio and hydraulic diameter) and the operational parameters (Reynolds number, Prandtl number, and Eckert number) for microconduits, and cannot be ignored in any analysis of this type of flow. This work was followed by an analysis of the effect of the surface roughness on the heat transfer performance of microtubes and channels (Koo and Kleinstreuer 2005a). The surface roughness was incorporated in a porous medium layer (PML) model. The surface roughness effects, in terms of the thermal conductivity ratio of the PML and the Darcy number, on the dimensionless temperature profile and the Nusselt number were investigated. It was shown that, depending on these parameters, the Nusselt number could be higher or lower than predicted by conventional theory, and that the influence of these parameters was more pronounced in microchannels than in microtubes. Koo and Kleinstreuer (2005b) also presented a simulation and analysis of the flow of nanofluids (water and ethylene glycol with small volume factions of copper oxide nanospheres) in microchannel heat sinks. New models for the effective thermal conductivity and dynamic viscosity of these fluids, which includes the effect of particle Brownian motion, were used in this model. The analysis was performed using the concentration of the nanofluid as a parameter, along with the aspect ratio. Performance enhancement was reported for large Prandtl number carrier fluids, in which nanoparticles with higher thermal conductivity ratios, i.e., the conductivity of the particle was high with respect to the carrier fluid, at volume concentrations of approximately 4%, in microchannels with high aspect ratios. In this investigation, the walls were treated to avoid nanoparticle accumulation.

Horiuchi and Dutta (2004) obtained an analytical solution to the problem of electro-osmotic flow in two-dimensional microchannels, based on an infinitesimal EDL theory, using a separation of variables technique for solving the modified Navier–Stokes equation and the normalized energy equation. Constant surface heat flux and constant surface temperature cases were considered in the investigation, and the inertial, diffusive, and Joule heating terms were all included in the analysis. Heat transfer characteristics were presented for the low Re regime, where the viscous and electric field terms are dominant. It was found that the Nusselt number is independent of the Peclet number, except in the thermally developing region of the channel. The analytical results showed very good agreement with the existing heat transfer results available in the literature.

Azimian and Sefid (2004) reported results of a three-dimensional computational analysis of the flow of Newtonian and non-Newtonian fluids in microchannels. The three-dimensional Navier–Stokes equations were solved using a finite difference models, with the use of the SIMPLE algorithm (Patankar 1980). The shear stress was incorporated in the governing equations through a power law to study the potential of using non-Newtonian fluids. It was found that the pressure drop was significantly reduced and the Nusselt number was increased for shear thinning fluids. The reverse was true for shear thickening fluids. Variations of the pressure drop, thermal entrance length, and Nusselt numbers were presented for various values of the power-law index assumed in the computation.

Morini (2004) presented an analysis of existing experimental results of heat transfer and pressure drop in microchannels and validated these against the Obot-Jones model for conventional noncircular channels. The results indicated that the conventional theory needed to be revised to accurately predict the behavior in channels with hydraulic diameters smaller than 1 mm. It was also observed that flow transition in mirochannels took place at smaller Reynolds numbers than expected in conventional channels.

Maynes and Webb (2003a,b) analyzed the fully developed electro-osmotically driven flow in microchannels, imposed by a voltage gradient along the length of the channel. Analytical solutions with constant wall heat flux and constant wall temperature boundary conditions, for combined pressure and electro-osmotically driven flows in microchannels, were obtained. Exact solutions of the fully developed dimensionless temperature gradient and Nusselt number were found to depend on the relative duct radius. The fully developed Nusselt number was found to depend on the duct radius to Debye length ratio, the dimensionless volumetric source, and a dimensionless parameter that characterizes the relative strengths of the two driving mechanisms—the pressure and the electro-osmotic forces. Maynes and Webb (2004) also reported theoretical investigations to study the effect of viscous dissipation on thermally fully developed electro-osmotically driven flows, where the flow is established due to an imposed voltage (potential) gradient along the length of the tube. The velocity distribution for the fully developed electro-osmotically driven flow was used in the energy equation, which was solved numerically to obtain the temperature distribution and the Nusselt number. Flow in parallel plate microchannels and circular microtubes was studied for constant wall temperature and constant heat flux boundary conditions. It was found that the effect of the viscous dissipation is significant only for low values of the relative duct radius with respect to the thickness of the EDL characterized by the Debye length, and the effect is negligible in most of the practical applications of electro-osmotically driven flows. Ng and Tan (2004) described finite volume solutions of a three-dimensional numerical model for developing flow in microchannels, introducing the EDL effect. The model was based on Navier–Stokes equations and the Poisson–Boltzmann equation to simulate the electric potential distribution. Comparisons of the performance of the microchannels under different conditions of Reynolds number and electric potentials were presented. It was concluded that the EDL effect could decrease the effectiveness in microchannel flows. Tardu (2004) presented an analysis of the effect of the EDL on the stability of the microchannel flow, by performing a parametric investigation of the two-dimensional governing equations for channel flow under the EDL effect. It was found that when the EDL is thick, the inviscid instability mechanism becomes dominant in microchannels. Chen et al. (2004) presented a numerical investigation of the hydrodynamically and thermally developing flow of a liquid in microchannels, with electrokinetic effects. The effects of EDL on flow and heat transfer between two parallel plates were considered, using the modified Navier–Stokes equations and the Poisson equation to describe the electrical potential distribution. Solution was obtained using a finite volume approach with the SIMPLER algorithm (Patankar 1980).

It was found that the friction factor and Nusselt number were greatly influenced by the effect of the electrical double layer effects.

Nonconventional modeling methods and philosophies in the analysis of microchannel flows were also introduced. These did not depend on the Navier–Stokes equations for the mathematical modeling of the problem.

Chen and Cheng (2002) presented and discussed a new design of fractal tree-like network applied to microchannel cooling of electronic chips, inspired from biological circulatory and respiratory networks. Analysis was performed to study the heat transfer and fluid flow behavior in these fractal microchannel nets. It was concluded that these networks have a better heat transfer capability and lower pumping power, compared to traditional parallel nets of microchannels.

Li and Kwok (2003) presented a Lattice–Boltzmann model to describe the flow of an electrolytic solution in a microchannel, which is termed electrokinetic flow. This model compensates for the difficulties in the use of the modified Navier–Stokes equation for the solution of this class of flow problems, and is based on a Poisson–Boltzmann equation. The results were found to be in excellent agreement with experimental results for pressure driven microchannel flow of electrolyte solutions. The most significant results of this work were presented as comparisons of the pressure drops and the Reynolds number of the flow.

Fabrication and experimental facilities, as well as measurement methods and instrumentation for microchannels, have progressed considerably in the past few years. With the advent of MEMS technology, probes and sensors can be effectively incorporated into the substrate and individual channels to obtain in situ measurements at the inner wall of the microchannels. This has resulted in more accurate observations of microchannel flows, when compared to the early experiments reported in the literature, which were primarily based on external measurements. A majority of the experimental results reported recently have utilized carefully fabricated test specimens, coupled with measured entrance and exit pressure losses in channels, and specific data measured on the inside of individual channels. This approach has resulted in new information, which clarifies the performance of microchannels, and supports the use of conventional analyses for estimating the behavior of microchannels with practical cross-sectional dimensions and shapes.

The experiments of Wu and Cheng (2003) concluded that the aspect ratio has a significant influence of the friction factor. Trapezoidal microchannels of hydraulic diameters ranging from 25.9 to 291.0 μm were investigated. Using the resulting experimental data, a correlation for the friction factor in fully developed laminar flow was developed as a function of the aspect ratio as follows:

$$f\mathrm{Re} = 11.43 + 0.80\exp\left(2.67\frac{W_b}{W_t}\right) \qquad (4.10)$$

where W_b and W_t are, respectively, the bottom width and top width of the trapezoidal microchannel. As shown in Figure 4.11, the experimental data were found to correlate well with the analytical solution developed by Ma and Peterson (1997a) for fully developed laminar flow, indicating that the Navier–Stokes equations are

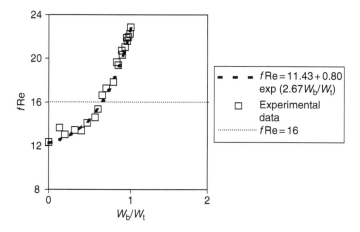

FIGURE 4.11 Comparison of experimental data with correlation equation for fully developed laminar flow. (Adapted from Wu, H.Y. and Cheng, P., *Int. J. Heat Mass Transfer*, 46, 2519, 2003. With permission.)

valid in microchannels of hydraulic diameters as low as 25.9 μm. In the larger channels analyzed, transition to turbulent flow was observed at Reynolds numbers in the range of 1500–2000.

Cheng and Wu (2003a), in subsequent experimental investigations of laminar convective heat transfer and pressure drop for water flowing in trapezoidal microchannels, indicated that the value of the Nusselt number and the apparent friction factor were dependent on a number of geometric parameters, increasing with both the surface roughness and the surface hydrophilic property. This increase was found to be more significant at larger Reynolds numbers. Dimensionless correlations were obtained for the Nusselt number and the apparent friction constant for the flow of water in trapezoidal microchannels, when the geometric dimensions, surface roughness, and hydrophilic properties were all known.

Garimella and Singhal (2004) presented experimental results and discussed the fluid flow in microchannels, focusing on the pumping capabilities and pumping power optimization for electronics cooling applications. The experimental data indicated that the conventional correlations can be used to accurately predict the fluid flow and heat transfer performance of microchannel flows, in channels with hydraulic diameters as small as 250 μm. On the basis of an analysis of the pumping requirements of microchannel heat sinks, the microchannel size was optimized to minimize the pumping power. Flow visualization results for laminar, transition, and turbulent flow in microchannels were also presented. Liu and Garimella (2004) presented experimental and numerical investigations for deionized water flowing in rectangular microchannels fabricated in Plexiglass to allow optical access for flow visualization studies. The hydraulic diameters of the microchannels evaluated were in the range of 250–1000 μm. To avoid external perturbations which can cause instabilities, a gas–pressure driven flow was used, with the help of a reservoir pressurized with nitrogen gas. Pressure drops in the longer channels were measured at locations remote from the entrance and the exit regions so that the entrance and exit effects could be avoided and would not adversely affect measured values. For the short microchannels, pressure measurements were made at the entrance and exit sections, and the entrance and exit losses were estimated and included in the calculations to obtain the pressure drop in the channels. Friction factors were determined for laminar and turbulent regimes and the experimental results were compared with simulations using a finite volume based CFD software package. It was found that the conventional theories accurately predict the flow characteristics in microchannels to a Reynolds number of approximately 2000. The onset of transition to turbulence was visualized, and it was observed that the development of fully turbulent flow was somewhat retarded in microchannels. The investigation indicated that reduced microchannel size may have a significant impact on the development of turbulence.

Lee et al. (2005) reported experimental results on single-phase flow of deionized water in copper microchannels ranging in width from 194 to 534 μm, maintaining an almost constant width to depth ratio of 5. These experiments covered the laminar range, and also extended into the turbulent flow range. The accuracy of a number of correlations was compared with the experimental results, with parameters such as the channel geometry, and whether the flow was developing or fully developed. The experimental results revealed that numerical predictions based on classical, continuum modeling accurately describe the flow and heat transfer in the range of microchannel dimensions evaluated, provided the entrance and boundary conditions are appropriately incorporated in the numerical models.

Favre-Marinet et al. (2004) conducted experimental investigations on fluid flow and heat transfer in parallel plate microchannels, and verified the experimentally determined optimal spacing between the channels to optimization results presented in the literature. The results of this comparison demonstrated that the classical prediction techniques were valid for channels with spacing greater than 0.4 mm, but that the Nusselt numbers were overpredicted, when the channel spacing was less than 0.4 mm. It was also noted that the flow transition is not affected by the channel size. Judy et al. (2002) reviewed the pressure driven liquid flow in microtubes and square microchannels fabricated from fused silica and stainless steel. Experiments were performed for channel hydraulic diameters in the range 15–150 μm, with water, methanol, and isopropanol to determine the effect of differences in the viscosity. The experimental results were found to match well with predicted values of the friction factor as determined from the Stokes flow theory. Lelea et al. (2004) presented experimental and numerical investigations of heat transfer and fluid flow in microtubes with various diameters in the developing laminar flow regime. A stainless steel-distilled water system was studied and comparisons of the experimental heat transfer and fluid flow characteristics with numerical and theoretical results for conventional tubes were presented. It was concluded that classical and conventional theories are applicable for water flow in microchannels, provided the entrance effects are properly included.

Ren et al. (2001) presented an experimental and theoretical investigation to determine the correlation between the pressure drop and flow rate of pure water and aqueous KCl solutions in microchannels. A strong dependence of the pressure drop on the channel size and the ionic concentration of the liquids were observed. The experimental data were also compared with the results of an electrokinetic flow model, signifying the effect of the electrical double layer and the electro-viscous effects on the pressure drop in pure water and the dilute ionic solutions studied.

Jiang et al. (2001) presented experimental investigations of heat exchangers with microchannels and porous media. In addition to comparing the relative performance, a numerical investigation was performed to analyze the influence of the microchannel size on the heat transfer performance of the exchangers. The heat exchangers analyzed consisted of deep and shallow microchannels and porous media. It was observed that, considering the heat transfer and pressure drop aspects, the deep microchannel heat exchanger offered better performance than the other two heat exchangers. Yin et al. (2002) reported measurements of the single-phase pressure drop in microchannel heat exchangers with complex headers and parallel circuits. The measurements were done to identify internal manufacturing defects. The results confirmed that the Moody chart is applicable for friction in submillimeter size channels.

Xu et al. (2005) developed a microchannel heat sink with a number of longitudinal and transverse microchannels, dividing the flow domain into several independent zones. This arrangement produced interruption and reattachment of the boundary layer, thus enhancing the heat transfer. In this investigation, the pressure drop was reduced, when compared to conventional longitudinal microchannels. In the experiments, the local chip temperature and heat fluxes were measured using a high resolution infrared temperature measuring system. An appropriate thermal resistance and a dimensionless pressure drop were both defined. In addition, a dimensionless parameter, which controls the heat transfer in each independent zone, was obtained as follows:

$$L_{h,s}^{+} = \frac{L_{h,s}}{D_h \text{Re} \ \text{Pr}} \tag{4.11}$$

where
$L_{h,s}^{+}$ is the nondimensional heating length
$L_{h,s}$ is the length of each independent zone

The variation of the overall Nusselt number with respect to this parameter was presented, and comparisons were made between the experimental results from two silicon microchannel heat sinks, one with, and one without the transverse channels, which indicated an enhancement in heat transfer and a reduction in the pressure drop for the new design.

Steinke and Kandlikar (2004a) reported experimental investigations on laminar single- and two-phase flow, during flow boiling of water in trapezoidal microchannels with a hydraulic diameter of 207 μm. It was found that under adiabatic conditions, the single-phase laminar flow friction factor could be accurately predicted using conventional correlations.

Though most of the above-cited references provided evidence to justify the use of conventional correlations and analyses for microchannels, there have been a number of investigations reported in the literature that identified deviations from conventional theory under certain conditions. Celata (2004) provided an overview of the research in single-phase heat transfer and fluid flow in capillary micro heat pipes, and presented experimental results of the friction factor and heat transfer in the laminar flow regime. It was found from the experimental investigations that the friction factor is in close agreement with the Hagen–Poiseuille law below Reynolds numbers of approximately 800. The transition from laminar to turbulent flow was found to occur in the Reynolds number range of 1800–2500. It was also noted that conventional heat transfer correlations are inadequate for the prediction of heat transfer rates in microtubes.

Some of the publications reviewed and discussed the nature of microchannel flow in general and micro heat exchangers and heat sinks in particular, and provided guidelines for research as well as the application of these devices. Eason et al. (2005a) presented a direct comparison of the pressure drop character-istics for microchannels, produced using different techniques (wet-etching, deep reactive ion etching, and precision sawing) and having different cross-sectional shapes (trapezoidal, rectangular, and near rectangular) and dimensions, in both silicon and plastic. The manufacturing process and the measurement of the channel dimensions were also presented. Experimental investigations of the microchannel system described by Eason et al. (2005a) were presented by Eason et al. (2005b), describing the experimental facility and the flow friction measurements. An error analysis and methods for accounting for the inaccuracies in the measurements were also presented. Comparisons of the experimental and predicted pressure drop and friction factor characteristics were presented and discussed for the various cases investigated.

Newport et al. (2004) reviewed the state of the art techniques in the measurement of the temperature distribution in microchannels, and suggested interferometric measurements for obtaining the temperature field in the fluid flow. The accuracy of extracting temperature data from phase difference intensity maps was discussed in the paper. Optical experiments to estimate the effect of the noise level on the phase evaluation were also described. Simulated interferograms and a two-dimensional intensity plot of the noise level in the measurements were presented.

Nishio (2004) discussed the heat transfer aspects of single-phase laminar flow, two-phase flow, and two-phase self-exciting, oscillating flow in microchannels. Detailed discussions of the thermal fluctuations in microchannels were presented, including the effect on evaporation and the thermo acoustic effects. The paper also discussed experimental results, which indicated that for tubes larger than 0.1 mm in diameter, the experimental results are in good agreement with conventional analyses. Kandlikar and Grande (2004) presented a roadmap of single-phase cooling technol-ogy to identify research opportunities to meet the cooling demands of future IC chips. An overview of the current technologies for chip cooling was presented, along with design considerations for direct water-cooling of microelectronics. The paper presented fabrication methods such as deep reactive etching technology, for micro-channels and microstructures within channels. It also presented discussions on key design concerns such as pressure drop, fouling, and cost considerations.

Kockmann et al. (2005) presented discussions on a numerical investigation of mixing and mass transfer in sharp 90° bends in microchannels. Heat transfer in "T" joints was also analyzed. The paper presented discussions of laminar flow pressure distributions, fluid forces, and transport processes (heat and mass transfer) in different configurations of bends and joints.

4.2.2.2   Gas Flow

The flow of gases and resulting heat transfer in microchannels has also received attention from various investigators. Rarefied flow conditions have been analyzed, and solutions obtained by applying slip and temperature jump boundary conditions. The following section summarizes the recent theoretical and experimental investigations of these flows.

Li et al. (2000) reported on a theoretical investigation of gas flow in microchannels. The wall effects on heat transfer characteristics were investigated for laminar flow, using a model, which utilized the kinetic theory for thermal conductivity variations. Analytical expressions for temperature distributions and heat transfer coefficients were derived for fully developed laminar flow in microtubes and microchannels. The investigation demonstrates that the variation of gas thermal conductivity in the wall-adjacent layer would influence the heat transfer significantly in gas flow in small dimension channels.

Yu and Ameel (2001, 2002) investigated laminar slip flow forced convection in the thermal entry length in a microchannel analytically, using a modified integral transform technique. Two dimensionless variables were identified, which when coupled with the Knudsen number, determines whether the heat transfer will increase, decrease, or remain unchanged, with respect to the nonslip conditions. These dimensionless variables include the effects of the channel size, rarefaction, and the fluid–wall interaction. The transition point which separates the effect, indicated by an enhancement or reduction of heat transfer, was obtained for different aspect ratios and mixed mean temperatures and Nusselt numbers were evaluated. The heat transfer was found to depend on two dimensionless variables that include the effect of rarefaction and the fluid–wall interaction. For different aspect ratios, the transition point between heat transfer enhancement and reduction was identified based on these dimensionless variables.

Analysis of convection heat transfer in the flow of rarefied gas through rectangular microchannels was presented by Tunc and Bayazitoglu (2002). The governing equations were solved using an integral transformation technique, for constant axial and peripheral heat fluxes and slip-boundary conditions, for various values of the Knudsen number. The results showed behavior similar to that occurring in microtubes and rectangular microchannels.

Sun and Faghri (2003) used a DSMC technique to analyze nitrogen flow in microchannels, modeling the surface roughness as an array of rectangular modules placed on the wall of a parallel plate channel. The application of the DSMC technique was checked against the friction coefficients using continuum theory and validated against the results obtained for rarefied flow. The variations of the friction coefficient with respect to surface roughness, and with the Knudsen number for rough channels, were presented. It was found that the roughness distribution has a significant effect on the friction factor. Colin et al. (2004) presented an analysis of gaseous flow in rectangular microchannels utilizing an analytical model with second order slip flow and temperature jump boundary conditions, taking into account the three-dimensional effects in the channel. Experimental measurements were performed, and a method to reduce the uncertainty was proposed. It was found that the proposed model predicts experimental results well up to Knudsen numbers of about 0.25. Helium and nitrogen gases were used in the investigation under rarefied conditions. Raju and Roy (2005) analyzed fluid flow and heat transfer in gaseous flow in microchannels, with first-order slip/jump boundary conditions, using a finite elements model. Helium and nitrogen gas flows were studied. The heat transfer characteristics predicted by the model were found to compare well with the results using the DSMC technique, reported in the literature. It was found that the assumption of a constant viscosity limits the accuracy of the solution.

Chen and Kuo (2004) presented solutions of the compressible boundary layer equations to study the heat transfer characteristics of gaseous laminar and turbulent flows in mini and microtubes, for uniform wall heat flux and isothermal boundary conditions. It was found that the numerical results agree well with the conventional correlations for large diameters and small Reynolds numbers, whereas for large Reynolds numbers and small channel diameters, they were found to vary considerably. The reasons for the difference were explained based on the compressibility effects of the gas. Figure 4.12 illustrates typical results, where the Nusselt number is plotted as a function of the turbulent Reynolds number, under a uniform wall heat flux condition.

Asako and Toriyama (2005) analyzed the heat transfer characteristics of gaseous flows in parallel plate microchannels using the arbitrary Lagrangian–Eulerian (ALE) method. The computations were performed for adiabatic and isothermal wall conditions, with channel heights ranging from 10 to 100 μm. Comparisons were presented for incompressible flow in conventional sized, parallel plate channels. Choi et al.

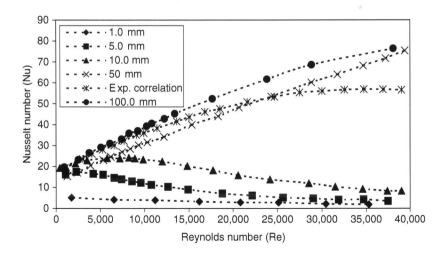

FIGURE 4.12 The Nusselt number as a function of Reynolds number for turbulent tube flows subject to a uniform wall heat flux. (Adapted from Chen, C.S. and Kuo, W.J., *Numerical Heat Transfer A*, 46, 497, 2004. With permission.)

(2003) analyzed gaseous slip flow in microchannels using a numerical model which combined the Navier–Stokes equations with a slip flow model called the Langmuir slip flow model, which is analogous to the Maxwell slip flow condition. The method was found to be an efficient and accurate tool for predicting microchannel slip flows.

Tso and Mahulikar (2000a) presented a numerical analysis of the interaction of radiation in an annular enclosure and an evaporating meniscus driven convection in a microchannel. The results of this analysis revealed the importance of the radiative heat transfer interaction in the coupled heat and fluid flow in microchannels.

Hsieh et al. (2004) presented an experimental and theoretical investigation of compressible flow of nitrogen in a microfabricated microchannel on a silicon substrate. The results on the friction factors were found to be in good agreement with the predicted values from a 2D analytical model with slip-boundary conditions, and a new momentum accommodation coefficient proposed based on the experimental investigation. Turner et al. (2004) presented experimental investigations of gas flow in microchannels. Etched microchannels on silicon substrates with almost rectangular cross sections were studied. The effect of the surface roughness, the Knudsen number, and the Mach number on the friction factor was investigated, using local pressure measurements along the microchannels. Microchannels varying in hydraulic diameters from 5 to 96 μm were studied. For low Mach and Knudsen numbers, the friction factor results showed close agreement with those predicted from continuum theory. An increase in the Mach number resulted in an increase in the friction factor, due to compressibility, whereas an increase in the Knudsen number showed a significant decrease in the friction factor. The influence of surface roughness was shown to be insignificant for both continuum and slip flow regimes. Kohl et al. (2005) conducted experiments on incompressible and compressible flows in microchannels with integrated pressure sensors, for a range of channel hydraulic diameters from 25 to 100 μm. Etched microchannels in silicon wafers were used in the experiment, and air and water were used as the test fluids. It was found that friction factors for microchannels could be accurately predicted using data for standard large channels. They suggested that the large inconsistencies in the previously reported data could be due to instrumentation errors or improper accounting for compressibility effects.

Murakami and Mikic (2003) investigated the effect of viscous dissipation on the performance of air cooled microchannel heat sinks. An optimization method is introduced, on the basis of dimensional groups, and applied to yield a design with the minimum pumping work, for given geometric dimensions and thermal load. The performance was evaluated and parametric influences on thermal load were calculated. The calculations revealed the importance of considering the effects of viscous dissipation in obtaining optimal designs for microchannel heat sinks. An in-depth review of the research related to heat transfer for gas flow in microchannels has been presented by Rostami et al. (2002).

4.2.2.3 Phase Change and Two-Phase Flow

Theoretical and experimental investigations have been reported on evaporation and boiling, condensation, and two-phase flow in microchannels and microtubes.

A majority of these investigations were directed at achieving a better understanding of the characteristics of phase change through measurements of flow and temperature, and by the application of flow visualization techniques. In the case of two-phase flows, comparisons with conventional theoretical models were performed. Information about two-phase flow regimes in microchannels was also brought out through careful experimentation.

On the basis of the classical theory of nucleation kinetics and bubble nucleation, Li and Cheng (2004) presented a numerical investigation of bubble cavitation in a microchannel, for water flowing in microchannels with different corner angles. A three-dimensional model was developed, and numerical computations were performed to understand the effects of mass flow rate and heat flux on bubble nucleation in microchannels. The effects of the contact angle, dissolved gases, microcavities, and shape of the microchannels (e.g., corner angles) on the nucleation/cavitation temperature were estimated. It was concluded that the parameters governing bubble nucleation in microchannels are the hydraulic diameter and length of the microchannel, the mass flow rate, the heat flux on the heater surface, the nucleation temperature, the inlet temperature of the fluid, and the top wall temperature of the microchannel.

Peng et al. (2000) hypothesized about what was referred to as "fictitious boiling" in microchannels, where bubble nucleation is apparently absent, and employed the principles of cluster dynamics, in an attempt to develop an understanding of this phenomenon. The role of perturbations on the dynamics of clusters was examined, and a pressure fluctuation model was proposed to provide a criterion for the occurrence of this fictitious boiling. Peng et al. (2002) also presented arguments on the thermodynamic and dynamic characteristics of bubble nucleation in microchannels. The cluster model was used to explain the density fluctuations in superheated liquids, and the kinematics of the clusters were analyzed using the Brownian motion phenomenon. The analysis explored the role of the scales of the microstructure in cluster kinematics and boiling nucleation in microchannels. Peng et al. (2001a) have also discussed the process of the growth of bubble embryo in detail and considered the effects of perturbations on embryo dynamics. It was concluded that the perturbations could alter the dynamic characteristics of bubble embryos, and that external perturbations have the possibility to either enhance or retard the generation of bubble embryos. A statistical mechanics approach was used to model the bubble embryos developed during boiling in microchannels (Peng et al. 2001b), and the bubble nucleation temperature of the fluid flowing in the microchannel was theoretically derived. Experimental measurements were performed to determine the nucleation temperatures for four working fluids, using a platinum wire heater inside a capillary tube. The results indicated that the bubble nucleation temperature increased as the microchannel size was reduced. Micro photographic observations suggested vigorous oscillations in the liquid, close to bubble nucleation temperature, which were attributed to the production of tiny bubble embryos.

Jacobi and Thome (2002) proposed a model for heat transfer in the elongated bubble flow regime in microchannels, with the hypothesis that thin film evaporation is the governing process in microchannel evaporation. Two-phase flow and energy balance equations that represent the characteristics as a function of the position were

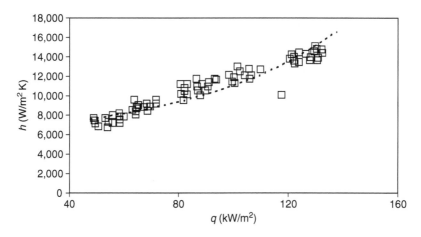

FIGURE 4.13 Comparison of the proposed model for elongated bubble flow regime with available experimental data. (Adapted from Jacobi, A.M. and Thome, J.R., *J. Heat Transfer*, 124, 1131, 2002. With permission.)

developed. Model results were compared with experimental data, and the model was found to be capable of representing the data with a reasonable degree of accuracy as shown in Figure 4.13, thus indicating that the thin-film evaporation mechanism could explain two-phase heat transfer in microchannels, as opposed to the conventional macroscale nucleate boiling. Thome et al. (2004) presented a three-zone evaporation model to predict the local dynamic heat transfer coefficient at fixed locations along a microchannel during flow and evaporation of an elongated bubble. A constant, uniform heat flux boundary condition was used in the analysis. The model was based on an energy balance, and the assumption that bubble nucleation occurs at the location where the fluid first attains the saturation temperature. This resulted in a time averaged local heat transfer coefficient. Dupont et al. (2004) in a subsequent paper compared the results from the three-zone flow boiling model with the existing experimental results, and conducted a parametric investigation. The strong dependence of heat transfer on the bubble frequency, the minimum liquid film thickness at dry-out, and the liquid film thickness were all apparent in the results of this investigation. The importance of the cyclic variation of the heat transfer coefficient was also illustrated using the validated model.

Nilson et al. (2004) obtained analytical solutions for evaporating flow in a tapered microchannel of uniform depth, using the one-dimensional energy equation and the Laplace–Young equation. The flow domain was divided into two regions, the entry domain where the meniscus is attached to the top corners of the channel and the recess domain where the meniscus retreats along the walls to the bottom of the channel, and the domains are matched at the interface in the analysis. It was found that tapered channels performed better than straight channels in terms of the cooling capacity.

Chakraborty and Som (2005) presented a theoretical investigation to analyze the flow of a thin evaporating film along the walls of a microchannel under the combined effects of gravity and surface tension, solving the conservation equations with coupled heat and mass transfer boundary conditions at the interface. A dimensionless number was suggested, which was found to influence the average Nusselt number, apart from the liquid film thickness, while the local Nusselt number was found to be influenced only by the film thickness.

Hao and Tao (2004) presented a numerical simulation of laminar hydrodynamic and heat transfer characteristics of the flow of a fluid containing a suspension of micro/nano sized phase change materials, in microchannels. Two-phase conservation equations were solved to investigate heat transfer enhancement mechanisms and their limits. Propagation of the melting region and local heat transfer coefficients for constant wall heat flux and constant wall temperature boundary conditions were obtained. Effects of nonthermal equilibrium melting of the phase change material, and the existence of a particle depleted layer, and particle–wall interaction were also observed. It was noted that the introduction of the phase change material strongly enhances the heat transfer in the melting region.

Du and Zhao (2003) presented a theoretical analysis of the film condensation process in a vertical microtube with a thin metal wire attached to its inner surface. A two-dimensional analysis was performed to obtain the radial and axial distributions of the condensate liquid along the tube wall and in the meniscus zone. The governing equations were developed based on the principle of minimum energy in the two-phase flow system. The influence of the contact angle and the wire diameter on the condensate distribution and the heat transfer characteristics were studied. The system under analysis was found to provide better condensation heat transfer performance than a simple round tube.

In addition to the analytical and numerical investigations, there have been a number of experimental investigations of phase change and two-phase flows in microchannels, using both measured values and flow visualization. Kandlikar and Balasubramanian (2004) explored and extended the applicability of the flow boiling heat transfer coefficient correlations developed for channels ranging from relatively large diameter channels to mini and microchannels. Extended correlations for laminar and transition flow regimes in both mini- and microchannels were presented based on data analysis, accounting for the low Reynolds number values encountered in these geometries. Steinke and Kandlikar (2004a) reported on experimental investigations of single- and two-phase flows, during flow boiling of water in trapezoidal microchannels. The results indicated a two-phase flow reversal under certain conditions. The changes occurring in the interface under dry-out conditions were visually observed and documented in this work. Steinke and Kandlikar (2004b) have also reported experimental investigations designed to investigate the effects of dissolved gases on the heat transfer and pressure drop in water flowing through microchannels. Experiments were conducted in a microchannel array, fabricated in a copper block substrate. The experimental test facility was developed to obtain the desired level of dissolved air in the test fluid (deionized water). The surface temperatures in the channels were deduced based on a heat flux calculation, using measured temperatures at different depths in the copper block, and the heat transfer coefficients were estimated based on the logarithmic mean temperature difference. It was observed that the appearance of nucleation and the values of the heat transfer coefficients are both dependent on the quantity of dissolved air. In the experimental case with relatively large dissolved gas quantities (8.0 ppm), a reduction in the heat transfer due to the presence of an insulating bubble layer over the surface was observed.

Kandlikar (2004) discussed the heat transfer mechanisms during flow boiling in microchannels, and identified the nondimensional groups which are relevant to microchannels. Two new nondimensional groups were proposed, which represent the surface tension forces around the contact line region, the momentum change forces due to evaporation at the interface, and the inertial forces. A correlation obtained from the combination of these groups was found to represent the boiling CHF data. The dominance of the nucleation effect of flow boiling in microchannels was also discussed based on high-speed flow visualization experiments. Balasubramanian and Kandlikar (2005) have also performed experimental investigations on flow boiling of water in parallel microchannels with hydraulic diameters of 333 μm. A set of six parallel rectangular microchannels were used in the investigation. Pressure fluctuations were recorded, and the dominant frequencies and amplitudes were obtained through signal analysis. Visualization investigations for observing the flow patterns were conducted using a high-speed camera, and the pressure fluctuations were mapped on to the flow patterns. The paper also discussed the conditions of maldistribution and reverse flow in the channel array, with the help of the visualization investigation.

Wu and Cheng (2004) reported experimental investigations on flow boiling of water in parallel silicon microchannels of two hydraulic diameters. Visualization studies and measurements were carried out. The visualization investigation indicated the alternate appearance of single- and multiphase flow in these microchannels once boiling heat transfer had been established. The experimental results confirmed that large-amplitude and long-periodic fluctuations characterized boiling when the fluctuations of pressure drop and mass flux had phase differences. Flow patterns in the microchannels were observed using a microscope and high-speed camera, and were presented. Experimental investigations are also reported in Cheng and Wu (2003b). Trapezoidal microchannels with a hydraulic diameter of 186 μm and a length of 30 mm were used in the experimental investigation with various wall heat fluxes and mass fluxes. Three kinds of flow patterns were observed: liquid–two-phase alternating flow (at low heat and mass fluxes), continuous two-phase flow (at medium heat and mass fluxes), and liquid/two-phase/vapor alternating flow (at high heat and mass fluxes). The investigation also focused on the amplitude and frequency of the oscillations for all cases. It was found that bubbly flow is the most dominant flow pattern during two-phase flow in alternating flow. Wu and Cheng (2005) also reported visualization and measurements of condensing flow of steam through silicon microchannels. Flow patterns such as droplet flow, annular, injection, and slug flow were all observed. It was noted that, due to the bubble growth and detachment, termed as vapor injection flow, the two- and single-phase flows appeared alternately in the microchannels, causing large fluctuations in the temperature and flow measurements. Visualization results for vapor injection flow and the associated instabilities were reported for the first time.

Ghiaasiaan and Chedester (2002) presented an analysis of the available data on the boiling incipience of water in microtubes in the diameter range of 0.1–1 mm. It was observed that macroscale models and correlation appear to under-predict the heat flux required for boiling incipience in microtubes. It was suggested that the boiling incipience in microchannels might be controlled by the thermocapillary forces that tend to suppress the formation of microbubbles on the wall cavities. A semiempirical method was suggested for predicting the onset of nucleate boiling in microtubes, the results of which were shown earlier in Figure 1.3 (Chapter 1), compared with the experimental data. The paper also discussed the effects of the turbulent characteristics of the microtube.

Piasecka and Poniewski (2004) investigated boiling incipience in a narrow rectangular vertical channel of 1 mm depth, measuring the temperature distribution using crystal thermography. The impact of various factors such as pressure, inlet liquid subcooling, and flow velocity on boiling incipience was studied. The occurrence of a nucleation hysteresis associated with boiling incipience was also observed and discussed.

Hetsroni et al. (2004) performed experimental investigations of convective boiling of a water–surfactant mixture flowing in parallel triangular microchannels. Temperature and flow patterns were measured using infrared radiometry and high-speed digital video imaging. Pressure and temperature instabilities in the channels were also investigated. It was found that boiling of surfactant solutions in microchannels may be used to provide a nearly isothermal heat sink. In addition, this paper presents discussions on incipience and growth of bubbles, and gives a number of flow visualization results. Hetsroni et al. (2002) have also presented investigations on the cooling of a heat sink for electronic applications, using the boiling of a dielectric liquid flowing through parallel triangular microchannels with a base dimension of 250 μm. The experiments were based on the measurement of temperature and pressure and utilized infrared radiometry and high-speed digital imaging to obtain the results. Comparison of the results for two- and single-phase flow was also presented.

Experimental investigations of bubble dynamics in a single trapezoidal microchannel with a hydraulic diameter of 41.3 μm and fabricated using micromachining and anodic bonding processes on silicon were presented by Lee et al. (2004). The bubble dynamic processes were visualized using a high-speed digital camera. The experiments indicated that bubble nucleation could be predicted using the classical model for microsized cavities. The bubble departure from the microchannel walls was found to be governed by the surface tension and frictional drag resulting from the bulk flow. The bubble frequency was found to be comparable with that of an ordinary sized channel, and the conventional form of frequency-departure diameter was found to be nonexistent in the microchannel studied. Li et al. (2004a) extended this experimental investigation to two parallel microchannels, and the results indicated that the bubble growth and departure behavior were similar to those observed in the single microchannel. Bubble dynamics in slug flow were also investigated.

Zhang et al. (2005) reported on an experimental investigation of micromachined silicon microchannels with hydraulic diameters ranging from 27 to 171 μm and varying surface roughnesses. Bubble nucleation, flow and temperature patterns, and transient pressure fluctuations were all observed and studied. Typical nucleate boiling and eruption boiling were noticed in these channels, and large pressure fluctuations were recorded due to bubble nucleation. The boiling mechanism was found to be strongly dependent on the wall surface roughness. Explanations of boiling in channels with small hydraulic diameters were provided, based on comparisons with theoretical models present in the literature (Hsu 1962).

Brutin et al. (2003) conducted an experimental investigation of unsteady convective boiling in a mini channel. Rectangular channels with a hydraulic diameter of 889 μm were used in the investigation. Steady-state and nonstationary two-phase flow patterns were identified through pressure signal fluctuations, and were analyzed. Steady and unsteady behavior depending on the heat flux and mass-velocity were analyzed in this experimental investigation.

Wang et al. (2004) presented flow visualizations of boiling in a heated microchannel 270 μm wide and 97 μm deep. Fluorescent polystyrene spheres, 700 nm in diameter, were used as seed particles to visualize the flow. The nucleation and growth of vapor bubbles and the onset of partial dry out were apparent in the pictures presented. Son and Allen (2004) presented visualization results for two-phase flow in 300 μm diameter circular and 500 μm square microchannels, using high-speed microscopy. Coated and uncoated channels were used to obtain different wetting conditions and the results indicated that wetting conditions play an important role in the fluid flow resistance. The coated channels resulted in less stable flow and enhanced droplet formation.

Coleman (2004) presented experimental investigations of two-phase pressure drop for R134a in microchannel headers. Experimental data were compared with various theoretical models and indicated that many of the commonly used correlations for estimating two-phase pressure losses significantly underpredict the pressure losses in compact microchannel tube headers. Kawahara et al. (2005) also reported experimental investigations of two-phase flow to determine the void fraction in adiabatic two-phase flow through circular microchannels. The experiments were performed using water/nitrogen gas and ethanol–water/nitrogen gas combinations, with various ethanol concentrations in water to vary the surface tension and viscosity. The experimental results for tube diameters of 50, 75, 100, and 251 μm indicated that the void fraction data agreed very well with microchannel correlations, which implies that the boundary between micro and mini channels would lie between 100 and 251 μm. The void fraction data indicated that the liquid properties do not influence this significantly.

Bergles and Kandlikar (2005) presented an extensive review and discussions of the stability mechanisms and CHF for boiling in microchannels, with particular emphasis on bulk boiling. The investigation focused on the flow distribution in microchannel heat exchangers, and more specifically, boiling incipience and CHF experiments. It was observed that the CHF could be a result of upstream compressive volume instability or an excursive instability, rather than dry out, and that a proper inlet restriction can reduce or potentially eliminate instabilities. It was suggested that the next step in microchannel research should be focused on the fabrication and investigation of the effects of such orifices.

Garimella (2004) presented an overview of experimental investigations of the condensation flow patterns, pressure drop, and heat transfer in micro and mini channels for various cross-sectional shapes and hydraulic diameters. Flow visualization results were presented and discussed to provide an understanding of the flow mechanisms. These investigations indicated that transitions between the various flow regimes in microchannels occur at conditions different than those expected in conventional geometries. Experimentally validated models for the condensation pressure drop were also presented in this paper. Garimella et al. (2005) combined experimental results on pressure drop measurements during the condensation of refrigerant R134a in horizontal microchannels with existing pressure drop models and proposed a comprehensive model that addresses the entire progression of the condensation process, from the vapor phase to the liquid phase. Overlap and transition between the flow regimes were also addressed using an appropriate interpolation technique. A comparison of the calculated and measured pressure drops for the noncircular channels reported in their work was shown earlier in Figure 3.14.

4.2.2.4   Design Optimization

Several investigators have discussed the design optimization of microchannel heat sinks and heat exchangers, and a number of suggestions and processes on this aspect have been presented. Kim (2004) presented, discussed, and summarized three methods of optimization: the application of a fin model, a porous medium model, and an optimization method based on the three-dimensional numerical simulation of microchannel heat sinks to evaluate the thermal resistance. The porous medium model was found to predict the thermal performance of the microchannel heat sink more accurately than the fin model with a one-dimensional heat conduction assumption. Nakayama et al. (2004) presented discussions on the geometric uncertainties associated with microchannels incorporated on electronic devices. A summary of the sources of uncertainties was presented, along with a CFD simulation based methodology to assess situations involving geometric uncertainties. A model channel was defined and used to illustrate the proposed methodology. The results of the simulation indicated the stringent measures required in assembling microchannel-cooling devices. A hypothetical experiment for comparing the performance of a channel of optimum height, with that of a nonoptimal channel under a given pumping power, was performed.

Hrnjak (2004) presented an overview of the issue of misdistribution of flow in microchannel heat exchangers. Various options for distributing single- and two-phase fluid evenly in the channels were presented and discussed. Two-phase regimes in a horizontal adiabatic header with R134a as the working fluid were described using flow visualization maps for the two-phase flow.

4.2.2.5   Review Papers

As mentioned previously, several review articles that compare and analyze the reported literature on microchannel flow and heat transfer have been published in recent years. Sobhan and Garimella (2001) presented a compilation and analysis of the results of experimental and theoretical research on fluid flow and heat transfer in microchannels in the 1990s, with a special focus on quantitative observations and predictions. Anomalies and deviations reported in the behavior of microchannel flow and heat transfer were identified and compared with the results of conventional prediction techniques. The need for extensive investigations with careful design, fabrication, and instrumentation of the microchannels to examine and eliminate the resulting disparities in the observed results was emphasized in the paper. A critical review of the literature related to microscale heat transfer, and discussions on the wide discrepancies among reported results can also be found in a subsequent paper by Garimella and Sobhan (2003).

A review of the literature related to single- and two-phase flow in microchannels was presented by Palm (2001). The information presented identified a number of unanswered questions related to the reasons for observed deviations from conventional theory in the microchannel flow characteristics and behavior. Guo and Li (2003) presented a review and discussion of the size effects of microscale single-phase flow and heat transfer, classifying them into the gas rarefaction effect and the effect of the influence of the dominant factors, as the characteristic length decreases. This work discussed in detail the effects of surface roughness, variation of the dominant forces acting on the fluid flow, the axial conduction effects, and other important factors such as surface geometry and measurement accuracy, in microscale fluid flow and heat transfer data.

A review of research on heat transfer in gas flow in microchannels has been presented by Rostami et al. (2002). The potential applications of gas flow in microchannels were discussed, followed by a critical review of the literature on flow and heat transfer in gases flowing in microchannels. The results indicated that the flow and heat transfer characteristics cannot be adequately predicted by the theories and correlations developed for conventional sized channels.

Thome (2004) has presented a review of recent experimental and theoretical research on boiling in microchannels. The review focused on the heat transfer mechanisms in microchannels and flow boiling models for microchannels, and compared microscale and macroscale heat transfer with a special emphasis on boiling and two-phase flow. Flow visualization investigations and data on evaporation models were also discussed along with a three-zone flow boiling model for evaporation of elongated bubbles in microchannels.

4.3   ANALYSIS OF SYSTEMS FOR ENGINEERING APPLICATIONS

In the foregoing sections, review has been presented on the advances in heat and fluid flow in microchannels, separately treating the investigations reported in the early stages of development and in recent times. A few cases of theoretical analyses and experimental investigations of systems, relevant to direct applications to microscale engineering, are discussed in the following sections.

4.3.1   COMPUTATIONAL ANALYSIS OF MICROCHANNEL HEAT SINKS

As discussed earlier, investigations of fluid flow and heat transfer in microscale passages have indicated that conventional correlations for large channels may not be capable of accurately predicting the frictional pressure drop or heat transfer characteristics in either individual microchannels or their arrays. The rationale for these variations, however, is not well understood and many researchers believe that the dimensions of these microchannels are sufficient in size to justify the use of assumptions and models based upon the continuum theory. Though it would require extensive and careful experimental research to bring out conclusive results and observations regarding the suitability of the use of conventional correlations for predicting the fluid and thermal behavior in these channels, it is very instructive to perform modeling and theoretical investigations on channels with microscale dimensions using conventional governing equations for fluid flow for comparative purposes. Overall performance characteristics from such analyses could serve not only to provide information and trends for comparison with existing experimental data, but also to optimize the channel dimensions and spacing for practical design applications.

Investigations performed on laminar single-phase flow of water in rectangular microchannels, fabricated in practically useful substrates such as silicon, have been used to determine the resulting heat dissipation effects. Some of these investigations are presented in this section. The objective of these investigations is the calculation of local and average heat transfer coefficients and Nusselt numbers using the temperature distributions obtained through the solution of the nondimensional governing equations. Multidimensional approaches have been utilized in the modeling, some of which will be discussed below.

4.3.1.1   Analysis of Rectangular Microchannels

An interesting computational investigation of integrated microchannels has been presented in the literature by Weisberg et al. (1992). In this work, a nondimensional formulation has been utilized, based on a nondimensional temperature defined in terms of the bulk fluid temperature, to solve the energy equation, while the nondimensional velocity field was obtained from classical expressions for fully developed laminar flow. The two dimensional analysis presented in this work was used for an optimization study, which led to the determination of the conditions for minimizing the nondimensional thermal resistance. Extending this investigation, the analysis has been made more realistic by incorporating a pseudo three-dimensional marching in the numerical scheme, as well as by solving the momentum equation to obtain the appropriate nondimensional velocity distribution in the channel cross-sections, as presented below. An optimization investigation was also performed, using the nondimensional thermal resistance defined by Weisberg et al. (1992), following the computational analysis. Comparisons with existing experimental results (Lee and Garimella 2003) suggest that the continuum model utilized in the analysis predicts the channel performance reasonably well.

The arrangement of rectangular microchannels on a substrate, analyzed in the present study, is as shown earlier in Figure 1.7. In this configuration, the channels have a uniform rectangular cross section of width W_c, a height of H_c, and a length L. Further, in this arrangement, it is assumed that the heat transfer into the system is uniform and is applied to the top surface of the wafer while the bottom surface

TABLE 4.2

Details of the Baseline Case Analyzed

Parameter	Value in the Baseline Case
Geometric dimensions	$H_c = 0.3$ mm, $W_c = 0.18$ mm
	$L = 25$ mm
	$H = 0.43$ mm, $W = 0.36$ mm
Inlet fluid temperature	24°C
Reynolds number	500 for base line case, 500–2400 range
Heat input	50 W/m^2
Pressure drop	80 kPa

Thermophysical Properties	Silicon	Water
Thermal conductivity	148 W/m K	0.628 W/m K
Density	2330 kg/m^3	995 kg/m^3
Specific heat	712 kJ/kg K	4178 kJ/kg K
Thermal diffusivity	89.2×10^{-6} m^2/s	0.1511×10^{-6} m^2/s
Kinematic viscosity	—	0.657×10^{-6} m^2/s

is insulated. Ultra pure water is used as the cooling fluid, which is uniformly circulated through the channels. To better understand the physical nature of the numerical solution, a baseline case was studied, with the dimensions and operating conditions listed in Table 4.2. Further, parametric and optimization studies were also performed, changing the geometrical dimensions.

The domain of analysis for the mathematical equivalent of the problem is shown as the shaded portion in Figure 1.7 given in Chapter 1, which represents the channel and the substrate configuration. A nondimensional formulation with the channel height, H_c, as the characteristic length dimension was utilized, and the governing equations were derived accordingly. The flow is assumed to be steady, laminar, and hydrodynamically and thermally fully developed, as suggested by Tuckerman and Pease (1981) based on their experimental observations.

A quasi three-dimensional approach was used to analyze the problem, which solves the governing equations in two dimensions (cross section) and utilizes a marching technique for the third dimension (length). The mathematical formulation, solution method, and the salient results of the analysis are given below.

The computational domain for the analysis is shown in Figure 4.14. Neglecting frictional dissipation effects for steady, laminar flow with convective heat transfer in the channel, the governing momentum equation can be written as follows. The dominance of the viscous terms also pertains to general microchannel flow.

$$\frac{\partial^2 u}{\partial x^2} + \frac{\partial^2 u}{\partial y^2} = \frac{1}{\mu}\frac{dP}{dz} \tag{4.12}$$

$$\frac{dP}{dx} = \frac{dP}{dy} = 0 \tag{4.13}$$

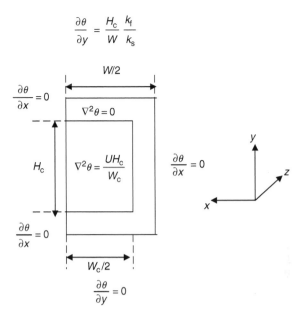

FIGURE 4.14 The domain of computational analysis with boundary conditions for the heat transfer problem.

As the flow becomes fully developed, the velocity "u" does not vary with the axial length, z, which implies that $\frac{dP}{dz}$ is a constant. Proper scaling of the momentum equation should include the parameters H_c, $\frac{dP}{dz}$, and μ.

The momentum equation can be nondimensionalized using the dimensionless variables given below:

$$X = \frac{x}{H_c}, \quad Y = \frac{y}{H_c}, \quad U = \frac{u}{\bar{u}}$$

where $\bar{u}$ is a bulk mean velocity given by

$$\bar{u} = \frac{1}{W_c H_c} \int_0^{W_c} \int_0^{H_c} u \, dy \, dx \tag{4.14}$$

To further simplify the momentum equation, a transformation can be applied as follows:

$$U^* = \frac{U\bar{u}}{\frac{H_c^2}{\mu}\left(-\frac{dP}{dz}\right)} \tag{4.15}$$

which allows the momentum equation to be expressed as

$$\frac{\partial^2 U^*}{\partial X^2} + \frac{\partial^2 U^*}{\partial Y^2} = -1 \tag{4.16}$$

subject to $U^* = 0$ at all of the boundaries of the channel cross section.

The actual nondimensional velocity U can be obtained from the solution for U^* as

$$U = \frac{u}{\bar{u}} = \frac{U^*}{\bar{U}^*} \tag{4.17}$$

$$\text{where} \quad \bar{U}^* = \frac{2H_c}{W_c} \int_0^{\frac{W_c}{2H_c}} \int_0^1 U^* \, dY \, dX \tag{4.18}$$

To nondimensionalize the energy equation utilizing the bulk fluid temperature, the approach suggested by Weisberg et al. (1992) was adopted. The method is outlined below:

The steady-state energy equation for laminar fully developed flow, neglecting viscous dissipation, can be written as

$$C_f \rho_f u \frac{\partial T_b}{\partial z} = k_f \left(\frac{\partial^2 T}{\partial x^2} + \frac{\partial^2 T}{\partial y^2} \right) \tag{4.19}$$

For the substrate, the two-dimensional conduction equation is given by

$$\frac{\partial^2 T}{\partial x^2} + \frac{\partial^2 T}{\partial y^2} = 0 \tag{4.20}$$

Using the expressions for the bulk fluid temperature rise, and the definition of the nondimensional temperature difference θ, as obtained by Weisberg et al. (1992), the above equations become

$$\frac{\partial^2 \theta}{\partial X^2} + \frac{\partial^2 \theta}{\partial Y^2} = \frac{U H_c}{W_c} \quad \text{(in the fluid)} \tag{4.21}$$

$$\frac{\partial^2 \theta}{\partial X^2} + \frac{\partial^2 \theta}{\partial Y^2} = 0 \quad \text{(in the substrate)} \tag{4.22}$$

where

$$\frac{dT_b}{dz} = \frac{QW}{\rho_f C_f \bar{u} W_c H_c} \tag{4.23}$$

$$\theta = \frac{k_f}{QW} (T - T_b) \tag{4.24}$$

The definition of the nondimensional temperature difference, θ, as expressed in Equation 4.24 implies that

$$
\int_0^{\frac{W_c}{H_c}} \int_0^1 U\theta \ \mathrm{d}Y \ \mathrm{d}X = 0 \tag{4.25}
$$

This condition is utilized to check the computational results.

The boundary conditions utilized for the fluid flow are the no-slip conditions on the solid walls. The boundary conditions utilized for solving the energy equation include symmetric boundary conditions in temperature, insulated boundary condition at the bottom boundary, as well as interfacial heat balance at the solid–liquid interfaces. The nondimensional governing equations and boundary conditions for the heat transfer problem are illustrated in Figure 4.14 showing the computational domain.

From the temperature distributions in the flow field, two important heat transfer parameters can be determined, namely the average Nusselt number and the overall thermal resistance. The Nusselt number is used for validation of the model by comparison with the existing experimental results. The thermal resistance is utilized in the optimization investigation to arrive at the optimal channel arrangement for a given operating condition. These parameters are further explained below.

The temperature distribution, deduced from the nondimensional results in the field, can be used to determine the surface heat fluxes and heat transfer coefficients on the surfaces at any location from the following equation:

$$
h_1 = -\left(\frac{k_f}{T_w - T_f}\right)\left(\frac{\partial T}{\partial n}\right)_w \tag{4.26}
$$

A peripheral mean heat transfer coefficient at a given axial location can be calculated as an integral average of the local coefficients along the channel periphery. The local temperature difference between the surface and the bulk fluid is used in defining this heat transfer coefficient. An integral average Nusselt number can be calculated for the entire channel, from the average Nusselt numbers at cross sections, through a numerical integration of the values obtained through repeated calculations at various axial locations. Such average Nusselt number values are used in representing the results of the computational analysis in graphical form to be discussed later.

Determination of the optimal channel arrangement requires that the resistance to heat flow from the substrate-fluid interface into the bulk fluid flowing through the channel be minimized. Following the analysis presented by Weisberg et al. (1992), the dimensionless thermal resistance is obtained as follows:

$$
r = \frac{W_T k_f L R}{H_c} = \left(\frac{W}{H_c}\right)\bar{\theta}_{\text{surface}} \tag{4.27}
$$

Minimization of this quantity leads to the optimum design of the heat sink.

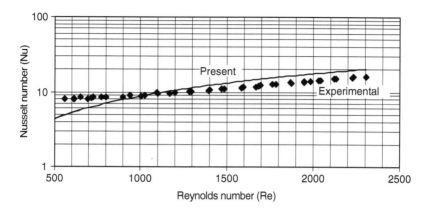

FIGURE 4.15 Comparison of the predicted average Nusselt number and the experimental data presented in the literature by Lee and Garimella (2003).

Utilizing a Finite Difference scheme for solution (Patankar 1980), it was possible to obtain convergence with a fairly coarse computational domain grid size (30×20). As the optimization investigation is based on averaging of surface temperatures, the use of coarse grids is justified, also taking into account the savings in computation time. However, the analysis can be carried out using much finer grid structures to produce more accurate local results.

To compare the overall performance of the channel and validate the numerical model, the experimental results of Lee and Garimella (2003) for channel hydraulic diameters of 318 μm were used in Figure 4.15. As shown, the correlation between the predicted and measured values is fairly good at Reynolds numbers above 800 with a gradually increasing deviation with increasing Reynolds numbers. At Reynolds numbers below 800, the computed Nusselt number decreases rather drastically with decreasing Reynolds numbers. The deviation at lower Reynolds numbers may, in part, be due to the constraints in measurement at relatively low flow rates.

The cross-sectional distributions of the velocity and temperature were obtained and used in the computation of the local and average Nusselt numbers and thermal resistances. Figure 4.16 illustrates the variation in the dimensionless thermal resistance with respect to the ratio of the channel width to the channel pitch, W_c/W, for various aspect ratios (H_c/W_c). The operating conditions and design constraints under which this plot was constructed are described in the later section on optimization. As shown, for a constant value of aspect ratio, the dimensionless thermal resistance decreases with increases in the width to pitch ratio. This is expected, as the larger the surface area for heat transfer, the more closely the bulk fluid temperature will approach the surface temperature, thus reducing the thermal resistance. The variation of the thermal resistance is steeper for smaller values of the aspect ratios.

On the basis of the field solution and computation of the thermal resistance previously presented, an investigation was performed to obtain the optimal arrangement of channels on the substrate. The design constraints selected to demonstrate the

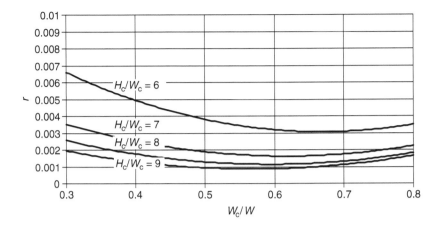

FIGURE 4.16 Variation of the dimensionless thermal resistance with respect to the width to pitch ratio of the channel for different aspect ratios of the channel.

procedure are a maximum rise in bulk temperature of the fluid of 40°C, an allowable maximum pressure drop in the channel of 80 kPa, and a maximum channel depth of 300 μm.

The optimization procedure involves arriving at the minimum thermal resistance subject to the given constraints. Optimization is done by minimizing the dimensionless thermal resistance with respect to the width to pitch ratio and the height to width ratio. The combination of these parameters that will produce the minimum thermal resistance can be determined, by performing a parametric investigation of the computational model. The optimization procedure for a typical case with the operating conditions and design constraints given in Table 4.3 can be best understood by examining Figures 4.16 and 4.17.

TABLE 4.3

Feasible Solutions: Dimensionless Thermal Resistances and Geometric Parameters

Design Constraints

Maximum rise in bulk temperature = 40°C
Allowable pressure drop in the channel = 80 kPa
Maximum channel depth = 300 μm

H_c/W_c	Width to Pitch Ratio (W_c/W)	Dimensionless Resistance (r)
6	0.372	0.0053
7	0.5	0.00185
8	0.644	0.00131
9	0.805	0.00149

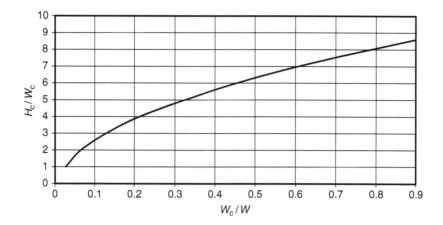

FIGURE 4.17 Comparison of the height to width ratio (H_c/W_c) and the width to pitch ratio (W_c/W).

The steps involved in selecting the optimal channel arrangement are given below:

1. A plot of the dimensionless thermal resistance with respect to the width to pitch ratio (W_c/W) is made, for various values of aspect ratios as given in Figure 4.16.
2. A plot of H_c/W_c for various values of the width to pitch ratio (W_c/W) is made, for specified values of pressure drop and bulk mean temperature rise (Figure 4.17).
3. The value of (W_c/W) is obtained for various values of H_c/W_c from Step 2.
4. The value of the dimensionless thermal resistance for each aspect ratio is noted, corresponding to the value of H_c/W_c, from the plot in Figure 4.16.
5. The channel dimensions and arrangement corresponding to the minimum value of the dimensionless thermal resistance are selected.

Table 4.3 lists the dimensionless thermal resistances for various aspect ratios and width to pitch ratios from the optimization investigation as performed on the design constraints. From Table 4.3, the solution corresponding to the minimum thermal resistance involves an aspect ratio of 8 and a width to pitch ratio of 0.644 for this case. Using this, the design geometric parameters can be calculated and are shown in Table 4.4.

A detailed numerical simulation of the heat transfer occurring in silicon-based microchannel heat sinks has been conducted (Li et al. 2004b, Li and Peterson 2006) to optimize the geometric structure using a simplified, three-dimensional conjugate heat transfer model (2D fluid flow and 3D heat transfer). The micro heat sink modeled in this investigation consists of a 10 mm length silicon substrate with rectangular microchannels fabricated with different geometries. The rectangular microchannels had bottom widths ranging from 20 to 350 μm and a depth ranging

TABLE 4.4

**Selected Microchannel Arrangement
from Optimization Study**

Design Parameter	Optimal Value
Channel height (H_c)	300 μm (constraint)
Channel width (W_c)	37.5 μm
Pitch (W)	57.87 μm
Aspect ratio (H_c/W_c)	8
Number of channels per mm	17

from 100 to 400 μm. Assuming fully developed laminar flow, the analysis involved the solution of the three-dimensional conservation equations of momentum and energy in the fluid and the heat conduction equation in the substrate using a nondimensional formulation, and employing an implicit finite difference procedure. The model was validated by comparing the predicted results with previously published experimental results and theoretical analyses.

The results of the analysis indicated that both the physical geometry of the microchannel and the thermophysical properties of the substrate are important parameters in the design and optimization of these microchannel heat sinks. For the silicon–water micro heat sink in fully developed laminar flow, the optimal configuration for rectangular channel heat sinks was determined. From the numerical simulation, it was recommended that under any constraints, the preferred height of the channel is the largest possible value that can be achieved through the microfabrication processes used. This was again verified by the optimal analysis for two different channel heights, 180 and 360 μm. The major contributions resulting from this investigation were: (1) using a full 3D simulation, the assumptions resulting from the 2D numerical fin calculations can be removed, resulting in more accurate results; (2) an insight into the problem was obtained which indicates that while some results obtained in previous investigations are correct, others may not be completely accurate; and (3) the 3D analysis performed provided a full understanding of the effects of the geometry of the channel on the heat transfer capacity of the heat sink.

A typical nondimensional temperature distribution obtained from the analysis and a typical optimization result from the investigation are given in Figures 4.18 and 4.19. A complete discussion on the formulation, solution methodology, and extensive interpretations of the solution and optimization results is presented in references (Li et al. 2004b, Li and Peterson 2006).

4.3.1.2   Micro Fin Arrays in the Slip Flow Domain

The application and utility of modified continuum solutions through the introduction of slip-boundary conditions will be demonstrated in this section, by analyzing the problem of a two-dimensional micro fin array in laminar natural convection.

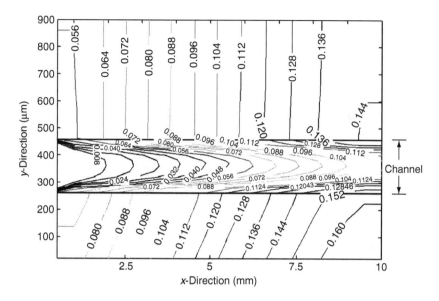

FIGURE 4.18 A typical nondimensional temperature distribution in the cross section of the heat sink at a particular longitudinal location. (From Li, J. and Peterson, G.P., *IEEE Trans. Comp. Packaging Technol.*, 29, 145, 2006. With permission.)

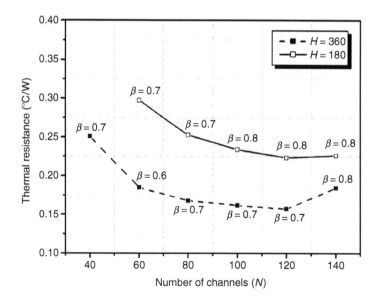

FIGURE 4.19 Variation of the thermal resistance with the number of channels for $H = 360$ and 180 µm. The corresponding width to pitch ratios (β) are also shown in the figure. (From Li, J. and Peterson, G.P., *IEEE Trans. Comp. Packaging Technol.*, 29, 145, 2006. With permission.)

Conventional analysis of fins using numerical techniques can be found in the literature, such as the one presented by Sunil et al. (1996), for realistic fin materials. The analysis essentially consists of solving the governing equations to get the velocity and temperature fields, and proceeding to obtain local and average heat transfer coefficients and Nusselt numbers. The results are further used for obtaining optimum geometric configurations in terms of the fin height, spacing in a fin array, and fin material, corresponding to given operating conditions such as the heat dissipation and temperature levels.

Miniaturization and MEMS designs have introduced scaled down systems, and miniature forms of fin arrays could find applications in heat dissipation from many such systems. The analysis of fin arrays, when extended to the micro-scale domain, would lead to modified or noncontinuum approaches, as the cases demand. As explained previously, the major parameter that decides the approach of the analysis will be the Knudsen number, and it is based on this parameter that a decision can be made whether to use the continuum, slip flow, or free molecular approach. In the following discussions, the case of a micro fin array which permits a two-dimensional approximation, but with dimensions so small that slip flow and temperature jump conditions are better suited at the solid boundaries, is analyzed. A relatively large length in the third dimension, compared to the cross-sectional dimensions of the channels formed by the fins, ensures that a two-dimensional analysis can be used, due to the expected "down and up flow pattern" (Sane and Sukhatme 1974) as shown in Figure 4.20a which shows the configuration of the micro fin array.

A long horizontal rectangular array with fins of microscale dimensions is considered for analysis. The fins are made of silicon, and the fin base is maintained at a constant temperature T_0. The fin array transfers heat to the surrounding air. The domain of analysis consists of half the channel spacing, as shown in Figure 4.20b. The calculation domain is extended as shown to specify ambient conditions on the top boundary.

The flow and temperature fields are coupled, and the analysis of the problem involves simultaneous solution of the continuity, momentum, and the energy equations with appropriate boundary conditions. The general two-dimensional governing equations for the flow field are given below:

Continuity equation:

$$\frac{\partial u}{\partial x} + \frac{\partial v}{\partial y} = 0 \qquad (4.28)$$

Momentum equations:

$$\frac{\partial u}{\partial t} + u\frac{\partial u}{\partial x} + v\frac{\partial u}{\partial y} = -\frac{1}{\rho}\frac{\partial p}{\partial x} + v\left(\frac{\partial^2 u}{\partial x^2} + \frac{\partial^2 u}{\partial y^2}\right) + g\beta(T - T_\infty) \qquad (4.29)$$

$$\frac{\partial v}{\partial t} + u\frac{\partial v}{\partial x} + v\frac{\partial v}{\partial y} = -\frac{1}{\rho}\frac{\partial p}{\partial y} + v\left(\frac{\partial^2 v}{\partial x^2} + \frac{\partial^2 v}{\partial y^2}\right) \qquad (4.30)$$

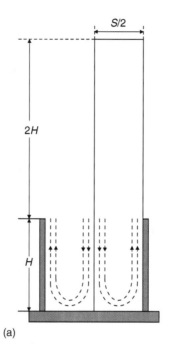

(a)

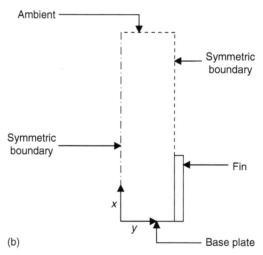

(b)

FIGURE 4.20 (a) The configuration of the micro fin array with the "down and up" flow pattern. (b) The domain of analysis for the natural convection problem.

Energy equation:

$$\rho c\left(\frac{\partial T}{\partial t} + u\frac{\partial T}{\partial x} + v\frac{\partial T}{\partial y}\right) = k\left(\frac{\partial^2 T}{\partial x^2} + \frac{\partial^2 T}{\partial y^2}\right) \qquad (4.31)$$

To eliminate the pressure gradients, a vorticity–stream function formulation is used in the present analysis (White 2003). The equations are rewritten using the following nondimensional variables:

$$X = \frac{x}{H}; \quad Y = \frac{y}{H};$$

$$U = \frac{u \cdot H}{\nu}; \quad V = \frac{\nu H}{\nu} \tag{4.32}$$

$$\theta = \frac{T - T_\infty}{T_0 - T_\infty}; \quad \tau = \frac{t\nu}{h^2}$$

$$U = \frac{\partial \psi}{\partial Y}; \quad V = -\frac{\partial \psi}{\partial X}; \quad \xi = \frac{\partial V}{\partial X} - \frac{\partial U}{\partial Y} \tag{4.33}$$

The dimensionless vorticity and stream function (ψ and ξ, respectively) automatically satisfy the continuity equation. The pressure gradient terms in the momentum equations get eliminated by the use of the vorticity–stream function equations. The governing equations in terms of the nondimensional variables are given below:

Vorticity equation:

$$\frac{\partial \xi}{\partial t} + U \frac{\partial \xi}{\partial X} + V \frac{\partial \xi}{\partial Y} = \frac{\partial^2 \xi}{\partial X^2} + \frac{\partial^2 \xi}{\partial Y^2} - Gr \frac{\partial \theta}{\partial Y} \tag{4.34}$$

Stream function equation:

$$-\xi = \frac{\partial^2 \psi}{\partial X^2} + \frac{\partial^2 \psi}{\partial Y^2} \tag{4.35}$$

Energy equation:

$$\frac{\partial \theta}{\partial \tau} + U \frac{\partial \theta}{\partial X} + V \frac{\partial \theta}{\partial Y} = \frac{1}{Pr} \left(\frac{\partial^2 \theta}{\partial X^2} + \frac{\partial^2 \theta}{\partial Y^2} \right) \tag{4.36}$$

There are two types of boundaries for the domain under analysis: the solid walls and the open boundaries. The open boundaries can be further classified into inflow and outflow boundaries. Momentum and temperature boundary conditions are specified for each of these boundaries. The boundary conditions are explained below.

Both the velocity components, U and V, are zero on the solid walls. The stream function is a constant on the solid walls, which is taken as zero (reference). Therefore,

$$\text{At } X = 0,\ 0 \leq Y \leq \left(\frac{S}{2H} \right); \quad \psi = 0$$

$$Y = \left(\frac{S}{2H} \right),\ 0 \leq X \leq 1; \quad \psi = 0$$

Some assumptions are made regarding the stream function: It is assumed that the stream function equation is satisfied on the walls, and that the gradients of the stream function (that is, the components of velocities) perpendicular to the walls are negligible in comparison with those parallel to the walls. By invoking such an assumption, the stream function equation reduces to the following, at the walls:

Vertical walls:

$$\frac{\partial^2 \psi}{\partial Y^2} = -\xi_v \tag{4.37}$$

Base:

$$\frac{\partial^2 \psi}{\partial X^2} = -\xi_H \tag{4.38}$$

The numerical equivalents of the vorticity boundary conditions at the vertical wall and the horizontal base are derived by expanding the stream function at the computational node adjacent to the boundary, using Taylor series and utilizing Equations 4.37 and 4.38. The numerical equivalents of these boundary conditions are obtained as follows:

For the vertical surface:

$$\text{At } Y = (S/2H),\ 0 \le X \le 1; \quad \xi_v = \frac{2}{(\Delta X)^2}\psi_n$$

Base:

$$\text{At } X = 0,\ 0 \le Y \le \left(\frac{S}{2H}\right); \quad \xi_H = -\frac{2}{(\Delta Y)^2}\psi_n$$

The parent surface is at a constant and uniform temperature. Therefore,

$$\text{At } X = 0,\ 0 \le Y \le \left(\frac{S}{2H}\right); \quad T = T_0 \text{ or } \theta = 1$$

The temperature profile along the fin satisfies the following fin equation:

$$\frac{\partial^2 T}{\partial x^2} - \frac{2K_a}{K_f t_1}\frac{\partial T}{\partial y} = \frac{1}{\alpha_f}\frac{\partial T}{\partial t} \tag{4.39}$$

where
t_1 is the thickness of the fin
K_a is the thermal conductivity of air
K_f is the thermal conductivity of the fin
$\alpha_f = \frac{K_f}{\rho C_p}$ is the thermal diffusivity of the fin

The top end of the domain of analysis is fixed at a fairly large distance from the base. Trials were done reducing the height of the domain successively, and it was found that the top boundary can be fixed at three times the fin height, from the base, without affecting the results. Ambient conditions are not prescribed at this boundary. It is assumed that the fluid enters and exits normally, and so the vorticity is zero at the open boundary.

$$\text{That is, at } X = 3, \, 0 \leq Y \leq \left(\frac{S}{2H}\right); \quad \xi = 0$$

Regarding the stream function, an assumption is made that the component of velocity normal to the opening is significant; the component parallel to the opening has a negligible effect. The assumption is justified by the physical nature of the flow. So the derivative of the stream function with respect the normal direction at the opening is zero.

$$\text{Thus, at } X = 3, \, 0 \leq Y \leq \left(\frac{S}{2H}\right); \quad \frac{\partial \psi}{\partial X} = 0$$

The above equation is valid for the inflow and outflow boundaries.
At the top end of the domain, the ambient conditions are assumed to prevail. Thus,

$$\text{At } X = 3, \, 0 \leq Y \leq \left(\frac{S}{2H}\right); \quad T = T_\infty \text{ or } \theta = 0$$

The velocity gradients normal to the planes of symmetry are zero. Therefore,

$$\text{At } Y = 0, \, 0 \leq X \leq 3; \quad \frac{\partial U}{\partial Y} = 0; \quad \frac{\partial^2 \psi}{\partial Y^2} = 0; \quad \frac{\partial \xi}{\partial Y} = 0$$

$$\text{At } Y = \left(\frac{S}{2H}\right), \, 1 \leq X \leq 3; \quad \frac{\partial U}{\partial Y} = 0; \quad \frac{\partial^2 \psi}{\partial Y^2} = 0; \quad \frac{\partial \xi}{\partial Y} = 0$$

The characteristic length of the micro fin array (fin spacing) considered is typically of the order of a few microns. When air is the working fluid, the mean free path is about 10–100 nm, resulting in a Knudsen number of about 0.05. Thus, the flow is considered to be in the slip regime, characterized by $0.001 < \text{Kn} < 0.1$. The effect is modeled through Maxwell's velocity slip and Smoluchowski's temperature jump boundary conditions (Zohar 2006).

$$U_s - U_w = \frac{2 - \sigma_u}{\sigma_u} \Lambda \frac{\partial U}{\partial n}\bigg|_w \tag{4.40}$$

$$T_j - T_w = \frac{2 - \sigma_t}{\sigma_t} \frac{2\gamma}{\gamma + 1} \frac{k}{\mu C_p} \Lambda \frac{\partial T}{\partial n}\bigg|_w \tag{4.41}$$

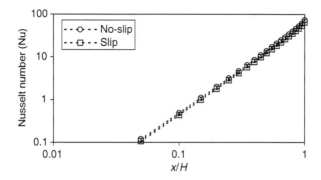

FIGURE 4.21 Variation of the local Nusselt number along the height of a micro fin in a fin array. Comparison is shown between the results from the application of no-slip and slip-boundary conditions.

U_w and T_w are the wall velocity and temperature, respectively; U_s and T_j are the gas flow velocity and the temperature at the boundary; and n is the direction normal to the solid boundary. Also, γ is the ratio of specific heats; and σ_u and σ_t are the momentum and energy accommodation coefficients, respectively. Experiments with gases over various surfaces have shown that both coefficients are approximately 1.0.

The governing differential equations presented above can be solved numerically, together with the boundary conditions specified. A fully implicit finite difference scheme with under-relaxation has been used to obtain the results (Patankar 1980). The primary solution yields the velocity and temperature distribution in the domain. The temperature distribution is used to calculate the local surface heat transfer coefficients and Nusselt numbers. The variation of the Nusselt number over the fin surface is shown in Figure 4.21. The deviation of the results from the solution using a conventional continuum approach (with no-slip boundary conditions) is obvious in the graph. The result shown corresponds to the following conditions: fin height: 125 μm; fin spacing: 50 μm; base temperature: 60°C; and ambient temperature: 23°C. Correspondingly, the Knudsen number is 0.136.

4.3.2 OPTICAL MEASUREMENTS

The major difficulty encountered in making measurements in microscale systems is the introduction of probes. This becomes extremely difficult in the case of channels. As probes intrude with the flow in the channels, the measurements will not give a picture of the actual flow situation, as the measurement process itself might change the nature of the physical problem considerably. Though MEMS probes integrated into experimental specimens have been used recently in many investigations, these also are basically invasive measurement methods which produce inaccuracies due to interaction with the flow field. Noninvasive methods such as interferometric methods are a good option, if appropriate experimental techniques are selected depending on the size of the channels, so that optical phenomena can be used as the measurement tool. Development of effective optical methods for local measurements in channels could contribute tremendously in providing explanations for the differences between conventional flows and microchannel flows, if such differences exist in reality.

Some of the pioneering work in the application of interferometric measurements in mini and microchannels has been undertaken by Dr. David Newport's research team at the Stokes Research Institute, University of Limerick, Ireland. This work focused on concentration measurements in mini and microchannels, producing what could possibly be the first ever results in interferometric measurements in microchannels. Credit is also due to the efforts of Professor Maurice Whelan, Head of Photonics at the Institute for Health and Consumer Protection (IHCP), at the Joint Research Centre (JRC) of the European Commission, Ispra, Italy, and his research team.

In conventional interferometric measurements using the Mach–Zehnder principle (Dyson 1970), two coherent beams are produced from a beam of monochromatic light or laser, which pass through the test section and a reference section, respectively, so that their physical length of travel is the same, thus producing a definite phase lag in them. On recombination, fringes are formed due to the difference in field variables (such as temperature or concentration) between the two sections, which produce differences in density and the refractive index. Thus, the fringe patterns can represent either isotherms or lines of constant concentration in the test section depending on whether the field is a heat transfer field or a diffusion field. The fringe map from which measurements are made has the handicap that the field variable can be measured only at locations where the fringes are formed. This can be avoided by using phase stepping methods, where a continuous phase map is the end product of the interferometric measurement, which can be analyzed to get the distribution of the field variable, using digital image processing. In one type of phase stepping, the interferograms are recorded either after introducing discrete incremental phase steps in one of the beams by changing the position of a mirror in the reference beam of the interferometer, or the alignment of a glass plate introduced in the beam. In an alternative method, interferograms are recorded while the phase is shifted continuously, and the intensity detected continuously. This is called an integrating-bucket method (Newport et al. 2008).

Another technique called the heterodyne technique (Garvey 2005) is more suitable for measurements in microchannels, as it gives much higher spatial accuracy. This method utilizes an electromechanical scanning device, such as an acousto-optic modulator (AOM) with the conventional Mach–Zehnder interferometric configuration. In the heterodyne technique, a known frequency shift is given to one of the laser beams in the interferometer, which produces a frequency difference between the two laser beams (called the carrier frequency). The phase value of a detected pixel is obtained in this case, by comparing its intensity signal to a reference carrier frequency signal. The heterodyne technique results in higher resolution than usual phase stepping interferometry, because this introduces a continuous phase shift without mechanical parts or reset of optics, just by introducing a frequency shift in the used light beam. The phase stepping and heterodyne techniques have been described elaborately by investigators (Garvey 2005, Newport et al. 2008) and will not be discussed in detail here.

4.3.2.1 Concentration and Diffusion Measurements

Optical measurements using phase stepping interferometry and the heterodyne technique have been tried out and reported in the literature, for obtaining concentration gradients and diffusion coefficients in fluids flowing through small channels including typical microchannels. The phase stepping technique based on the Mach–Zehnder interferometer has been used for the measurement of mixing by diffusion in a 4 mm mini channel, and to obtain quantitatively the diffusion coefficient using the concentration map (Garvey 2005, Newport et al. 2008). Phase stepping interferometry has been utilized to make concentration measurements in mini channels and to obtain diffusion coefficients. Heterodyne interferometry has been utilized in microchannels of size 500 μm, in another breakthrough application of optical measurement in microchannels (Garvey 2005), which could possibly pave the way for a lot of interesting future experimental investigations on fluid flow and heat transfer in microchannels.

The test specimen consisted of a 4 mm mini channel mixer as shown in Figure 4.22. Two-fluid streams, deionized water and 0.2 gmol/l NaCl solution, are passed through the channel in parallel. The flow rate to both inlets was controlled via Becton Dickinson plastic syringes using a Harvard Apparatus PHD 22/2000 Syringe Pump. The syringes were connected to the inlet connectors through 1/32″ internal diameter FEP Teflon tubing. The two flow rates used were 0.1 and 0.01 ml/min corresponding to Reynolds numbers of 0.848 and 0.0848, respectively. Optical access for interferometric measurement of the diffusion process was obtained through two glass windows forming two sidewalls of the device. The device was placed in the Mach–Zehnder interferometer. Phase stepping was implemented using a PZT Piezomechanik Amplifier, Model SVR 150/1, with a power output range of 0–150 V. The amplifier was connected to a PC and controlled by a software analog output device. A CCD camera, with an array size of 811×508, acquired the interferograms for processing. All processing and control was implemented digitally via ESPI Test, an in-house software package. Measurements were obtained at the locations specified in Figure 4.23.

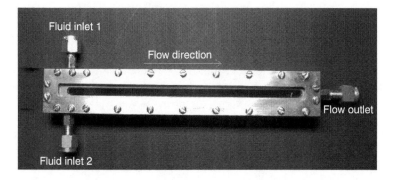

FIGURE 4.22 The mini channel test section. (From Newport, D., Sobhan, C.B., and Garvey, J., *Heat Mass Transfer*, 44, 535, 2008. With permission.)

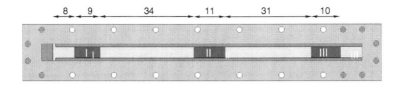

FIGURE 4.23 Schematic of the mini channel showing the three measurement sections. (From Newport, D., Sobhan, C.B., and Garvey, J., *Heat Mass Transfer*, 44, 535, 2008. With permission.)

Figure 4.24 shows the concentration distribution map resulting from the interferometric measurement (using phase stepping interferometry), corresponding to a typical section. The diffusion coefficient determined for each of the three sections is shown in Table 4.5 and excellent agreement with the literature is achieved (Welty et al. 2000).

Due to its high accuracy, the heterodyne technique has been utilized to make measurements of concentration gradients in microchannels by the same research group, which could be the first attempt to reach into microchannels by a totally noninvasive technique. This utilized AOM to introduce the required frequency shift, and image sensing by CMOS camera and a microscopic image detection system (Garvey 2005) to give the best results, apart from attempts with CCD and photodiode detection systems. The method was used to measure the concentration profiles in a binary fluid system, similar to the mini channel fluid system described above, but with a channel size of 500 μm. Steady-state and transient measurements were undertaken. The concentration profiles measured in the channel at various flow rates with the binary fluid system of water and NaCl solution are shown in Figure 4.25.

4.3.2.2 Temperature Field and Heat Transfer in Mini Channels

Investigations using optical measurements in small channels are being undertaken in the area of temperature measurements also. Attempts are in progress to obtain temperature distributions in mini channels using Mach–Zehnder interferometry at

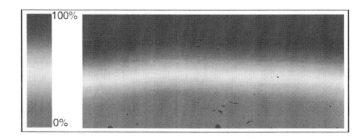

FIGURE 4.24 Phase map at Section II at a Reynolds number of 0.848, shown as a percentage of the NaCl solution. (From Newport, D., Sobhan, C.B., and Garvey, J., *Heat Mass Transfer*, 44, 535, 2008. With permission.)

TABLE 4.5

Diffusion Coefficient Values Measured Using Residual Phase Subtraction

Literature	$D = 1.21 \times 10^{-9}$ m^2/s (Welty et al. 2000)	
Section 1	$D = 1.194 \times 10^{-9}$ m^2/s	Error 1.322%
Section 2	$D = 1.238 \times 10^{-9}$ m^2/s	Error 2.314%
Section 3	$D = 1.201 \times 10^{-9}$ m^2/s	Error 0.924%

National Institute of Technology, Calicut, India. As an experimental case, laminar convective heat transfer in mini channels heated from one of its sides has been studied. The configuration of the setup with two identical specimen SL used in the test section and the reference section of the interferometer is shown in Figure 4.26.

For given heat fluxes representative of the values encountered in electronics cooling, and with appropriate flow rates for keeping the dissipating surface temperature at safe values for substrates in electronic devices (around 60°C), interferograms were recorded. By appropriate fringe interpolation for water, the temperature values corresponding to the fringe pattern were identified. A typical interference fringe pattern is shown in Figure 4.27.

The Reynolds number corresponding to Figure 4.27 is 1345, and it is seen from the fringe pattern that the flow has started turning into turbulent, indicating an early transition similar to what has been noticed in earlier investigations on microchannels using overall measurements. However, extensive experimentation would be necessary before conclusive remarks can be made in this regard.

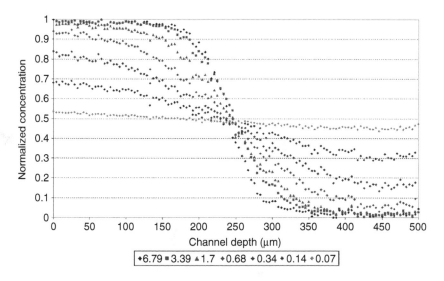

FIGURE 4.25 Normalized concentration profiles across the channel, at 487 µm from the inlet, with various Reynolds numbers.

FIGURE 4.26 Configuration of the interferometer setup for temperature measurement in mini channels.

The temperature distribution is used to calculate the surface heat transfer coefficients and the local Nusselt number values, for a constant heat flux application at the heater. The variation of the local Nusselt number on the heated surface of a channel with laminar flow conditions is shown in Figure 4.28.

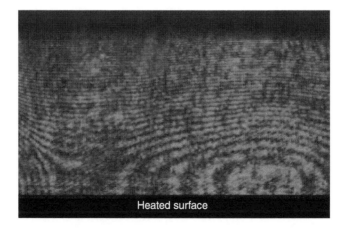

FIGURE 4.27 Typical interference fringe pattern in a longitudinal region, for steady-state convective flow of water through a mini channel of hydraulic diameter 8.8 mm and length 150 mm. Re = 1345, Pr = 5.134. The input flux to the heated surface is 6.3 W. The fringe patterns correspond to a variation of temperature in the range 23°C–42°C in the fluid.

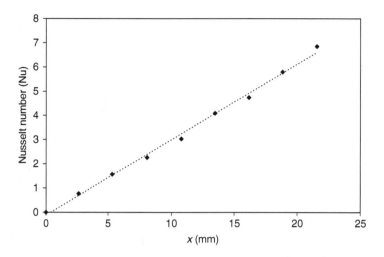

FIGURE 4.28 The local Nusselt number variation along certain length of the heated surface, for a case with fully laminar flow.

4.3.3 Other Visualization Methods

A few results of visualization experiments in microchannels have been reported in the literature. Richter et al. (1997) conducted microscopic measurements in microchannels with flow rates in the range 0.01–1000 μL/min. This work was related to minute discharge and precise flow control in liquid dosing. V grooves fabricated in silicon, covered with Pyrex glass by anodic bonding, constituted the microchannels under the investigation.

PIV was utilized for measurements in microchannel flow by Meinhart et al. (1999). The application of the method and its accuracy were demonstrated by measuring known flow fields in a 30×300 μm microchannel, and by comparison with analytical flow solutions in rectangular channels. The method of PIV and its application to determine the flow entrance length in microchannels have been explained in detail by Lee et al. (2006). Two different cross-sectional dimensions were used—one was fabricated in transparent acrylic by precision sawing and had the cross-sectional dimensions of 252×694 μm, whereas the other test specimen, fabricated in silicon using photolithography had a cross section of 104.6×38.6 μm. The measured velocity distributions were used to identify the entrance length in the channels. The variations of the entrance lengths with respect to the Reynolds number in the two channels were presented, and a comparison of the experimental results was also done with correlations available in the literature as shown in Figure 4.29.

The most remarkable feature of optical measurements in microchannels is their noninvasive nature. It is a matter of concern that the measurement reported in microscale convective flow and heat transfer are mostly based on external measurement and extrapolation, or by techniques which interfere with the flow stream, which have always been inadequate to make conclusive interpretations. Optical noninvasive

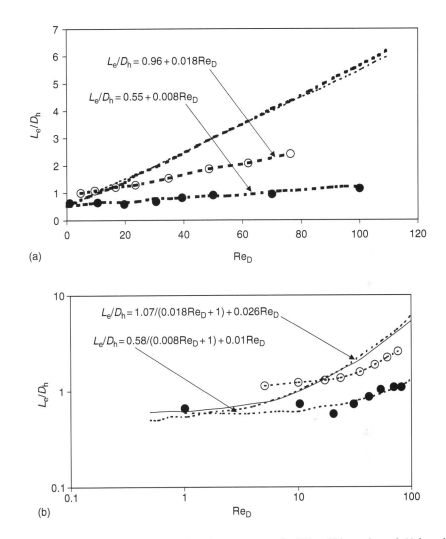

FIGURE 4.29 (a) Results on entrance length measurement for $252 \times 694 \, \mu m$ channel. (Adapted from Lee, S.Y., Jang, J., and Wereley, S.T., *The MEMS Handbook. 2nd ed. Design and Fabrication*, Mohammed Gad-el Hak (Ed.), Taylor & Francis, Boca Raton, 2006. With permission.) (b) Results on entrance length measurement for $104.6 \times 38.6 \, \mu m$ channel. (Adapted from Lee, S.Y., Jang, J., and Wereley, S.T., *The MEMS Handbook. 2nd ed. Design and Fabrication*, Mohammed Gad-el Hak (Ed.), Taylor & Francis, Boca Raton, 2006. With permission.)

measurements are expected to have a profound impact on understanding the physical phenomena associated with microscale convective flows and help in the design of effective engineering systems in MEMS and other microminiaturization applications.

4.3.4 MICRO HEAT PIPES AND MICRO HEAT SPREADERS

The heat pipe is a passive two-phase heat transfer device that utilizes the liquid–vapor phase change processes occurring in a working fluid to transfer heat, and to pump the working fluid by capillary action. In a conventional heat pipe, the required capillary action or "wicking" is obtained by the use of a capillary wick, which could be a metallic screen wick, or a porous wick. Many kinds of wick structures have been tested and utilized in the conventional heat pipe. The heat added externally at the evaporator section of the heat pipe vaporizes the liquid inside the heat pipe, which is at the saturation temperature. The vapor moves to the condenser section as more and more liquid evaporates along the evaporator section. At the condenser section, due to external heat transfer from the heat pipe body, the vapor condenses and the liquid is circulated back to the evaporator section through the capillary wick. Thus, the working fluid undergoes a thermodynamic cycle, and the physical processes involved in the working of a heat pipe are heat transfer, phase change, and fluid flow by capillary action. The effective thermal conductivity of the heat pipe is normally many times that of the wall material, as most of the heat is transferred by the thermodynamic cycle that the working fluid undergoes. For heat pipes operating in steady-state, there are a number of fundamental mechanisms that limit the maximum heat transfer. These have been summarized by Marto and Peterson (1988) and include the capillary wicking limit, viscous limit, sonic limit, and entrainment and boiling limits. The first two mechanisms deal with the pressure drops occurring in the liquid and vapor phases. The sonic limit results from pressure gradient induced vapor velocities that may result in choked vapor flow, while the entrainment limit focuses on the entrainment of liquid droplets in the vapor stream, which inhibits the return of the liquid to the evaporator and ultimately leads to dry out. Unlike these limits, which depend upon the axial transport, the boiling limit is reached when the heat flux applied in the evaporator portion is high enough that nucleate boiling occurs in the evaporator wick, creating vapor bubbles that partially block the return of fluid.

As mentioned in Chapter 1, the micro heat pipe is a compact, passive heat removal device, with extremely high effective thermal conductance, suitable for thermal management of microelectronic devices. Though the basic working principle of both micro heat pipes and larger, conventional heat pipes are very similar, there is an essential difference between their operations, in that the micro heat pipe typically does not employ a wicking structure for the circulation of the working fluid, but rather depends upon small liquid arteries. The micro heat pipe is essentially a channel of a polygonal cross section, which contains a small, predetermined quantity of saturated working fluid. Heat added to the evaporator section of the micro heat pipe results in the vaporization of a portion of the working fluid. The vapor then flows through the central portion of the channel cross section. The return flow of the liquid formed in the condenser is accomplished by utilizing the capillary action at the narrow corner regions of the passage. Thus, in the micro heat pipe, "wicking" is provided by the corners of the passage, thus avoiding the need for a wick structure for liquid recirculation. The vapor and liquid flow in the micro heat pipe are also characterized by the varying cross-sectional areas of the two-fluid paths, unlike the flow of the vapor and liquid confined to the core and the wick regions of the conventional heat pipe (Peterson and Sobhan 2006).

The required condition for micro heat pipe operation is that the average radius of the liquid–vapor meniscus formed at the corners of the channel is comparable in magnitude with the reciprocal of the hydraulic radius, i.e., the characteristic dimension, of the total flow channel (Cotter 1984). For additional information on the theory and working principle of heat pipes and the fundamental phenomena which cause the working limitations of heat pipes, readers are referred to the books by Tien (1975), Chi (1976), Dunn and Reay (1982), Peterson (1994), and Faghri (1995).

Since the initial introduction of the micro heat pipe concept, the investigation of microscale heat transfer has grown enormously and has encompassed not only phase change heat transfer, but also the entire field of heat transfer, fluid flow and in particular, a large number of fundamental investigations in thin film behavior. While the division between micro- and macroscale phase-change behavior is virtually indistinguishable, in applications involving phase change heat transfer devices, such as micro heat pipes and micro heat spreaders; it can best be described by applying a dimensionless expression developed by Babin et al. (1990), which related the capillary radius of the interface and the hydraulic radius of the passage and provided a good indicator of when the forces particular to the microscale began to dominate.

A number of previous reviews have summarized the literature published prior to 2000 (Peterson and Ortega 1990, Peterson 1992, Cao et al. 1993, Peterson et al. 1998, Faghri 2001, Garimella and Sobhan 2001). However, significant advances have been made over the past few years, particularly in the development of a better understanding of the thin film behavior that governs the operation of these devices.

The earliest embodiments of micro heat pipes typically consisted of a long thin tube with one or more small noncircular channels that utilized the sharp angled corner regions as liquid arteries. While initially quite novel in size (Figure 4.30), it was soon apparent that devices with characteristic diameters of approximately 1 mm functioned in nearly the same manner as larger, more conventional liquid artery heat pipes. Heat applied to one end of the heat pipe vaporizes the liquid in that region and forces it to move to the cooler end where it condenses and gives up the latent heat of vaporization. This vaporization and condensation process causes the liquid–vapor interface in the liquid arteries to change continually along the pipe, as illustrated in Figure 4.31 and results in a capillary pressure difference between the evaporator and condenser regions. This capillary pressure difference promotes the flow of the working fluid from the condenser back to the evaporator through the triangular-shaped corner regions. These corner regions serve as liquid arteries, thus no wicking

FIGURE 4.30 Micro heat pipe cooled ceramic chip carrier. (Adapted from Peterson, G.P. and Sobhan, C.B., *The MEMS Handbook. 2nd ed. Applications*, Mohammed Gad-el Hak (Ed.), Taylor & Francis, Boca Raton, 2006. With permission.)

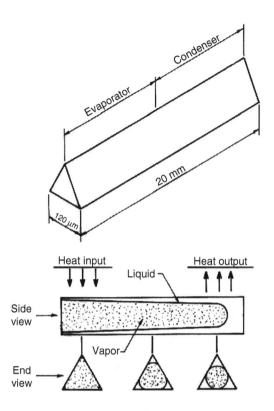

FIGURE 4.31 Micro heat pipe operation. (Adapted from Peterson, G.P. and Sobhan, C.B., *The MEMS Handbook. 2nd ed. Applications*, Mohammed Gad-el Hak (Ed.), Taylor & Francis, Boca Raton, 2006. With permission.)

structure is required (Peterson 1990, 1994). The following sections present a summary of the analytical and experimental investigations conducted on individual micro heat pipes, arrays of micro heat pipes, flat plate microscale heat spreaders, and the latest advances in the development of highly conductive, flexible phase-change heat spreaders.

4.3.4.1 Modeling of Conventional Micro Heat Pipes

The early steady-state analytical models of micro heat pipes utilized pressure balances in the flow streams inside the heat pipe, and provided a mechanism by which the steady-state and transient performance characteristics of micro heat pipes could be determined. These indicated that while the operation was similar to that observed in larger more conventional heat pipes, the relative importance of many of the parameters is quite different. Perhaps the most significant difference was the relative sensitivity of the micro heat pipes to the amount of working fluid present. These steady-state models later led to the development of both transient numerical models and 3D numerical models of the velocity, temperature, and pressure distribution within individual micro heat pipes (Peterson 1992, 1994, Longtin et al. 1994, Peterson et al. 1998, Sobhan et al. 2000, Sobhan and Peterson 2004).

The first steady-state model specifically designed for use in modeling of micro heat pipes was developed by Cotter (1984). Starting with the momentum equation, and assuming uniform cross-sectional area and no-slip conditions at the boundaries, this expression was solved for both the liquid and vapor pressure differential and then combined with the continuity expression. The result was a first-order ordinary differential equation, which related the radius of curvature of the liquid–vapor interface to the axial position along the pipe. Building upon this model, Peterson (1988a) and Babin et al. (1990) developed a steady-state model for a trapezoidal micro heat pipe, using the conventional steady-state modeling techniques outlined by Chi (1976). The resulting model demonstrated that the capillary pumping pressure governed the maximum heat transport capacity of these devices.

The performance limitations resulting from the model presented by Cotter (1984) and by Babin et al. (1990) were compared, and indicated that significant differences in the capillary limit predicted by the two models existed. These differences have been analyzed and found to be the result of specific assumptions made in the initial formulation of the models (Peterson 1992).

A comparative analysis of these two early models was performed by Gerner et al. (1992) who indicated that the most important contributions of Babin et al. (1990) were the inclusion of the gravitational body force and the recognition of the significance of the vapor pressure losses. In addition, the assumption that the pressure gradient in the liquid flow passages was similar to that occurring in Hagen–Poiseuille flow was questioned and a new scaling argument for the liquid pressure drop was presented. In this development, it was assumed that the average film thickness was approximately one-fourth the hydraulic radius, resulting in a modified expression for the capillary limitation.

A significant contribution made by Gerner et al. (1992) was the recognition that the capillary limit may never actually be reached due to instabilities occurring at the liquid–vapor interface. The decision as to whether to use the traditional capillary limit proposed by Babin et al. (1990) or the interfacial instability limit proposed by Gerner et al. (1992) should be governed by evaluating the shape and physical dimensions of the specific micro heat pipe being considered. Khrustalev and Faghri (1994) presented a detailed mathematical model of the heat and mass transfer processes in micro heat pipes, which described the distribution of the liquid, and the thermal characteristics as a function of the liquid charge. The liquid flow in the triangular-shaped corners of a micro heat pipe with polygonal cross-section was considered by accounting for the variation of the curvature of the free liquid surface and the interfacial shear stresses due to the liquid–vapor interaction. The predicted results were compared with the experimental data obtained by Wu and Peterson (1991) and Wu et al. (1991), and indicated the importance of the liquid charge, the contact angle, and the shear stresses at the liquid–vapor interface in predicting the maximum heat transfer capacity and thermal resistance of these devices.

Longtin et al. (1994) developed a one-dimensional, steady-state model for the evaporator section of the micro heat pipe. The governing equations were solved, assuming a uniform temperature along the heat pipe. The solution indicated that the maximum heat transport capacity varied with respect to the cube of the hydraulic diameter of the channel.

An analytical model for the etched triangular micro heat pipe was developed by Duncan and Peterson (1995), which is capable of calculating the curvature of the liquid–vapor meniscus in the evaporator. This model was used to predict the capillary limit of operation of the heat pipe and to arrive at the optimal value of the liquid charge. In a subsequent work, a hydraulic diameter was defined, incorporating the frictional effects of the liquid and the vapor, and was used in a model for predicting the minimum meniscus radius and the maximum heat transport in triangular grooves (Peterson and Ma 1996b). The major parameters influencing the heat transport capacity of the micro heat pipe were found to be the apex angle of the liquid arteries, the contact angle, heat pipe length, vapor velocity, and the tilt angle. Ma and Peterson (1997b) presented analytical expressions for the minimum meniscus radius and the maximum capillary heat transport limit in micro heat pipes, which were validated with experimental data.

A detailed steady-state mathematical model for predicting the heat transport capability of a micro heat pipe, and the temperature gradients that contribute to the overall axial temperature drop as a function of the heat transfer, was developed by Peterson and Ma (1999). The unique nature of this model was that it considered the governing equation for fluid flow and heat transfer in the evaporating thin film region. The model also consisted of an analytical solution of the two-dimensional heat conduction in the macro evaporating regions in the triangular corners. The effects of the vapor and liquid flows in the passage, the flow and condensation of the thin film caused by the surface tension in the condenser, and the capillary flow along the axial direction of the micro heat pipe were considered in this model. The predicted axial temperature distribution was compared with experimental data, with very good agreement. The model was capable of calculating the heat transfer distribution through the thin film region and the heat transfer-operating temperature dependence of the micro heat pipe. It was concluded from the investigation that the evaporator temperature drop was considerably larger than that at the condenser, and that the temperature drops increased with an increase in input power, when the condenser is kept at a constant temperature.

The maximum heat transfer capacity of copper–water micro heat pipes was also explored by Hopkins et al. (1999) using a one-dimensional model for predicting the capillary limitation. In this analysis, the liquid–vapor meniscus was divided into two regions depending on whether the contact angle can be treated as a constant at the evaporator or as a variable along the adiabatic and condenser sections.

As heat pipes diminish in size, the transient nature becomes of increasing interest. The ability to respond to rapid changes in heat flux coupled with the need to maintain constant evaporator temperature in modern high-powered electronics necessitates a complete understanding of the temporal behavior of these devices. The first reported transient investigation of micro heat pipes was conducted by Wu and Peterson (1991). This initial analysis utilized the relationship developed by Collier (1981) and used later by Colwell and Chang (1984) to determine the free molecular flow mass flux of evaporation. The most interesting result from this model was the observations that reverse liquid flow occurred during the startup of micro heat pipes. As explained in the original reference (Wu et al. 1991), this reverse liquid flow is the result of an imbalance in the total pressure drop and occurs because the evaporation rate does not provide an adequate change in the liquid–vapor interfacial curvature to compensate for the pressure drop. As a result, the increased pressure in the evaporator causes the meniscus to recede into the corner regions forcing liquid out of the evaporator and into the condenser. During start-up, the pressures of both the liquid and vapor are higher in the evaporator and gradually decrease with position, promoting flow away from the evaporator. Once the heat input reaches full load, the reverse liquid flow disappears and the liquid mass flow rate into the evaporator gradually increases until a steady-state condition is reached. At this time the change in the liquid mass flow rate is equal to the change in the vapor mass flow rate for any given section (Wu and Peterson 1991).

Several, more detailed transient models have been developed. Badran et al. (1993) developed a conjugate model to account for the transport of heat within the heat pipe and conduction within the heat pipe case. This model indicated that the specific thermal conductivity of micro heat pipes (effective thermal conductivity divided by the density) could be as high as 200 times that of copper.

Ma et al. (1996) developed a closed mathematical model of the liquid friction factor for flow occurring in triangular grooves. This model, which built upon the earlier work of Ma et al. (1994), considered the interfacial shear stresses due to liquid–vapor frictional interactions for countercurrent flow. Using a coordinate transformation and an iteration scheme, the importance of the liquid–vapor interactions on the operational characteristics of micro heat pipes and other small phase change devices was demonstrated. The solution resulted in a method by which the velocity distribution for countercurrent liquid–vapor flow could be determined, and allowed the governing liquid flow equations to be solved for cases where the liquid surface is strongly influenced by the vapor flow direction and velocity. The results of the analysis were verified using an experimental test facility constructed with channel angles of 20°, 40°, and 60°. The experimental and predicted results were compared and found to be in good agreement (Ma and Peterson 1996a,b, Peterson and Ma 1996a).

A transient model for a triangular micro heat pipe with an evaporator and condenser section was presented by Sobhan and Peterson (2004). The energy equation as well as the fluid flow equations were solved numerically, incorporating the longitudinal variation of the cross-sectional areas of the vapor and liquid flows, to yield the velocity, pressure, and temperature distributions. The effective thermal conductivity was computed, and characterized with respect to the heat input and the cooling rate, under steady and transient operation of the heat pipe. The salient features of this model and the important results are discussed below.

A flat micro heat pipe heat sink consisting of an array of micro heat pipe channels, used to form a compact heat dissipation device to remove heat from electronic chips, was analyzed. Each channel in the array served as an independent heat transport device. The analysis presented here examined an individual channel in such an array. The individual micro heat pipe channel analyzed had a triangular cross-section. The channel was fabricated on a copper substrate and the working fluid used was ultra pure water. The external view and constructional details of the device are shown in Figure 4.32.

The micro heat pipe consisted of an externally heated evaporator section and a condenser section subjected to forced convective cooling. A one-dimensional model was sufficient for the analysis, as the variations in the field variables were significant only in the axial direction, due to the geometry of the channels. A transient model which proceeded until steady-state was utilized to analyze the problem completely. In this problem, the flow and heat transfer processes are governed by the continuity, momentum, and energy equations for the liquid and vapor phases. A nonconservative formulation can be utilized, as the problem deals with low velocity flows. As phase change occurs, the local mass rates of the individual liquid and vapor phases are coupled through a mass balance at the liquid–vapor interface. The cross-sectional

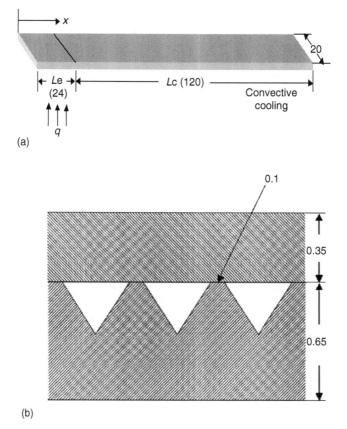

FIGURE 4.32 (a) External view of the micro heat pipe heat sink. Dimensions are in mm. (From Sobhan, C.B., Xiaoyang, H., and Liu, C.Y., *AIAA J. Thermophys. Heat Transfer*, 14,161, 2000. With permission.) (b) Construction details of the micro heat pipe channels, fabricated on a copper substrate. The side of triangle is 300 μm. Dimensions are in mm.

areas of the vapor and liquid regions and the interfacial area vary along the axial length, due to the progressive phase change occurring as the fluid flows along the channel. These variations in the area can be incorporated into the model through the use of suitable geometric area coefficients, as described in Longtin et al. (1994) and Peterson and Sobhan (2006). The local meniscus radii at the liquid–vapor interface are calculated using the Laplace–Young equation. The friction factor, which appears in the momentum and energy equations, is incorporated through appropriate models for fluid friction in varying area channels, as described in the literature.

The governing differential equations can be described as follows:

Laplace–Young equation:

$$P_v - P_1 = \frac{\sigma}{r} \tag{4.42}$$

Vapor continuity equation: evaporator section

$$\left(\frac{\sqrt{3}}{4}d^2 - \beta_1 r^2\right)\frac{\partial u_v}{\partial x} - 2\beta_1 u_v r \frac{\partial r}{\partial x} + \beta_i \frac{\rho_1}{\rho_v} r V_{il} = 0 \tag{4.43}$$

Vapor continuity equation: condenser section

$$\left(\frac{\sqrt{3}}{4}d^2 - \beta_1 r^2\right)\frac{\partial u_v}{\partial x} - 2\beta_1 u_v r \frac{\partial r}{\partial x} - \beta_i \frac{\rho_1}{\rho_v} r V_{il} = 0 \tag{4.44}$$

It should be noted that the vapor continuity equation incorporates the interfacial mass balance equation:

$$\rho_1 V_{il} = \rho_v V_{iv} \tag{4.45}$$

Vapor momentum equation:

$$\rho_v \left(\frac{\sqrt{3}}{4}d^2 - \beta_1 r^2\right)\frac{\partial u_v}{\partial t} = 2\rho_v \left(\frac{\sqrt{3}}{4}d^2 - \beta_1 r^2\right)u_v \frac{\partial u_v}{\partial x} - 2\rho_v \beta_1 r u_v^2 \frac{\partial r}{\partial x}$$
$$+ \left(\frac{\sqrt{3}}{4}d^2 - \beta_1 r^2\right)\frac{\partial P_v}{\partial x} - \frac{1}{2}\rho_v u_v^2 f_{vw}(3d - \beta_{1w} r) - \frac{1}{2}\rho_v^2 u_v^2 f_{vi}\beta_1 r \tag{4.46}$$

Vapor energy equation: evaporator section

$$\left(\frac{\sqrt{3}}{4}d^2 - \beta_1 r^2\right)\frac{\partial E_v}{\partial t} + \frac{\partial}{\partial x}\left(u_v \left(\frac{\sqrt{3}}{4}d^2 - \beta_1 r^2\right)(E_v + P_v)\right)$$
$$= \frac{\partial}{\partial x}\left\{\frac{4}{3}\mu_v \left(\frac{\sqrt{3}}{4}d^2 - \beta_1 r^2\right)u_v \frac{\partial u_v}{\partial x} + K_v \left(\frac{\sqrt{3}}{4}d^2 - \beta_1 r^2\right)\frac{\partial T_v}{\partial x}\right\}$$
$$+ q(3d - \beta_{1w} r) + h_{fg} V_{il}\rho_1 \beta_1 r + \frac{1}{2}\rho_v u_v (3d - \beta_{1w} r) + \frac{1}{2}\rho_v u_v^2 f_{vi} u_v \beta_1 r \tag{4.47}$$

Vapor energy equation: condenser section

$$\left(\frac{\sqrt{3}}{4}d^2 - \beta_1 r^2\right)\frac{\partial E_v}{\partial t} + \frac{\partial}{\partial x}\left[u_v\left(\frac{\sqrt{3}}{4}d^2 - \beta_1 r^2\right)(E_v + P_v)\right]$$

$$= \frac{\partial}{\partial x}\left[\frac{4}{3}\mu_v\left(\frac{\sqrt{3}}{4}d^2 - \beta_1 r^2\right)u_v\frac{\partial u_v}{\partial x} + k_v\left(\frac{\sqrt{3}}{4}d^2 - \beta_1 r^2\right)\frac{\partial T_v}{\partial x}\right]$$

$$+ h_{fg}V_{il}\rho_1\beta_1 r - h_o(3d - \beta_{1w}r)\Delta T + \frac{1}{2}\rho_v u_v^2 f_{vw} u_v(3d - \beta_{1w}r) + \frac{1}{2}\rho_v u_v^2 f_{vi} u_v\beta_i r$$

$$(4.48)$$

Liquid continuity equation: evaporator section

$$r\frac{\partial u_1}{\partial x} + 2u_1\frac{\partial r}{\partial x} - \frac{\beta_i}{\beta_1}V_{il} = 0 \tag{4.49}$$

Liquid continuity equation: condenser section

$$r\frac{\partial u_1}{\partial x} + 2u_1\frac{\partial r}{\partial x} + \frac{\beta_i}{\beta_1}V_{il} = 0 \tag{4.50}$$

Liquid momentum equation:

$$\rho_1 r\frac{\partial u_1}{\partial t} = -2\rho_1\left(ru_1\frac{\partial u_1}{\partial x} + u_1^2\frac{\partial r}{\partial x}\right) - r\frac{\partial P_1}{\partial x} - \frac{1}{2}\rho_1 u_1^2 f_{lw}\frac{\beta_{1w}}{\beta_1} - \frac{1}{2}\rho_1 u_1^2 f_{li}\frac{\beta_i}{\beta_1} \tag{4.51}$$

Liquid energy equation: evaporator section

$$\beta_1 r^2\frac{\partial E_1}{\partial t} + \frac{\partial}{\partial x}\left(u_1\beta_1 r^2(E_1 + P_1)\right) = \frac{\partial}{\partial x}\left(\frac{4}{3}\mu_1 u_1\beta_1 r^2\frac{\partial u_1}{\partial x} + k\beta_1 r^2\frac{\partial T_1}{\partial x}\right) + q\beta_{1w}r$$

$$- h_{fg}V_{il}\rho_1\beta_i r + \frac{1}{2}\rho_1 u_1^2 f_{lw} u_1\beta_{1w} r + \frac{1}{2}\rho_1 u_1^2 f_{li} u_1\beta_i r$$

$$(4.52)$$

Liquid energy equation: condenser section

$$\beta_1 r^2\frac{\partial E_1}{\partial t} + \frac{\partial}{\partial x}\left(u_1\beta_1 r^2[E_1 + P_1]\right) = \frac{\partial}{\partial x}\left(\frac{4}{3}\mu_1 u_1\beta_1 r^2\frac{\partial u_1}{\partial x} + k\beta_1 r^2\frac{\partial T_l}{\partial x}\right)$$

$$+ h_{fg}V_{il}\rho_1\beta_i r - h_o\beta_{1w}r\Delta T$$

$$+ \frac{1}{2}\rho_1 u_1^2 f_{lw} u_1\beta_{1w} r + \frac{1}{2}\rho_1 u_1^2 f_{li} u_1\beta_i r \quad (4.53)$$

The vapor and liquid pressures can be computed as follows:

1. Ideal gas equation of state is utilized for computing the pressure in the vapor. Because the vapor is either saturated or super heated, the ideal gas state equation is reasonably correct and is used extensively in the analysis.
2. For the liquid phase, the Hagen–Poiseuille equation is used as a first approximation, with the local hydraulic diameter for the wetted portion of the liquid-filled region adjacent to the corners. The values of pressure obtained from this first approximation are substituted into the momentum equations and iterated for spatial convergence.

Equation of state for the vapor:

$$P_v = \rho_v R_v T_v \tag{4.54}$$

The Hagen–Poiseuille equation as a first approximation, for the liquid flow

$$\frac{\partial P_l}{\partial x} = -\frac{8\mu_l u_l}{\left(\frac{D_H^2}{4}\right)} \tag{4.55}$$

The boundary conditions are
at $x=0$ and $x=L$

$$u_l = 0; \quad u_v = 0; \quad \frac{\partial T}{\partial x} = 0$$

The initial conditions are
at $t=0$ and for all x

$$P_l = P_v = P_{sat}; \quad T_l = T_v = T_{amb}$$

At $x=0$

$$P_v - P_l = \frac{\sigma}{r_o}$$

The value of r_o, the initial radius of curvature of the interface meniscus for the copper–water system, was adopted from the literature.

Figure 4.33 illustrates the geometric configuration of the vapor and liquid flow in the cross section of the micro heat pipe, along with the meniscus idealized as an arc of a circle at any longitudinal location. The definitions of the area coefficients, as derived for this configuration, are given below:

Referring to Figure 4.33,

The cross section is an equilateral triangle with $\phi = \frac{\pi}{3} - \alpha$, and $\eta = r \sin \phi$.

The total area of the liquid in the cross section is

$$A_l = \beta_l r^2 \tag{4.56}$$

where $\beta_l = 3\left[\sqrt{3}\sin^2\left(\frac{\pi}{3} - \alpha\right) + 0.5 \sin 2\left(\frac{\pi}{3} - \alpha\right) - \left(\frac{\pi}{3} - \alpha\right)\right]$

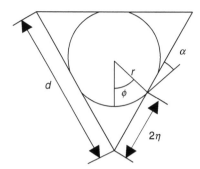

FIGURE 4.33 Cross-sectional geometry of the triangular micro heat pipe for determining the area coefficients. (Adapted from Peterson, G.P. and Sobhan, C.B., *The MEMS Handbook. 2nd ed. Applications*, Mohammed Gad-el Hak (Ed.), Taylor & Francis, Boca Raton, 2006. With permission.)

The total area of interface in length dx is

$$A_i = \beta_i \, r \, dx, \quad \text{where } \beta_i = 6\left(\frac{\pi}{3} - \alpha\right) \qquad (4.57)$$

The total wetted perimeter for the three corners is $\beta_{lw} r$,

$$\text{where } \beta_{lw} = \pi \sin\left(\frac{\pi}{3} - \alpha\right) \qquad (4.58)$$

A numerical procedure based on the finite difference method was used to solve the above system of equations to obtain the transient behavior and field distributions in the micro heat pipe, and the results of the computation have been discussed in the literature (Sobhan and Peterson 2004). Parametric studies were also presented in this paper. The results of the analysis are discussed below.

The primary results obtained from the numerical simulation were the instantaneous local velocity, the temperature, and the pressure distribution throughout the domain. To make a comparative analysis, an effective thermal conductivity was defined using a one-dimensional heat conduction analogy in a solid rod.

$$k_{\text{eff}} = \frac{Q_{\text{in}}}{A_c[(T_e - T_c)/L]} \qquad (4.59)$$

In this expression, Q_{in} represents the heat input in the evaporator section and A_c is the overall cross-sectional area of the heat sink. A dimensionless effective thermal conductivity ratio (k_{eff}^*) was also defined as the ratio of the effective thermal conductivity to the thermal conductivity of copper.

The first step in computing the variations of the field variables is the calculation of the radius curvature of the liquid–vapor meniscus as a function of the longitudinal distance along the micro heat pipe channel. As described above, the Laplace–Young equation was used for this purpose and was solved along with the other governing

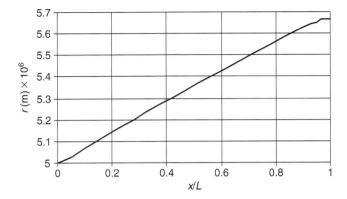

FIGURE 4.34 Distribution of the radius of curvature of the liquid–vapor interface meniscus in the micro heat pipe.

equations incorporating the computed pressures from the model at each step. The results were then stepped forward in time to yield the steady-state distribution of the meniscus radius of curvature. The resulting distribution of the meniscus radius of curvature is illustrated in Figure 4.34. The nature of the distribution is in accordance with those presented previously for copper–water systems by Khrustalev and Faghri (1994).

The numerical solution procedure involved marching in the time domain to obtain the steady-state results as characterized by constant temperature values along the length of the heat pipe. The vapor temperature in the evaporator was then used as the indicator for the attainment of steady-state and the computational process was terminated when the value of the midpoint temperature did not vary more than 0.1% over a period of 20 s. A further check for steady-state was performed by ensuring that the input heat at the evaporator matched the heat transfer at the condenser. A typical transient variation of the vapor temperature at the evaporator midpoint is shown in Figure 4.35.

Typical steady-state distributions of the normalized vapor and liquid velocity are shown in Figures 4.36 and 4.37. The velocity distribution depicts the mass addition and depletion along the evaporator and condenser sections of the micro heat pipe. The liquid velocity distribution also signifies the impact of the area variation of the liquid flow stream as it moves along the heat pipe adjacent to the corners of the triangular channel. This is evident from the nonlinear nature of the longitudinal distribution in the condenser section. The typical steady-state temperature profiles shown in Figure 4.38 correspond to three different heat flux rates at the surface of the 40-channel heat sink and a heat transfer coefficient of 160 W/m² at the condenser. The temperature variation is much smaller along the heat pipe than would be expected in a solid copper rod of the same dimensions, assigning a high value for the effective thermal conductance. Similar trends are observed for all of the cases analyzed.

The overall effect of the temperature distribution was consolidated in terms of the effective thermal conductivity of the heat pipe, using an analogy with heat conduction in a solid rod with the same dimensions as the heat sink, as illustrated

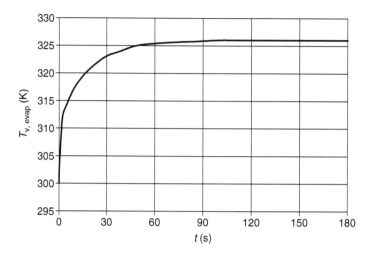

FIGURE 4.35 Transient variation of the vapor temperature at the midpoint of the evaporator section, showing the attainment of steady-state, corresponding to a heat input of 3.5 W/cm^2 and a condenser heat transfer coefficient of 120 W/m^2 K.

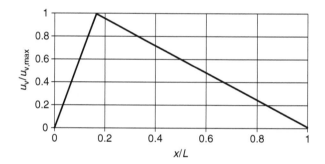

FIGURE 4.36 The normalized vapor velocity distribution along the micro heat pipe corresponding to a heat input of 2.5 W/cm^2 and a condenser heat transfer coefficient of 120 W/m^2 K.

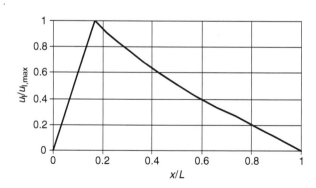

FIGURE 4.37 The normalized liquid velocity distribution in the micro heat pipe for the case corresponding to that presented in Figure 4.36.

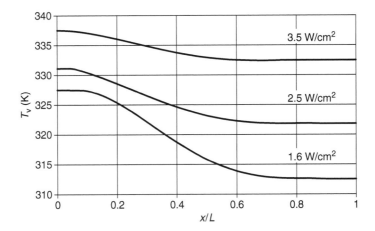

FIGURE 4.38 Vapor temperature distributions in the micro heat pipe corresponding to various input heat fluxes at the evaporator section of the heat sink. The condenser heat transfer coefficient is 120 W/m² K for all cases shown.

in Equation 4.59. The effective thermal conductivity is presented in the form of a normalized value, k^*, with respect to the thermal conductivity of copper, which can be used as a performance index for the system analyzed. The influence of two major operating parameters of the heat pipe, namely the input heat and the heat transfer coefficient at the condenser, is studied and quantified below.

The variation of the effective thermal conductivity ratio with respect to the heat flux at the evaporator section is shown in Figure 4.39, and the increasing trend is expected until the heat transfer limit is reached. In this case, the heat input is increased and the heat transfer coefficient at the condenser is held constant. These

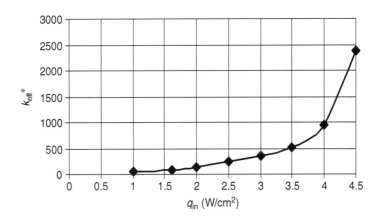

FIGURE 4.39 Variation of the effective thermal conductivity ratio with respect to the input heat flux to the heat sink. The case corresponds to a heat transfer coefficient of 160 W/m² K at the condenser section.

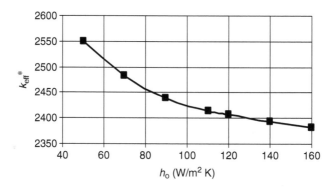

FIGURE 4.40 Variation of the effective thermal conductivity ratio with respect to the heat transfer coefficient at the condenser section of the heat sink. The input heat flux is 4.5 W/cm^2.

results clearly indicate the function of the heat pipe as a phase change device, where the effects of the circulation of the vapor and liquid on heat transport improve as the device operates at higher power levels and hence higher operating temperatures. The increase in the performance index of the heat pipe with an increase in the heat input is expected to continue until the operating limit is reached, which indicates that operating conditions close to the maximum operating limits are best characterized by the optimal performance of the heat pipe.

The effect of variations in the heat transfer coefficient at the condenser section on the effective thermal conductivity ratio is shown in Figure 4.40. In the present investigation, increases in the condenser heat transfer coefficient, with constant coolant temperature, yield a larger temperature drop between the evaporator and the condenser, for a given heat input. Thus, the calculated values of the effective thermal conductivity as defined in Equation 4.59 decrease with respect to increases in the condenser heat transfer coefficients. The rate of the decrease in the effective thermal conductivity ratio becomes smaller as the heat transfer coefficient increases as shown in Figure 4.40.

4.3.4.2   Wire-Sandwiched (Wire-Bonded) Micro Heat Pipes

One of the designs that has been developed and evaluated for use in both conventional electronic applications and also for advanced spacecraft applications consists of a flexible micro heat pipe array, fabricated by sintering an array of aluminum wires between two thin aluminum sheets as shown in Figure 4.41. In this design, the sharp corner regions formed by the junction of the plate and the wires act as the liquid arteries. When made of aluminum with ammonia or acetone as the working fluid, these devices become excellent candidates for use as flexible radiator panels for long-term spacecraft missions and can have a thermal conductivity that greatly exceeds the conductivity of an equivalent thickness of any known material.

A numerical model, which combined both conduction and radiation effects, was established to predict the heat transfer performance and temperature distribution of these types of radiator fins in a simulated space environment has been developed

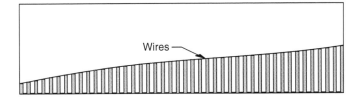

FIGURE 4.41 Flexible wire-bonded micro heat pipe. (Adapted from Peterson, G.P. and Sobhan, C.B., *The MEMS Handbook. 2nd ed. Applications*, Mohammed Gad-el Hak (Ed.), Taylor & Francis, Boca Raton, 2006. With permission.)

(Wang et al. 2001). Three different configurations were analyzed, experimentally evaluated and the results compared. Each of the three configurations was modeled both with, and without, a working fluid charge to determine the reduction in the maximum temperature, mean temperature, and temperature gradient on the radiator surface. Acetone was used as the working fluid in both the modeling effort and also in the actual experimental tests. The flexible radiator with the array of micro heat pipes was found to have an effective thermal conductivity of more than 20 times that of the uncharged version, and 10 times that of a solid material.

The results of the preliminary tests conducted on these configurations are shown in Figure 4.42. As indicated, the heat transport was proportional to the temperature

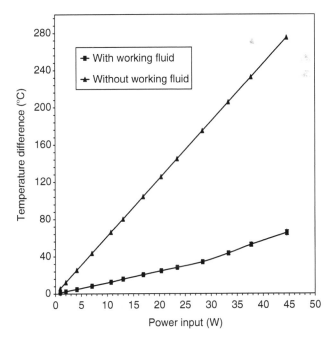

FIGURE 4.42 Temperature difference of micro heat pipe arrays with or without working fluid. (Adapted from Peterson, G.P. and Sobhan, C.B., *The MEMS Handbook. 2nd ed. Applications*, Mohammed Gad-el Hak (Ed.), Taylor & Francis, Boca Raton, 2006. With permission.)

difference between the evaporator and condenser, i.e., the effective thermal conductivity of the micro heat pipe array was constant with respect to the temperature. From the temperature difference and the heat transport obtained as shown in Figure 4.42 the effective conductivity was calculated. As illustrated in Figure 4.43, the effective thermal conductivities of micro heat pipe arrays No. 1, 2, and 3 were 1446.2, 521.3, and 3023.1 W/K m, respectively. For the micro heat pipe arrays without any working fluid, the effective conductivities in the x-direction were 126.3, 113.0, and 136.2 W/K m, respectively. Comparison of the predicted and experimental results indicated these flexible radiators with the arrays of micro heat pipes have an effective thermal conductivity between 15 and 20 times that of the uncharged version. This results in a more uniform temperature distribution, which could significantly improve the overall radiation effectiveness and reduce the overall size, and meet or exceed the baseline design requirements for long-term space missions.

Wang and Peterson (2002a) presented an analysis of wire-bonded micro heat pipe arrays, using a one-dimensional steady-state analytical model, which incorporated the effects of the liquid–vapor phase interactions and the variation in the cross-section area. The model was used to predict the heat transfer performance and

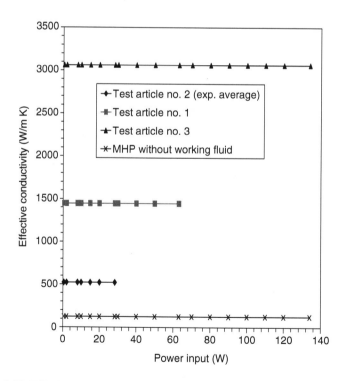

FIGURE 4.43 Effective thermal conductivity of micro heat pipe arrays. (Adapted from Peterson, G.P. and Sobhan, C.B., *The MEMS Handbook. 2nd ed. Applications*, Mohammed Gad-el Hak (Ed.), Taylor & Francis, Boca Raton, 2006. With permission.)

optimum design parameters. An experimental facility was fabricated and tests were conducted to verify the concept, as well as to validate the proposed model. The results indicated that the maximum heat transport capacity increased with increases in wire diameter and that the overall value was proportional to the square of the wire diameter. The numerical model indicated that the maximum heat transport capacity increased with increases in the wire spacing, and predicted the existence of an optimal configuration for the maximum heat transfer capacity. Further optimization studies on a wire-bonded micro heat pipe radiator in a radiation environment were reported in Wang and Peterson (2002b). A combined numerical and experimental investigation was performed to optimize the heat transfer performance of the radiator. The optimal charge volume was found to decrease with increasing heat flux. The overall maximum heat transport capacity of the radiator is found to be strongly governed by the spacing of the wires, the length of the radiator, and the radiation capacity of the radiator surface. The numerical results were consistent with experimental results, which indicated that the uniformity of the temperature distribution and the radiation efficiency both increased with increasing wire diameter. The maximum heat transport capacity of 15.2 W was found to exist for radiators utilizing a wire diameter of 0.635 mm, among the specimens tested. Comparison of the proposed micro heat pipe radiators with solid conductors and uncharged versions indicated significant improvements in the temperature uniformity and overall radiation efficiency. Aluminum–acetone systems of wire-bonded micro heat pipes were tested in this investigation.

A flat heat pipe thermal module, consisting of a wire-bonded heat pipe and a fin structure to dissipate heat, for use as a cooling device for mobile computers, was analyzed by Peterson and Wang (2003). The temperature and heat flux distributions were calculated, and a performance analysis was done using a resistance model. Effects of the wire diameter, mesh number of the wire configuration, and the tilt angle of the heat pipe, on the maximum heat transport capacity were investigated. The effect of the air flow rate on the thermal resistance and the influence of the operating temperature and air flow velocities on the heat dissipation capacity were also studied. Larger wire diameters were found to lead to a significant increase in the maximum heat transport capacity.

4.3.4.2.1 Analysis of Wire-Sandwiched Micro Heat Pipes
The analysis of an individual heat pipe channel in a wire-sandwiched micro heat pipe array will be discussed in this section. The geometry of the wire-sandwiched micro heat pipe channel is shown in Figure 4.44. This micro heat pipe consists of an externally heated evaporator section and a condenser section subjected to forced convection cooling. In the present analysis a one-dimensional model will be used because the major variation of the field variables namely the velocity, pressure, and temperature is in the direction of flow. It will also be assumed that the thermophysical properties of the working fluid both in liquid phase and vapor phase are constant.

The methodology of the mathematical formulation is the same as that used for the analysis of triangular micro heat pipes, as presented before. The governing differential equations are derived for a varying area domain, pertaining to the channel

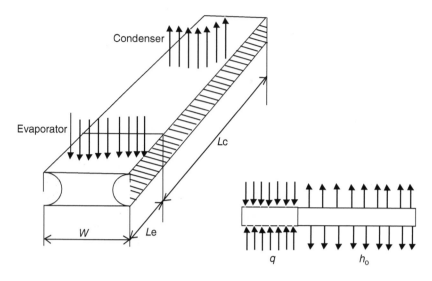

FIGURE 4.44 Individual wire-sandwiched micro heat pipe in an array.

formed in the wire-sandwiched micro heat pipe construction. The shape of the vapor and liquid regions and the mathematical representations of the areas occupied by them will be clear from the following description.

The geometric configuration of the vapor and liquid flow in the cross section of the micro heat pipe is shown in Figure 4.45. The areas of the liquid and the vapor cross sections vary continuously along the length due to progressive phase change as the fluid flows along the channel. These areas are expressed in terms of interfacial meniscus radius as derived geometrically from Figure 4.45.

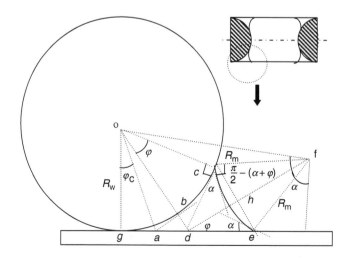

FIGURE 4.45 Geometry of the liquid meniscus in the wire-sandwiched micro heat pipe.

If ($\varphi > \varphi_c$) then area of liquid

$$A_l = 4 \left[R_m^2 \left(\frac{\cos^2(\alpha + \varphi)}{\tan \varphi} + \frac{\sin 2(\alpha + \varphi)}{2} - \frac{\pi}{2} + (\alpha + \varphi) \right) \right.$$
$$\left. + R_w^2 (\tan \varphi - \varphi + \varphi_c - \tan \varphi_c) \right] \qquad (4.60)$$

If ($\varphi < \varphi_c$) then

$$A_l = 4 R_m^2 \left[\left(\frac{\cos^2(\varphi_c + \alpha)}{\tan \varphi_c} \right) - \left(\frac{\pi}{2} - \varphi_c - \alpha - \frac{\sin 2(\varphi_c + \alpha)}{2} \right) \right] \qquad (4.61)$$

$$\text{Area of vapor } A_v = R_w^2 [4 - \pi - 4(\tan \varphi_c - \varphi_c)] + 2rl - A_l \qquad (4.62)$$

$$\text{Liquid–vapor interface area } A_i = 4[\pi - 2(\varphi - \alpha)]R \, dx = \beta_i R \, dx \qquad (4.63)$$

$$\text{Liquid–wall contact area } A_{lw} = 4(2R_w \tan \varphi + 2R_w \varphi - R_w \varphi_c - R_w \tan \varphi_c) \, dx \qquad (4.64)$$

$$\text{Vapor–wall contact area } A_{vw} = [2\pi R_w + 2(l + R_w)] \, dx - A_{lw} \qquad (4.65)$$

Relation between α and φ:

$$\tan \varphi = -\frac{1}{2} \times \frac{R_m}{R_w} \sin \alpha \pm \left[\left(\frac{R_m}{4R_w} \right)^2 \sin^2 \alpha + \frac{R_m}{R_w} \cos \alpha \right]^{\frac{1}{2}} \qquad (4.66)$$

The constructional details of the micro heat pipe considered in the baseline analysis discussed here and the properties of the working fluid used are given in Table 4.6.

The results obtained from an analysis along the same lines as described for the triangular micro heat pipe are discussed briefly below. A comparison of the effective thermal conductivities for the cases of the triangular and wire-sandwiched micro heat pipes, with similar heat inputs, is also made here.

To benchmark the results of the present computation, the values of the meniscus radius were compared with those reported in literature for the case of a trapezoidal micro heat pipe (Cotter 1984). The nature of the variations agrees as shown in Figure 4.46. The radius of curvature of the meniscus increases from the evaporator to the condenser section. With an increase in the heat load the meniscus radius at the end of the evaporator decreases continuously and reaches a minimum when the heat transport capacity reaches a maximum.

The distributions of the velocities and temperatures in the liquid and the vapor were obtained from the computational analysis (Reddy and Sobhan 2006). These distributions are found to be qualitatively similar to those for the triangular cross-section micro heat pipes. Using the temperature drop, a normalized effective thermal conductivity is calculated in this case also, as described previously (Equation 4.59).

TABLE 4.6

Details of the Wire-Sandwiched Micro Heat Pipe

Solid material	Copper
Working fluid	Water
Length of the heat pipe	140 mm
Radius of wire	0.165 mm
Distance between wires	1.46 mm
Length of the evaporator	26 mm
Length of the condenser	114 mm
Density of vapor	0.5542 kg/m^3
Density of liquid	998.2 kg/m^3
Specific heat of vapor	4186 J/kg K
Specific heat of liquid	2014 J/kg K
Thermal conductivity of liquid	0.6 W/m K
Thermal conductivity of vapor	0.0261 W/m K
Viscosity of liquid	1.003×10^{-3} N s/m
Viscosity of vapor	1.34×10^{-5} N s/m
Surface tension	0.0718 N/m

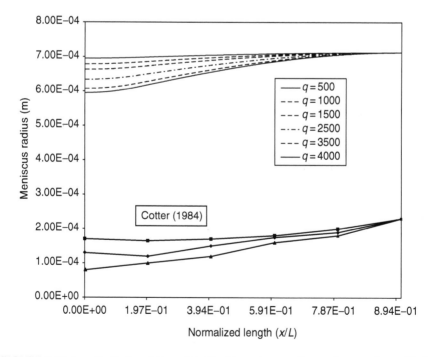

FIGURE 4.46 Longitudinal variation of the liquid meniscus radius, and comparison with the case of a trapezoidal micro heat pipe.

TABLE 4.7

Comparison of Performance of a Wire-Sandwiched Micro Heat Pipe and a Triangular Micro Heat Pipe

Heat Input, W	Effective Thermal Conductivity Ratio, k^*	
	Triangular	Wire Sandwiched
0.1	125	88.359
0.15	210	130.58
0.25	400	276.8
0.35	900	514.85

As the heat input is increased, the effective conductivity is also found to be increasing, making the heat pipe operate closer to an isothermal device, as observed in the case of the triangular cross section micro heat pipe.

A comparison is presented in Table 4.7 for the calculated values of the effective thermal conductivities for the wire-sandwiched micro heat pipe and the triangular micro heat pipe (Sobhan et al. 2000), for the same heat input values. It is to be noted that while the values of effective thermal conductivities are larger for the triangular cross section indicating better performance, the ease of fabrication of the wire-sandwiched micro heat pipe makes the use of this device attractive to a large extent.

4.3.4.3 Flat Plate Micro Heat Spreaders

While arrays of micro heat pipes have the ability to significantly improve the effective thermal conductivity of silicon wafers and other conventional heat spreaders they are of limited value in that they only provide heat transfer along the axial direction of the individual heat pipes. To overcome this problem, flat plate heat spreaders capable of distributing heat over a large two-dimensional surface have been proposed by Peterson (1992, 1994). In this application, a wicking structure is fabricated in silicon MCM substrates to promote the distribution of the fluid and the vaporization of the working fluid. This wick structure is the key element in these devices and several methods for wick manufacture have been considered (Peterson et al. 1998).

In the most comprehensive investigation of these devices to date, a flat plate micro heat pipe similar to that described by Peterson et al. (1998) was fabricated in silicon MCM substrates 5×5 mm^2 (Benson et al. 1996a,b). These devices, as illustrated earlier in Figure 1.9, utilized two separate silicon wafers. On one of the two wafers, the wick pattern was fabricated leaving a small region around the perimeter of the wafer unpatterned to allow the package to be hermetically sealed. The other silicon wafer was etched in such a manner that a shallow well was formed that corresponded to the wick area. The two pieces were then wafer bonded together along the seal ring. Upon completion of the fabrication, the flat plate micro heat pipe was filled through a small laser drilled port located in one corner of the wafer. Because the entire wicking area was interconnected, the volume of the liquid required to charge was sufficient that conventional charging techniques could be utilized (Benson et al. 1996a).

4.3.4.3.1 Modeling of Micro Heat Spreaders

Analytical investigations of the performance of micro heat spreaders or flat plate heat pipes have been under way for some time (Benson et al. 1996a,b) and Peterson (1996) has summarized the results. These investigations have demonstrated that these devices can provide an effective mechanism for distributing the thermal load in semiconductor devices and reducing the localized hot spots resulting from active chip sites (Peterson 1996). The models indicate that the performance of these devices is excellent. In addition, because these devices can be made from silicon, Kovar or a wide variety of other materials, an excellent match between the coefficient of thermal expansion (CTE), can be achieved, while keeping the material and fabrication costs very low. A number of different wicking structures have been considered. Among these are wicks fabricated using a silicon dicing saw (Figure 4.47), wicks fabricated using conventional anisotropic etching techniques (Figure 4.48), and wicks fabricated using a deep plasma etching technique (Figure 4.49). Recent modeling has focused on the development of optimized wicking structures that could be fabricated directly into the wafer and provide maximum capillary pumping while optimizing the thin film region of the meniscus to maximize the heat flux (Wayner et al. 1976, Peterson and Ma 1996b, 1999).

The results of these optimization efforts have demonstrated that these microscale flat plate heat spreaders allow the heat to be dissipated in any direction across the wafer surface, thereby vastly improving performance. The resulting effective thermal conductivities can approach and perhaps exceed that of diamond coatings of equivalent thicknesses. A relative comparison of these flat plate heat pipes and other types

FIGURE 4.47 Wick pattern prepared with bidirectional saw cuts on a silicon wafer. (Adapted from Peterson, G.P. and Sobhan, C.B., *The MEMS Handbook. 2nd ed. Applications*, Mohammed Gad-el Hak (Ed.), Taylor & Francis, Boca Raton, 2006. With permission.)

FIGURE 4.48 Chemically etched orthogonal, triangular groove wick. (Adapted from Peterson, G.P. and Sobhan, C.B., *The MEMS Handbook. 2nd ed. Applications*, Mohammed Gad-el Hak (Ed.), Taylor & Francis, Boca Raton, 2006. With permission.)

of materials traditionally utilized in the electronics industry for heat spreading, as conducted by Benson et al. (1998) illustrated that the ideal heat spreader would have the thermal conductivity of diamond, a CTE of silicon, and a cost comparable to aluminum. Flat plate heat pipes fabricated in either silicon or kovar compare very favorably with diamond in terms of thermal conductivity, have a close CTE of silicon relatively (or exactly in the case of silicon), and a projected cost that is quite low. On the basis of this comparison, it would appear that these flat plate heat pipes have tremendous commercial potential.

As described by Benson et al. (1998), a number of different flat plate micro heat pipe test articles have been evaluated using an IR camera to determine the spatially resolved temperature distribution. Using this method, a series of micro heat spreaders were evaluated experimentally. The results indicated that an effective thermal conductivity between 10 and 20 W/cm K was possible over a fairly broad temperature range. These values of thermal conductivity approach that of polycrystalline diamond substrates or more than five times that of a solid silicon substrate even at

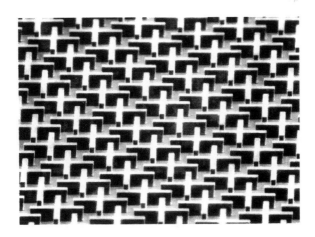

FIGURE 4.49 Wick pattern on silicon prepared by a photomask and deep plasma etching technique—wick features are 25 μm wide and 50 μm deep wafer. (Adapted from Peterson, G.P. and Sobhan, C.B., *The MEMS Handbook. 2nd ed. Applications*, Mohammed Gad-el Hak (Ed.), Taylor & Francis, Boca Raton, 2006. With permission.)

elevated temperatures (50°C) and power levels (15 W/cm^2). The cost of such advanced silicon substrates is estimated at \$60/cm^2. Any other inexpensive material with a CTE close to that of the chip may also be a potential option for the heat pipe case material. For example, many alloys in the Fe/Ni/Co family have CTEs closely matching those of semiconductor materials (Benson et al. 1996a,b).

As noted by Peterson (1992), several aspects of the technology remain to be examined before flat plate micro heat spreaders can come into widespread use, but it is clear from the results of these early experimental tests that spreaders such as the ones discussed here, fabricated as an integral part of silicon chips, present a feasible alternative cooling scheme that merits serious consideration for a number of heat transfer applications.

4.3.4.4 Other Innovative Designs of Micro Heat Pipes

In addition to the designs described above, several new designs are currently being developed and evaluated for use in conventional electronic applications, advanced spacecraft applications, and biomedical applications. In electronic applications, the function of the heat pipe design may entail the collection of heat from a microprocessor and transporting it to a conventional heat spreader or to a more readily available heat sink, such as the screen of a laptop computer. In the advanced spacecraft applications, these devices may be used to fabricate highly flexible radiator fin structures for use on long-term spacecraft missions.

A design currently being investigated consists of an array of flexible micro heat pipes fabricated in a polymer material is illustrated in Figure 4.50a. This material is extruded in such a fashion that it has a series of large rectangular grooves that serve

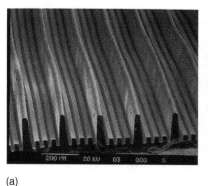
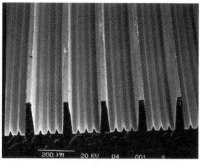

(a)   (b)

FIGURE 4.50 (a) Flexible polymer micro heat pipe: Rectangular grooves. (Adapted from Peterson, G.P. and Sobhan, C.B., *The MEMS Handbook. 2nd ed. Applications*, Mohammed Gad-el Hak (Ed.), Taylor & Francis, Boca Raton, 2006. With permission.) (b) Flexible polymer micro heat pipe: Trapezoidal grooves. (Adapted from Peterson, G.P. and Sobhan, C.B., *The MEMS Handbook. 2nd ed. Applications*, Mohammed Gad-el Hak (Ed.), Taylor & Francis, Boca Raton, 2006. With permission.)

as the actual heat pipes, each approximately 200 μm wide. Within each of these micro heat pipes is a series of smaller grooves which serve as the liquid arteries. As shown in Figure 4.50a, these grooves can be either rectangular in nature or trapezoidal, as shown in Figure 4.50b. In both cases, the material is polypropylene and has an internal dimension of the individual heat pipes of approximately 200 μm. The smaller grooves within each of the individual heat pipes are designed to transport the fluid from the evaporator to the condenser.

While only preliminary experimental test data are available, this design appears to hold great promise for both spacecraft radiator applications and also for flexible heat spreaders for use in earth based electronic applications.

To understand the heat transfer and fluid flow mechanisms in the micro wick structures of flexible micro membrane/thin film heat pipes, experimental and theoretical investigations were performed (Wang and Peterson 2002c). Experimental tests were conducted to evaluate the evaporation heat transfer limit in the polymer microfilm with 26 μm capillary grooves. The experiments indicated that the maximum heat transport capacity decreased significantly as the effective length of the polymer film increased. The experimental observations also indicated that the maximum liquid meniscus radius occurred in the microgrooves just prior to dry out.

An analytical model based on the Darcy law was used to obtain the pressure gradients, and the experimental results were validated. Two models for predicting the maximum heat transport capacity were developed—one assuming that the liquid only fills the microgrooves, and the other considering flooding of the space above the microgrooves, and the calculated results were compared with experimental values. It was found that the experimentally determined maximum capillary evaporation heat transfer agreed better with the second model, which took into account the flooding effect. Figure 4.51 shows the comparison of the experimental and analytical results.

The analytical model, based on parametric investigations, indicated that decreasing the bottom width of trapezoidal grooves very slightly can improve the evaporation heat transfer performance significantly. The analytical models were also used to determine the optimal half angle of the groove for the best heat transfer performance.

Investigations of polymer-based flexible micro heat pipes for application in spacecraft radiators have also been undertaken (McDaniels and Peterson 2001). Building upon the demonstrated effectiveness of micro heat pipe arrays as heat spreaders in electronic applications, the possibility of use of regions of micro heat pipe arrays in flexible radiators was tested. Analytical modeling suggested that a light weight polymeric material with embedded micro heat pipe arrays can meet heat dissipation requirements while contributing less mass than other flexible materials. The capillary pumping limit was estimated as a function of operating temperature, using the analytical model, with water and methanol as the working fluid. For water, the maximum heat transport was found to be 18 mW per channel, at around 160°C, while for methanol it was 2.2 mW per channel at 120°C. It was shown that the obtained radiator capacity in the range 6.0–12.2 kW, at source temperatures of 40°C or higher, met, or exceeded, the dissipation requirements of a reference spacecraft design.

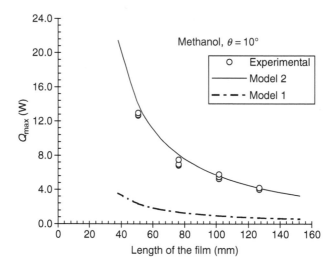

FIGURE 4.51 Comparison of the modeling and experimental results on microchanneled polymer films. (Adapted from Peterson, G.P. and Sobhan, C.B., *The MEMS Handbook. 2nd ed. Applications*, Mohammed Gad-el Hak (Ed.), Taylor & Francis, Boca Raton, 2006. With permission.)

The focus of this investigation consisted of micro heat pipe arrays which were made from a composite of two layers: an ungrooved metal foil and a grooved polymer film. A low heat bonding between a polymer coating of the foil and the raised points of the grooved film formed the micro heat pipe channels. The analysis was used to compute the capillary pumping pressure and the dynamic and frictional pressure drops in the liquid and the vapor. The results showing the variation of the capillary limit with respect to the temperature, for the two working fluids, are shown in Figure 4.52a,b. Selected results were used, with the Reynolds number as the criterion, to assess the validity of model simplifications regarding the liquid and vapor flow regimes, which assumed laminar flow for liquid and vapor.

Diverse uses for the micro heat pipe and micro heat spreader can be found in biomedical applications. One such application is in catheters which provide a hyperthermia or hypothermia source, which can be used in the treatment of tumors and cancers. U.S. patents have been granted for micro heat pipe catheters (Fletcher and Peterson 1993, 1995, 1997). In one of these designs, the micro heat pipe catheter enables the hypo or hyperthermic treatment of cancerous tumors or other diseased tissue. The heat pipe is about the size of a hypodermic needle, and is thermally insulated along a substantial portion of its length. The heat pipe includes a channel, partially charged with an appropriate working fluid. The device provides the delivery or removal of thermal energy directly to or from the tumor or diseased tissue site. In another design, the catheter uses a variety of passive heat pipe structures alone or in combination with feedback devices. This catheter is particularly useful in treating diseased tissue that cannot be removed by surgery,

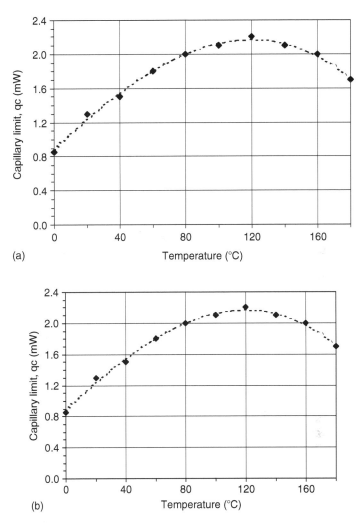

FIGURE 4.52 (a) Capillary limitation for the polymer micro heat pipe, for a single channel as of a function of temperature for the contact angle in the range 0°–20°. Methanol is the working fluid. (Adapted from Peterson, G.P. and Sobhan, C.B., *The MEMS Handbook. 2nd ed. Applications*, Mohammed Gad-el Hak (Ed.), Taylor & Francis, Boca Raton, 2006. With permission.) (b) Capillary limitation for the polymer micro heat pipe, for a single channel as of a function of temperature for the contact angle in the range 0°–20°. Water is the working fluid. (Adapted from Peterson, G.P. and Sobhan, C.B., *The MEMS Handbook. 2nd ed. Applications*, Mohammed Gad-el Hak (Ed.), Taylor & Francis, Boca Raton, 2006. With permission.)

such as a brain tumor. Another biomedical application under development is the polymer-based micro heat pipe heat spreader, which is being proposed for the treatment of neocortical seizures, by implanting a device that can provide localized cooling.

4.3.4.5 Comparative Study of Micro Heat Pipes

Attempts to miniaturize electronic packages and devices have led to extensive investigations on various geometric designs of micro heat pipes, suitable for different working conditions, and using different working fluid–material combinations. As a result, different kinds of micro heat pipes have been proposed in the past few years for use in the cooling of substrates.

Because the performance of micro heat pipes depends on a number of parameters, a comparison of performance is not easy. The geometric design and the substrate-working fluid combination are chosen based on the application, operating temperature, and heat transfer limits. Some of the investigations have used an effective thermal conductivity as the performance parameter. This parameter is useful in comparing micro heat pipe performance when the channel arrangement is in longitudinal arrays. The temperature drop in efficient micro heat pipes will normally be very small, thus making the effective conductivity values very large. Thermal conductance or thermal resistance of these types of heat pipes is also used as a performance indicator.

A more direct method of comparison could be in terms of the heat transport per unit area of cross section. In using various geometric designs for the cross section, this is an effective method of comparing the heat transfer capability of micro heat pipes, as the quantity per unit area will be independent of the number of channels used. However, this performance indicator will not directly give any information about the gradient of temperature along the heat pipe, unlike the effective thermal conductivity.

To compare the various heat pipe designs, the heat transport per unit area of cross section of the channel has been calculated for most of the experimental heat pipes discussed in the literature, wherever data are available. This performance indicator for the various cases studied is given along with the geometric parameters in Table 4.8. The values of the thermal resistances are also given in the table, wherever data are available. It is expected that Table 4.8 will provide a means to make a comparative assessment of the performance of micro heat pipe designs with different cross-sectional shapes.

A summary of the constructional details, material-working fluid combinations, and analysis methods of the micro heat pipe investigations presented in the literature is given in Table 4.9 for quick reference. This table also includes the important findings from the investigations on the common geometric designs.

4.3.5 INTEGRATION OF MICROCHANNEL HEAT SINKS TO SUBSTRATES

Various fabrication techniques used to integrate microchannels into silicon substrates have been discussed in the literature. Tuckerman and Pease (1982) describe the use of the etching process and precision machining methods for fabrication of microchannels and heat sinks. Anisotropic etching of $<110>$ silicon using KOH has originally been used in fabrication. The use of precision sawing (such as wafer dicing saw) has been used with excellent results, producing grooves as narrow as 30 μm. Very smooth finish has been obtained using 1 μm diamond grit. Precision

TABLE 4.8
Comparison of the Heat Transfer Performance of Various Micro Heat Pipe Designs

Reference	Cross-Sectional Shape	Operating Temperature/ Evaporator Temperature (°C)	Heat Input (W)	Cross-Sectional Area of a Channel (m²)	Heat Transfer per Unit Cross-Sectional Area (W/m²)	Temperature Drop (°C)	Length (m)	Thermal Resistance (°C/W)
Babin et al. (1990)	Trapezoidal	30	1.1×10^{-1}	1×10^{-6}	110,000	—	57×10^{-3}	—
		40	1.25×10^{-1}		125,000			
		50	1.35×0^{-1}		135,000			
		60	1.4×10^{-1}		140,000			
		70	1.45×10^{-1}		145,000			
		80	1.5×10^{-1}		150,000			
Wu and Peterson (1991)	Trapezoidal	—	0.2	1×10^{-6}	200,000	1.3	57×10^{-3}	—
			0.4		400,000	2.1		
			0.6		600,000	2.2		
			0.8		800,000	2.1		
			1.0		1,000,000	2.0		
			1.2		1,200,000	1.9		
			1.4		1,400,000	2.1		
			1.6		1,600,000	2.6		
			1.8		1,800,000	4.4		

(continued)

TABLE 4.8 (continued)
Comparison of the Heat Transfer Performance of Various Micro Heat Pipe Designs

Reference	Cross-Sectional Shape	Operating Temperature/ Evaporator Temperature (°C)	Heat Input (W)	Cross-Sectional Area of a Channel (m²)	Heat Transfer per Unit Cross-Sectional Area (W/m²)	Temperature Drop (°C)	Length (m)	Thermal Resistance (°C/W)
Peterson and Ma (1999)	Trapezoidal	35	0.225	1.4×10^{-6}	160,714	—	57×10^{-3}	—
		40	0.25		178,571			
		45	0.25		178,571			
		50	0.28		200,000			
		55	0.3125		223,214			
		60	0.35		250,000			
Peterson et al. (1993)	(a) Triangular	30	1.6	4800×10^{-12}	333,333,333.3	—	20×10^{-3}	—
		40	2.3		479,166,666.6			
		50	3.0		624,999,999.9			
		60	3.8		791,666,666.5			
	(b) Rectangular	30	1.3	3600×10^{-12}	361,111,110.9			
		40	2.0		555,555,555.2			
		50	2.75		763,888,888.4			
		60	3.4		944,444,443.8			
		70	4.0		11,111,111,110			

Moon et al. (2003)	(a) Curved-triangular		1	$1.767145868 \times 10^{-06}$	565,884.24	—	50×10^{-3}	—	18
			2		1131,768.484				10
			3		1697,652.726				6.25
			4		2263,536.968				5
			5		2829,421.211				3.5
	(b) Curved-rectangular		1	$1.767145868 \times 10^{-06}$	565,884.24				13.5
			2		1,131,768.484				7.5
			3		1,697,652.726				4.9
			4		2,263,536.968				2.4
Mallik and Peterson (1995)	(a) Triangular 34 MHP array	30	1.4	6.875×10^{-10}	2,036,363,636	—	20×10^{-3}	—	—
		40	1.95		2,836,363,636				
		50	2.35		3,418,181,818				
		60	3		4,363,636,364				
	(b) Triangular 66 Mhp array	30	1.6		2,327,272,727				
		40	2.2		3,200,000,000				
		50	2.7		2,909,090,909				
		60	3.2		4,654,545,455				
Ha and Peterson (1998b)	Triangular	40	0.15	8.65×10^{-09}	17,341,040.46	—	15×10^{-3}	—	—

TABLE 4.9

Summary of Investigations on Micro Heat Pipes of Various Designs

Trapezoidal Micro Heat Pipes

Heat Pipe Design	Nature of Work	Important Findings	Reference
Length = 57 mm, breadth = 1 mm, depth = 1 mm, water in copper and silver	Experimental and analytical investigations	Verification of the concept of micro heat pipes. An expression for capillary limit was derived	Babin et al. (1990)
Length = 8 cm, breadth = 2.54 mm, depth = 2.54 mm, water in copper and silver	Analytical investigations of transient characteristics	Reversal of flow occurs in the liquid arteries due to pressure imbalance during early transient period. Wetting angle is an important parameter contributing to transport capacity	Wu and Peterson (1991)
Length = 60 mm, breadth = 1 mm, depth = 2 mm, water in copper and silver	Experimental and analytical investigations	The maximum heat transport capacity in tapered micro heat pipe was obtained	Wu et al. (1991)
Length = 57 mm, breadth = 2 mm, depth = 0.7 mm, water in silver	Experimental and analytical investigation on temperature response	The temperature drop occurring in the evaporator is larger than in the condenser. With an increase in input power, there is an increase in the temperature drops in the evaporator and the condenser	Peterson and Ma (1999)
Length = 1000 μm, width = 2 μm, depth = 2 μm, height = 30 μm, ammonia	Analytical investigation of thermal performance	Variation of performance parameters incorporating the interfacial shear stress was obtained	Suh and Park (2003)
Length = 5.5 mm, breadth = 800 μm, depth = 0.6 mm, methanol in copper	Experimental investigation of metallic micro heat pipes	Characterized the cooling performance	Kang et al. (2004)

Triangular Micro Heat Pipes

Specifications	Type	Findings	Reference
Length = 20.48–20.58 mm, length of one triangular edge = 100 μm, water in silver and copper	One-dimensional modeling for steady-state operation	Demonstrated use of momentum differential equations to analyze the fluid flow in the evaporator section, and obtained velocity and pressure distributions. Compared results with experimental results of Babin et al. (1990)	Longtin et al. (1994)
Array of 40 channels, equilateral triangles, side = 0.3 mm, length = 144 mm, water in copper	Numerical investigations on the transient performance	The vapor velocity distributions in the evaporator and the condenser sections are linear. The pressure drop in the liquid is larger compared to that in the vapor. The instantaneous effective thermal conductivity increases in the transient period	Sobhan et al. (2000)
Array of 40 channels, equilateral triangles, side = 0.3 mm, length = 144 mm, water in copper	Modeling of the flow and heat transfer in micro heat pipes using implicit finite difference procedure	The effective thermal conductivity increases with increases in the heat input. An increase in the heat transfer coefficient with a constant coolant temperature reduces the effective thermal conductivity for a given heat input	Sobhan and Peterson (2004)
Array of 34 or 66 vapor deposited micro heat pipes. Length = 20 mm, width = 25 μm, depth 55 μm, methanol in silicon wafer	Experimental and analytical investigation for steady-state performance	The percentage of reductions in maximum temperature and maximum temperature difference are higher for a higher micro heat pipe density. The variations in the density of micro heat pipes are insignificant in changing effective thermal conductivity of the array. The increase in density of micro heat pipes reduces the effectiveness of the array	Mallik and Peterson (1995)

(continued)

TABLE 4.9 (continued)
Summary of Investigations on Micro Heat Pipes of Various Designs

Heat Pipe Design	Nature of Work	Important Findings	Reference
Array of 34 or 66 vapor deposited micro heat pipes. Length = 20 mm, width = 25 μm, depth = 55 μm, methanol in silicon wafer	Experimental and analytical investigation for the transient state	Array of heat pipes provides significant reductions in both the maximum temperature of the vapor and transient thermal gradients	Peterson and Mallik (1995)
Array of 59 channels, width = 120 μm, depth = 80 μm, methanol in silicon wafer	Experimental investigations and analytical modeling aimed at charge optimization	Analytical model predicted the capillary limit of operation, the curvature of the meniscus in the evaporator, and the optimal value of the liquid charge	Duncan and Peterson (1995)
Width at top = 64 μm, width at bottom = 90.9 μm, vertex angle of groove = 18.7°, methanol in silicon	Analytical investigation of heat transfer of evaporating thin films	Predicted the maximum heat transport capacity of micro heat pipe	Ha and Peterson (1996)
Different areas of cross sections, apex angle = 60°, methanol in silicon	Experimental investigation of maximum heat transport	Predicted the minimum meniscus radius and maximum heat transport. The heat transport capacity depended strongly on the apex channel angle, contact angle of liquid flow, length of the heat pipe, vapor flow velocity, and tilt angle	Peterson and Ma (1996b)
Length = 762 mm, width = 3.6, 7.4, 11.7 mm, apex angle = 20°, 40°, 60°, water and air in aluminum	Experimental investigation of countercurrent liquid–vapor interaction	Friction factor of the liquid flow increases with the air velocity. For grooves with larger channel angles, the air flow had a greater influence on the liquid friction factor	Ma et al. (1996)

Different areas of cross sections, apex angle = 60°, methanol in silicon	Experimental investigation of maximum heat transport in triangular grooves	There exists an optimum groove configuration which maximizes the capillary pumping capacity while minimizing the combined effects of capillary pumping pressure and the liquid viscous pressure losses	Ma and Peterson (1996c) Peterson and Ma (1996c)
Hydraulic diameter = 0.17–1.43 mm, apex angle = 60°, methanol in silicon	Analytical investigation	Existence of an optimum hydraulic radius. When the hydraulic radius is less than the optimal value, the groove dimension directly limits the capillary heat transport capability When the hydraulic radius is increased above the optimal value, no increase in capillary pumping occurs	Ma and Peterson (1997b)
Width = 0.38 mm, apex angle = 30°, ethanol or methanol in copper	Analytical and experimental investigations of capillary performance	The axial variation of the radius of curvature profile was determined	Peterson and Ha (1998)
Width = 0.38 mm, apex angle = 30°, ethanol or methanol in copper	Analytical investigation of capillary performance	Inclusion of the tilt angle variation makes the governing equation nonlinear	Ha and Peterson (1998a)
Length = 15 mm, side of triangle = 0.2 mm, water in copper and silver	Analytical investigation of heat transport capacity	The model incorporated the effects of the intrusion of the evaporator length into the adiabatic region Obtained an expression for liquid flow shape factor	Ha and Peterson (1998b)
19 channels in array, length = 120 mm, side = 0.7 mm, equilateral, spacing = 0.2 mm, ammonia in aluminum	Analytical investigation of the effect of interfacial phenomena	Major part of the heat input in the evaporator goes through the microregion. The apparent contact angle and the heat transfer rate in the microregion increase with an increase in the superheat The inner wall heat flux and temperature and the contact angle decrease along the evaporator section	Sartre et al. (2000)

(continued)

TABLE 4.9 (continued)
Summary of Investigations on Micro Heat Pipes of Various Designs

Heat Pipe Design	Nature of Work	Important Findings	Reference
Array of 55 pipes, MHP#1—length = 20 mm, hydraulic diameter = 120 μm, MHP#2—length = 20 mm, hydraulic diameter = 210 μm, ethanol in silicon wafer	Experimental investigations	MHP#2 with arteries for increasing the liquid flow cross-sectional area showed improvement ineffective thermal conductivity compared to the other	Launay et al. (2004)
Length = 2.5 cm, width = 0.2 mm, apex angle = 60°, pentane in silicon	Analytical transient investigation of microgrooved heat pipes	Predicted substrate temperature profiles	Suman et al. (2005)
Length = 2 cm, width = 0.1 mm, pentane in silicon	Analytical investigation for finding the variations in thermophysical properties and design parameters	Increase in apex angle, inclination, sharpness of the corner, length, and viscosity reduces the performance Increase in surface tension and contact angle increases the performance	Suman and Hoda (2005)
Rectangular and Triangular Micro Heat Pipes, and Comparisons			
Rectangular: length = 200 mm, width = 2.1 mm, depth = 10 mm, water in copper	Analytical investigation of fluid flow and heat transfer	Obtained the time evolution of the temperature, pressure, and velocity profiles along the heat pipe for different heat sources and sinks	Murer et al. (2005)
Rectangular: length = 20 mm, width = 45 μm, depth = 80 μm, triangular: length = 20 mm, width = 120 μm, depth = 80 μm, methanol in silicon	Experimental investigation of micro heat pipes fabricated in silicon wafers	The effective thermal conductivity is increased by 31% with rectangular and 81% with triangular cross-section micro heat pipes compared to ungrooved silicon wafer There was a reduction in maximum temperature when both rectangular and triangular heat pipes were used Due to better capillary pumping effect, the triangular micro heat pipes show better performance over rectangular	Peterson et al. (1993)

Rectangular: length = 20 mm, width = 800 μm, depth = 400 μm, triangular: length = 20 mm, side = 400 μm, equilateral pentane in silicon	Analytical investigation of capillary limit	The effect of body forces is small compared to pressure jump forces The transport capacity of triangular heat pipes is more than that of a rectangular one The dry-out length increases with increase in heat input and inclination	Suman et al. (2005)
Rectangular: length = 20 mm, width = 800 μm, depth = 400 μm, triangular: length = 20 mm, side = 400 μm, equilateral, pentane in silicon	Analytical investigation of fluid flow and heat transfer	The model calculated a performance factor which is useful in design	Suman and Kumar (2005)

Curved Rectangular and Curved Triangular Micro Heat Pipes

Comparison of curved rectangular and curved triangular designs: length = 50 mm, 100 mm, equivalent diameter = 1.5 mm, water in copper	Experimental study on the thermal performance	The effect of the inclination angle on the thermal performance was negligible. The effect of heat pipe length on thermal performance was predominant in the case of curved triangular heat pipes The heat transfer limit was more for curved triangular design The heat transfer limit increased as the operating temperature increased The overall heat transfer coefficient was enhanced by a reduction in total length, for the same heat input	Moon et al. (2003)

sawing has the advantage that hard substrates can be machined by appropriate selection of the cutting material, whether the substrate is crystalline or amorphous.

Fabrication of micropillars, obtained by various cross-cuts using precision sawing, has been discussed by Tuckerman and Pease (1982). Micropillars are much more effective than micro fins, as they are less susceptible to clogging. Also, they produce vortices at high Reynolds number, which interrupts the growth of the boundary layer and hence enhances heat transfer.

Methods of bonding silicon substrates with integrated microchannels to cover plates have also been discussed by Tuckerman and Pease (1982). The anodic bonding process involves heating the assembly of silicon and Pyrex glass to about 400°C and subsequently applying a high voltage of the order of 1 kV to permanently join them. In some applications, thermal stresses due to expansion hinder the use of anodic bonding, and epoxy resins are used for bonding silicon channel arrays with glass.

Various techniques have been utilized to fabricate microchannels as an integral part of silicon or gallium arsenide wafers, most of which have found applications as micro heat pipe channels (Gerner 1990, Peterson et al. 1991, Weichold et al. 1993). Fabrication techniques included those for the formation of parallel triangular groove arrays using directionally dependent etching (Peterson 1988a, Gerner 1990) or the vapor deposition process (Mallik et al. 1991).

Anisotropic or orientation-dependent etching techniques have long been used to produce microchannels in silicon wafers (Bean 1978, Kendall 1979). In the investigations by Peterson (1988b), Gerner (1990), and Peterson et al. (1991), a series of V-shaped grooves were etched using a string of processes consisting of oxidation, photolithography, anisotropic wet chemical etching, diffusion, and various bonding processes (Peterson 1991, Mallik and Peterson 1991, Ramadas et al. 1993, Ramadas 1993). The first step in the fabrication of micro heat pipe grooves is a sequence of dry–wet–dry oxidation to form silicon dioxide on the silicon wafer. Photolithography is then used to precisely align the heat pipe pattern edges with the [110] direction for anisotropic etching, such that the etched angle is 54.74° to the (100) surface, with walls formed by [111]-type planes which serve as etching barriers due to the atomic bond density in that direction. Anisotropic etchants such as potassium hydroxide (KOH), ethylene diamine pyrocatechol (EDP), and hydrazine were used to remove the exposed silicon without affecting the silicon dioxide, of which KOH was found to be the most preferred for both quality and safety of processing. Applying a coating of silicon dioxide on the silicon surfaces enhances the surface wetting characteristics as silicon dioxide is hydrophilic (and silicon is hydrophobic). In the case of micro heat pipes, the final process is the bonding of the cover plate, which is achieved through ionic bonding (silicon to silicon), ultraviolet bonding, or electrostatic bonding (silicon to glass). In the first successful demonstration of the operation of an array of micro heat pipes, Peterson et al. (1991) fabricated a series of 19 etched V-shaped channels with a depth of 80 μm and a width of approximately 84 μm. An array of 120 micro heat pipes were fabricated by etching V-shaped channels on silicon wafer by Gerner et al. (1994).

A dual source vapor deposition process was proposed by Weichold et al. (1993) to overcome problems associated with the direct contact of the working fluid with silicon. This method consisted of two stages. In the first stage, a series of rectangular or square grooves were machined or etched in a silicon wafer. In the second stage, a dual E-beam vapor deposition process was used to create triangular channels by depositing copper, which formed a thin layer separating the silicon surface from the working fluid (methanol) of the heat pipe, thus avoiding migration of methanol on the semiconductor material while in operation.

Fabrication of micro heat spreaders involves an extension of the methods used to fabricate individual channels of micro heat pipes. A number of different wicking structures have been utilized in this case, such as kovar, silicon, or gallium arsenide, and the methods include the use of conventional techniques such as machining, directionally dependent etching, plasma etching, or vapor deposition processes.

4.4 CONCLUDING REMARKS

As seen from the discussions presented in the previous sections, interest in flow and heat transfer in microchannels has been growing significantly in the past decade. A number of investigations have been conducted to better understand the fluid flow and heat transfer in microchannel heat sinks, particularly for electronics cooling applications, using theoretical modeling methods and experimental techniques. Comparison of the existing experimental results, however, indicates instances where disparities have been identified among the results from various investigations, and between the values predicted using classical macro techniques and the existing experimental data. Nevertheless, some of the previous investigations have indicated that the cross-sectional shape of the microchannel and other parameters can greatly influence the fluid flow behavior and the resulting heat transfer capability inside very small, noncircular microchannels.

REFERENCES

Adams, T.M., S.I. Abdel-Khalik, S.M. Jeter, and Z.H. Qureshi. 1998. An experimental investigation of single-phase forced convection in microchannels. *International Journal of Heat and Mass Transfer* 41:851–857.

Adams, T.M., M.F. Dowling, S.I. Abdel-Khalik, and S.M. Jeter. 1999. Applicability of traditional turbulent single phase forced convection correlations to non-circular microchannels. *International Journal of Heat and Mass Transfer* 42:4411–4415.

Aranyosi, A., L.M.R. Bolle, and H.A. Buyse. 1997. Compact air-cooled heat sinks for power packages. *IEEE Transactions on Components, Packaging, and Manufacturing Technology Part A* 20:442–451.

Arkilic, E.B., K.S. Breuer, and M.A. Schmidt. 1994. Gaseous flow in microchannels. Application of microfabrication to fluid mechanics. *ASME FED* 197:57–66.

Arkilic, E.B., M.A. Schmidt, and K.S. Breuer. 1997. Gaseous slip flow in long microchannels. *IEEE Journal of Microelectromechanical Systems* 6:167–178.

Asako, Y. and H. Toriyama. 2005. Heat transfer characteristics of gaseous flows in microchannels. *Microscale Thermophysical Engineering* 9:15–31.

Ayyaswamy, P.S., I. Catton, and D.K. Edwards. 1974. Capillary flow in triangular grooves. *ASME Journal of Applied Mechanics* 41:332–336.

Azimian, A.R. and M. Sefid. 2004. Performance of microchannel heat sinks with Newtonian and non-Newtonian fluids. *Heat Transfer Engineering* 25:17–27.

Babin, B.R., G.P. Peterson, and D. Wu. 1990. Steady-state modeling and testing of a micro heat pipe. *Journal of Heat Transfer* 112:595–601.

Badran, B., F.M. Gerner, P. Ramadas, H.T. Henderson, and K.W. Baker. 1993. Liquid metal micro heat pipes. *Proceedings of the 29th National Heat Transfer Conference,* Atlanta, Georgia HTD 236:71–85.

Balasubramanian, P. and S.G. Kandlikar. 2005. Experimental study of flow patterns, pressure drop, and flow instabilities in parallel rectangular minichannels. *Heat Transfer Engineering* 26:20–27.

Bau, H.H. 1998. Optimization of conduits' shape in microscale heat exchangers. *International Journal of Heat and Mass Transfer* 41:2717–2723.

Bean, K.E. 1978. Anisotropic etching of silicon. *IEEE Transactions on Electron Devices* 25:1185–1193.

Benson, D.A., R.T. Mitchell, M.R. Tuck, D.R. Adkins, and D.W. Palmer. 1996a. Micro-machined heat pipes in silicon MCM substrates. *Proceedings of the IEEE Multichip Module Conference*, Santa Clara, California.

Benson, D.A., D.R. Adkins, G.P. Peterson, R.T. Mitchell, M.R. Tuck, and D.W. Palmer. 1996b. Turning silicon substrates into diamond: Micromachining heat pipes. *Advances in Design, Materials and Processes for Thermal Spreaders and Heat sinks Workshop*, Vail, Colorado, April 19–21.

Benson, D.A., D.R. Adkins, R.T. Mitchell, M.R. Tuck, D.W. Palmer, and G.P. Peterson. 1998. Ultra high capacity micro machined heat spreaders. *Microscale Thermophysical Engineering* 2:21–29.

Bergles, A.E. and S.G. Kandlikar. 2005. On the nature of critical heat flux in microchannels. *Journal of Heat Transfer* 127:101–107.

Bowers, M.B. and I. Mudawar. 1994a. Two-phase electronic cooling using mini-channel and micro-channel heat sinks. Part 1: Design criteria and heat diffusion constraints. *ASME Journal of Electronic Packaging* 116:290–297.

Bowers, M.B. and I. Mudawar. 1994b. Two-phase electronic cooling using mini-channel and micro-channel heat sinks. Part 2: Flow rate and pressure drop constraints. *ASME Journal of Electronic Packaging* 116:298–305.

Bowers, M.B. and I. Mudawar. 1994c. High flux boiling in low flow rate, low pressure drop mini-channel and micro-channel heat sinks. *International Journal of Heat and Mass Transfer* 37:321–332.

Brutin, D., F. Topin, and L. Tadrist. 2003. Experimental study of unsteady convective boiling in heated minichannels. *International Journal of Heat and Mass Transfer* 46:2957–2965.

Cao, Y., A. Faghri, and E.T. Mahefkey. 1993. Micro/miniature heat pipes and operating limitations. *Heat Pipes and Capillary Pumped Loops, ASME HTD* 236:55–62.

Celata, G.P. 2004. Single-phase heat transfer and fluid flow in micropipes. *Heat Transfer Engineering* 25:13–22.

Celata, G.P., M. Cumo, and A. Mariani. 1993. Burnout in highly subcooled water flow boiling in small diameter tubes. *International Journal of Heat and Mass Transfer* 36:1269–1285.

Chakraborty, S. and S.K. Som. 2005. Heat transfer in an evaporating thin liquid film moving slowly along the walls of an inclined microchannel. *International Journal of Heat and Mass Transfer* 48:2801–2805.

Chen, C.S. and W.J. Kuo. 2004. Heat transfer characteristics of gaseous flow in long mini and microtubes. *Numerical Heat Transfer A* 46:497–514.

Chen, C.S., S.M. Lee, and J.D. Sheu. 1998. Numerical analysis of gas flow in microchannels. *Numerical Heat Transfer A* 33:749–762.

Chen, X.Y., K.C. Toh, C. Yang, and J.C. Chai. 2004. Numerical computation of hydrodynamically and thermally developing liquid flow in microchannels with electrokinetics effects. *Journal of Heat Transfer* 126:70–75.

Chen, Y. and P. Cheng. 2002. Heat transfer and pressure drop in fractal tree-like microchannel nets. *International Journal of Heat and Mass Transfer* 45:2643–2648.

Cheng, P. and H.Y. Wu. 2003a. An experimental study of convective heat transfer in silicon microchannels with different surface conditions. *International Journal of Heat and Mass Transfer* 46:2547–2556.

Cheng, P. and H.Y. Wu. 2003b. Liquid/two-phase/vapor alternating flow during boiling in microchannels at high heat flux. *International Communications in Heat and Mass Transfer* 30:295–302.

Chi, S.W. 1976. *Heat Pipe Theory and Practice*. New York: McGraw Hill.

Choi, H., D. Lee, and J.S. Maeng, 2003. Computation of slip flow in microchannels using Langmuir slip condition. *Numerical Heat Transfer Part A: Applications* 44:59–71.

Choi, S.B., R.F. Barron, and R.O. Warrington. 1991. Fluid flow and heat transfer in microtubes. *Micromechanical Sensors, Actuators and Systems ASME DSC* 32:123–134.

Coleman, J.W. 2004. An experimentally validated model for two-phase sudden contraction pressure drop in microchannel tube headers. *Heat Transfer Engineering* 25:69–77.

Coleman, J.W. and S. Garimella. 1999. Characterization of two-phase flow patterns in small diameter round and rectangular tubes. *International Journal of Heat and Mass Transfer* 42:2869–2881.

Colin, S., P. Lalonde, and R. Caen. 2004. Validation of a second-order slip flow model in rectangular microchannels. *Heat Transfer Engineering* 25:23–30.

Collier, J.G. 1981. *Convective Boiling and Condensation*. New York: McGraw Hill.

Colwell, G.T. and W.S. Chang. 1984. Measurements of the transient behavior of a capillary structure under heavy thermal loading. *International Journal of Heat and Mass Transfer* 27:541–551.

Cotter, T.P. 1984. Principles and prospects of micro heat pipes. *Proceedings of the 5th International Heat Pipe Conference*, Tsukuba, Japan, 328–335.

Cuta, J.M., C.E. McDonald, and A. Shekarriz. 1996. Forced convection heat transfer in parallel channel array microchannel heat exchanger. *Advances in Energy Efficiency, Heat/Mass Transfer Enhancement, ASME* PID-2/HTD-338:17–23.

Du, X.Z. and T.S. Zhao. 2003. Analysis of film condensation heat transfer inside a vertical micro tube with consideration of the meniscus draining effect. *International Journal of Heat and Mass Transfer* 46:4669–4679.

Duncan, A.B. and G.P. Peterson. 1995. Charge optimization for a triangular shaped etched micro heat pipe. *AIAA Journal of Thermophysics and Heat Transfer* 9:365–368.

Dunn, P.D. and D.A. Reay. 1982. *Heat Pipes. 3rd ed.* New York: Pergamon Press.

Dupont, V., J.R. Thome, and A.M. Jacobi. 2004. Heat transfer model for evaporation in microchannels. Part II: Comparison with the database. *International Journal of Heat and Mass Transfer* 47:3387–3401.

Dyson, J. 1970. *Interferometry as a Measuring Tool*. U.K.: Machinery Pub. Co.

Eason, C., T. Dalton, M. Davies, C. O'Mathuna, and O. Slattery. 2005a. Direct comparison between five different microchannels, part 1: Channel manufacture and measurement. *Heat Transfer Engineering* 26:79–88.

Eason, C., T. Dalton, M. Davies, C. O'Mathuna, and O. Slattery. 2005b. Direct comparison between five different microchannels, part 2: Experimental description and flow friction measurement. *Heat Transfer Engineering* 26:89–98.

Faghri, A. 1995. *Heat Pipe Science and Technology*. Washington, DC: Taylor & Francis.

Faghri, A. 2001. Advances and challenges in micro/miniature heat pipes. *Annual Review of Heat Transfer* 12:1–26.

Favre-Marinet, M., S. Le Person, and A. Bejan. 2004. Maximum heat transfer rate density in two-dimensional minichannels and microchannels. *Microscale Thermophysical Engineering* 8:225–237.

Fedorov, A.G. and R. Viskanta. 2000. Three-dimensional conjugate heat transfer in the microchannel heat sink for electronic packaging. *International Journal of Heat and Mass Transfer* 43:399–415.

Filonenko, G. 1954. Hydraulic resistance in pipes. *Teploenergetika* 1:40–44.

Fletcher, L.S. and G.P. Peterson. 1993. A micro heat pipe catheter for local tumor hyperthermia. U.S. Patent No. 5,190,539.

Fletcher, L.S. and G.P. Peterson. 1995. Temperature control mechanisms for a micro heat pipe catheter. U.S. Patent No. 5,417,686.

Fletcher, L.S. and G.P. Peterson. 1997. Treatment method using a micro heat pipe catheter. U.S. Patent No. 5,591,162.

Garimella, S. 2004. Condensation flow mechanisms in microchannels: Basis for pressure drop and heat transfer models. *Heat Transfer Engineering* 25:104–116.

Garimella, S., A. Agarwal, and J.D. Killion. 2005. Condensation pressure drop in circular microchannels. *Heat Transfer Engineering* 26:28–35.

Garimella, S.V. and C.B. Sobhan. 2001. Recent advances in the modeling and applications of nonconventional heat pipes. *Advances in Heat Transfer* 35:249–308.

Garimella, S.V. and C.B. Sobhan. 2003. Transport in microchannels—a critical review. *Annual Review of Heat Transfer* 13:1–50.

Garimella, S.V. and V. Singhal. 2004. Single-phase flow and heat transport and pumping considerations in microchannel heat sinks. *Heat Transfer Engineering* 25:15–25.

Garvey, J. 2005. *The Application of Interferometry to Microchannel Phase Measurement.* Ph.D dissertation, Department of Mechanical and Aeronautical Engineering, University of Limerick: Ireland.

Gerner, F.M. 1990. Micro heat pipes. AFSOR Final Report No. S-210-10MG-066, Wright-Patterson AFB, Dayton, Ohio.

Gerner, F.M., J.P. Longtin, P. Ramadas, T.H. Henderson, and W.S. Chang. 1992. Flow and heat transfer limitations in micro heat pipes. *Proceedings of the 28th National Heat Transfer Conference,* San Diego, California, August 9–12.

Gerner, F.M., B. Badran, H.T. Henderson, and P. Ramadas. 1994. Silicon wafer micro heat pipes. AFB, Dayton, Ohio.

Ghiaasiaan, S.M. and R.C. Chedester. 2002. Boiling incipience in microchannels. *International Journal of Heat and Mass Transfer* 45:4599–4606.

Gillot, C., C. Schaeffer, and A. Bricard. 1998. Integrated micro heat sink for power multichip module. *IEEE Industry Applications Soceity Annual Meet* 2:1046–1050.

Gillot, C., L. Meysenc, and C. Schaeffer. 1999. Integrated single and two-phase micro heat sinks under IGBT chips. *IEEE Transactions on Components and Packaging Technology* 22:384–389.

Gnielinski, V. 1976. New equations for heat and mass transfer in turbulent pipe and channel flow. *International Chemical Engineering* 16:359–368.

Guo, Z.Y. and X.B. Wu. 1997. Compressibility effect on the gas flow and heat transfer in a micro tube. *International Journal of Heat and Mass Transfer* 40:3251–3254.

Guo, Z.Y. and Z.X. Li. 2003. Size effect on microscale single-phase flow and heat transfer. *International Journal of Heat Mass and Transfer* 46:149–159.

Ha, J.M. and G.P. Peterson. 1996. The interline heat transfer of evaporating thin films along a micro grooved surface. *Journal of Heat Transfer* 118:747–754.

Ha, J.M. and G.P. Peterson. 1998a. Capillary performance of evaporating flow in microgrooves—an analytical approach for very small tilt angles. *Journal of Heat Transfer* 120:452–457.

Ha, J.M. and G.P. Peterson. 1998b. The heat transport capacity of micro heat pipes. *Journal of Heat transfer* 120:1064–1071.

Hao, Y.L. and Y.X. Tao. 2004. A numerical model for phase-change suspension flow in microchannels. *Numerical Heat Transfer A* 46:55–77.

Harms, T.M., M. Kazmierczak, F.M. Gerner, F.M., Holke, A., Henderson, H.T., Pilchowski, J., and Baber, K. 1997. Experimental investigation of heat transfer and pressure drop through deep microchannels in a (110) silicon substrate *Proceedings of ASME Heat Transfer Division ASME HTD* 351-1:347–357.

Hetsroni, G., A. Mosyak, Z. Segal, and G. Ziskind. 2002. A uniform temperature heat sink for cooling of electronic devices. *International Journal Heat and Mass Transfer* 45:3275–3286.

Hetsroni, G., D. Klein, A. Mosyak, Z. Segal, and E. Pogrebnyak. 2004. Convective boiling in parallel microchannels. *Microscale Thermophysical Engineering* 8:403–421.

Hoopman, T.L. 1990. Microchanneled structures. *Microstructures, Sensors and Actuators. ASME DSC* 108:171–174.

Hopkins, R., A. Faghri, and D. Khrustalev. 1999. Flat miniature heat pipes with micro capillary grooves. *Journal of Heat Transfer* 121:102–109.

Horiuchi, K. and P. Dutta. 2004. Joule heating effects in electroosmotically driven microchannel flows. *International Journal of Heat and Mass Transfer* 47:3085–3095.

Hrnjak, P. 2004. Developing adiabatic two-phase flow in headers—distribution issue in parallel flow microchannel heat exchangers. *Heat Transfer Engineering* 25:61–68.

Hsieh, S.S., H.H. Tsai, C.Y. Lin, C.F. Huang, and C. Chien. 2004. Gas flow in a long microchannel. *International Journal of Heat and Mass Transfer* 47:3877–3887.

Hsu, Y.Y. 1962. On the size range of active nucleation cavities on a heating surface. *Journal of Heat Transfer* 84:207–216.

Jacobi, A.M. and J.R. Thome. 2002. Heat transfer model for evaporation of elongated bubble flows in microchannels. *Journal of Heat Transfer* 124:1131–1136.

Jiang, P.X., M.H. Fan, G.S. Si, and Z.P. Ren. 2001. Thermal-hydraulic performance of small scale micro-channel and porous-media heat-exchangers. *International Journal of Heat and Mass Transfer* 44:1039–1051.

Judy, J., D. Maynes, and B.W. Webb. 2002. Characterization of frictional pressure drop for liquid flows through microchannels. *International Journal of Heat and Mass Transfer* 45:3477–3489.

Kandlikar, S.G. 2004. Heat transfer mechanisms during flow boiling in microchannels. *Journal of Heat Transfer* 126:8–16.

Kandlikar, S.G. and P. Balasubramanian. 2004. An extension of the flow boiling correlation to transition, laminar, and deep laminar flows and microchannels source. *Heat Transfer Engineering* 25:86–93.

Kandlikar, S.G. and W.J. Grande. 2004. Evaluation of single-phase flow in microchannels for high heat flux chip cooling-thermohydraulic performance enhancement and fabrication technology. *Heat Transfer Engineering* 25:5–16.

Kang, S.W., S.H. Tsai, and M.H. Ko. 2004. Metallic micro heat pipe heat spreader fabrication. *Applied Thermal Engineering* 24:299–309.

Kavehpour, H.P., M. Faghri, and Y. Asako. 1997. Effects of compressibility and rarefaction on gaseous flows in microchannels. *Numerical Heat Transfer A* 32:677–696.

Kawahara, A., M. Sadatomi, K. Okayama, M. Kawaji, and P.M. Chung. 2005. Effects of channel diameter and liquid properties on void fraction in adiabatic two-phase flow through microchannels. *Heat Transfer Engineering* 26:13–19.

Kendall, D.L. 1979. Vertical etching of silicon at very high aspect ratios. *Annual Review of Materials Science* 9:373–403.

Khrustalev, D. and A. Faghri. 1994. Thermal analysis of a micro heat pipe. *Journal of Heat Transfer* 116:189–198.

Kim, S.J. 2004. Methods for thermal optimization of microchannel heat sinks. *Heat Transfer Engineering* 25:37–49.

Kim, S.J., D. Kim, and D.Y. Lee. 2000. On the local thermal equilibrium in microchannel heat sinks. *International Journal of Heat Mass Transfer* 43:1735–1748.

Kleiner, M.B., S.A. Kuehn, and K. Haberger. 1995. High performance forced air cooling scheme employing microchannel heat exchangers. *IEEE Transactions on Components, Packaging and Manufacturing Technology A* 18:795–804.

Kockmann, N., M. Engler, D. Haller, and P. Woias. 2005. Fluid dynamics and transfer processes in bended microchannels. *Heat Transfer Engineering* 26:71–78.

Kohl, M.J., S.I. Abdel-Khalik, S.M. Jeter, and D.L. Sadowski. 2005. An experimental investigation of microchannel flow with internal pressure measurements. *International Journal of Heat and Mass Transfer* 48:1518–1533.

Koo, J. and C. Kleinstreuer. 2004. Viscous dissipation effects in microtubes and microchannels. *International Journal of Heat and Mass Transfer* 47:3159–3169.

Koo, J. and C. Kleinstreuer. 2005a. Analysis of surface roughness effects on heat transfer in micro-conduits. *International Journal of Heat and Mass Transfer* 48:2625–2634.

Koo, J. and C. Kleinstreuer. 2005b. Laminar nanofluid flow in microheat-sinks. *International Journal of Heat and Mass Transfer* 48:2652–2661.

Launay, S., V. Sartre, and M. Lallemand. 2004. Experimental study on silicon micro heat pipe arrays. *Applied Thermal Engineering* 24:233–243.

Lee, D.Y. and K. Vafai. 1999. Comparative analysis of jet impingement and microchannel cooling for high heat flux applications. *International Journal of Heat and Mass Transfer* 42:1555–1568.

Lee, P.C., F.G. Tseng, and C. Pan. 2004. Bubble dynamics in microchannels. Part I: Single microchannel. *International Journal of Heat and Mass Transfer* 47:5575–5589.

Lee, P.S. and S.V. Garimella. 2003. Experimental investigation of heat transfer in microchannels. *Proceedings of ASME Summer Heat Transfer Conference*, Las Vegas, Nevada, HT2003–472932003.

Lee, P.S., S.V. Garimella, and D. Liu. 2005. Investigation of heat transfer in rectangular microchannels. *International Journal of Heat and Mass Transfer* 48:1688–1704.

Lee, S.Y., J. Jang, and S.T. Wereley. 2006. Optical diagnostics to investigate the entrance length in microchannels. In Mohammed Gad-el Hak (Ed.), *The MEMS Handbook. 2nd ed. Design and Fabrication* (pp. 10:1–10:35). Boca Raton: Taylor & Francis.

Lelea, D., S. Nishio, and K. Takano. 2004. The experimental research on microtube heat transfer and fluid flow of distilled water. *International Journal of Heat and Mass Transfer* 47:2817–2830.

Li, B. and D.Y. Kwok. 2003. A lattice Boltzmann model for electrokinetic microchannel flow of electrolyte solution in the presence of external forces with the Poisson–Boltzmann equation. *International Journal of Heat and Mass Transfer* 46:4235–4244.

Li, H.Y., F.G. Tseng, and C. Pan. 2004a. Bubble dynamics in microchannels. Part II: Two parallel microchannels. *International Journal of Heat and Mass Transfer* 47:5591–5601.

Li, J. and G.P. Peterson. 2006. Geometric optimization of a micro heat sink with liquid flow. *IEEE Transactions on Components and Packaging Technologies* 29:145–154.

Li, J. and P. Cheng. 2004. Bubble cavitation in a microchannel. *International Journal of Heat and Mass Transfer* 47:2689–2698.

Li, J., G.P. Peterson, and P. Cheng. 2004b. Three-dimensional analysis of heat transfer in a micro heat sink with single-phase flow. *International Journal of Heat and Mass Transfer* 47:4215–4231.

Li, J.M., B.X. Wang, and X.F. Peng. 2000. Wall-adjacent layer analysis for developed-flow laminar heat transfer of gases in microchannels. *International Journal of Heat and Mass Transfer* 43:839–847.

Liu, D. and S.V. Garimella. 2004. Investigation of liquid flow in microchannels. *AIAA Journal of Thermophysics and Heat Transfer* 18:65–72.

Longtin, J.P., B. Badran, and F.M. Gerner. 1994. A one-dimensional model of a micro heat pipe during steady-state operation. *Journal of Heat Transfer* 116:709–715.

Ma, H.B. and G.P. Peterson. 1996a. Experimental investigation of the maximum heat transport in triangular grooves. *Journal of Heat Transfer* 118:740–746.

Ma, H.B. and G.P. Peterson. 1996b. Temperature variation and heat transfer in triangular grooves with an evaporating film. *AIAA Journal of Thermophysics and Heat Transfer* 11:90–98.

Ma, H.B. and G.P. Peterson. 1996c. Experimental investigation of the maximum heat transport in triangular grooves. *Journal of Heat Transfer* 118:740–746.

Ma, H.B. and G.P. Peterson. 1997a. Laminar friction factor in microscale ducts of irregular cross-section. *Microscale Thermophysical Engineering* 1:253–265.

Ma, H.B. and G.P. Peterson. 1997b. The minimum meniscus radius and capillary heat transport limit in micro heat pipes. *Microelectromechanical Systems. ASME* DSC 62/HTD 354:213–220.

Ma, H.B., G.P. Peterson, and X.J. Lu. 1994. Influence of vapor–liquid interactions on the liquid pressure drop in triangular microgrooves. *International Journal of Heat Mass Transfer* 37:2211–2219.

Ma, H.B., G.P. Peterson, and X.F. Peng. 1996. Experimental investigation of countercurrent liquid–vapor interactions and their effect on the friction factor. *Experimental Thermal and Fluid Science* 12:25–32.

Mahalingam, M. and J. Andrews. 1987. High performance air cooling for microelectronics. *Proceedings of the International Symposium on Cooling Technology for Electronic Equipment*, Honolulu, Hawaii, 608–625.

Mala, G.M., D. Li, and J.D. Dale. 1997a. Heat transfer and fluid flow in microchannels. *International Journal of Heat and Mass Transfer* 40:3079–3088.

Mala, G.M., D. Li, C. Werner, H.J. Jacobasch, and Y.B. Ning. 1997b. Flow characteristics of water through a microchannel between two parallel plates with electrokinetic effects. *International Journal of Heat Fluid Flow* 18:489–496.

Mallik, A.K. and G.P. Peterson. 1991. On the use of micro heat pipes as an integral part of semiconductors. *Proceedings of the 3rd ASME–JSME Thermal Engineering Joint Conference*, Reno, Nevada, 2:394–401.

Mallik, A.K. and G.P. Peterson. 1995. Steady-state investigation of vapor deposited micro heat pipe array. *ASME Journal of Electronic Packaging* 117:75–81.

Mallik, A.K., G.P. Peterson, and W. Weichold. 1991. Construction processes for vapor deposited micro heat pipes. *Proceedings of the 10th Symposium on Electronic Materials Processing and Characteristics*, Richardson, Texas, June 3–4.

Marto, P.J. and G.P. Peterson. 1988. Application of heat pipes to electronics cooling. In A. Bar Cohen and A.D. Kraus (Eds.), *Advances in Thermal Modeling of Electronic Components and Systems* (pp. 283–336). New York: Hemisphere Pub. Co.

Mavriplis, C., J.C. Ahn, and R. Goulard. 1997. Heat transfer and flow fields in short microchannels using direct simulation Monte Carlo. *AIAA Journal of Thermophysics and Heat Transfer* 11:489–496.

Maynes, D. and B.W. Webb. 2003a. Fully developed electro-osmotic heat transfer in micro-channels. *International Journal of Heat and Mass Transfer* 46:1359–1369.

Maynes, D. and B.W. Webb. 2003b. Fully-developed thermal transport in combined pressure and electro-osmotically driven flow in microchannels. *Journal of Heat Transfer* 125:889–895.

Maynes, D. and B.W. Webb. 2004. The effect of viscous dissipation in thermally fully-developed electro-osmotic heat transfer in microchannels. *International Journal of Heat and Mass Transfer* 47:987–999.

McDaniels, D. and G.P. Peterson. 2001. Investigation of polymer based micro heat pipes for flexible spacecraft radiators. *Proceedings of ASME Heat Transfer Division, HTD-142*, New York, November 11–16, 2001.

Meinhart. C.D., S.T. Wereley, and J.G. Santiago. 1999. PIV measurements of a microchannel flow. *Experiments in Fluids* 27:414–419.

Mertz, R., A. Wein, and M. Groll. 1996. Experimental investigation of flow boiling heat transfer in narrow channels. *Heat and Technology* 14:47–54.

Min, J.Y., S.P. Jang, and S.J. Kim. 2004. Effect of tip clearance on the cooling performance of a microchannel heat sink. *International Journal of Heat and Mass Transfer* 47: 1099–1103.

Moon, S.H., G. Hwang, S.C. Ko, and Y.T. Kim. 2003. Experimental study on the thermal performance of micro heat pipe with cross-section of polygon. *Microelectronics Reliability* 44:315–321.

Morini, G.L. 2004. Laminar-to-turbulent flow transition in microchannels. *Microscale Thermophysical Engineering* 8:15–30.

Murakami, Y. and B.B. Mikic. 2003. Parametric investigation of viscous dissipation effects on optimized air cooling microchanneled heat sinks. *Heat Transfer Engineering* 24:53–62.

Murer, S., P. Lybaert, L. Gleton, and A. Sturbois. 2005. Experimental and numerical analysis of the transient response of a miniature heat pipe. *Applied Thermal Engineering* 25:2566–2577.

Nakayama, W., Y. Ohba, and K. Maeda. 2004. The effects of geometric uncertainties on micro-channel heat transfer—A CFD study on the tilted-cover problem. *Heat Transfer Engineering* 25:50–60.

Newport, D., J. Garvey, T. Dalton, V. Egan, and M. Whelan. 2004. Development of inter-ferometric temperature measurement procedures for microfluid flow. *Microscale Thermophysical Engineering* 8:141–154.

Newport, D., C.B. Sobhan, and J. Garvey. 2008. Digital Interferometry: Techniques and trends for fluid measurement. *Heat and Mass Transfer* 44:535–546.

Ng, E.Y.K. and S.T. Tan. 2004. Computation of three-dimensional developing pressure-driven liquid flow in a microchannel with EDL effect. *Numerical Heat Transfer A* 45: 1013–1027.

Nilson, R.H., S.K. Griffiths, S.W. Tchikanda, and M.J. Martinez. 2004. Axially tapered microchannels of high aspect ratio for evaporative cooling devices. *Journal of Heat Transfer* 126:453–462.

Nishio, S. 2004. Single-phase laminar-flow heat transfer and two-phase oscillating-flow heat transport in microchannels. *Heat Transfer Engineering* 25:31–43.

Niu, Y.Y. 1999. Navier–Stokes analysis of gaseous slip flow in long grooves. *Numerical Heat Transfer A* 36:75–93.

Palm, B. 2001. Heat transfer in microchannels. *Microscale Thermophysical Engineering* 5:155–175.

Patankar, S.V. 1980. *Numerical Heat Transfer and Fluid Flow*. Washington DC: Hemisphere Pub. Co.

Peng, X.F. and B.X. Wang. 1993. Forced convection and flow boiling heat transfer for liquid flowing through microchannels. *International Journal of Heat and Mass Transfer* 36:3421–3427.

Peng, X.F. and G.P. Peterson. 1995. The effect of thermo fluid and geometrical parameters on convection of liquids through rectangular microchannels. *International Journal of Heat and Mass Transfer* 38:755–758.

Peng, X.F. and G.P. Peterson. 1996a. Forced convection heat transfer of single phase binary mixtures through microchannels. *Experimental Thermal and Fluid Science* 12:98–104.

Peng, X.F. and G.P. Peterson. 1996b. Convective heat transfer and flow friction for water flow in microchannel structures. *International Journal of Heat and Mass Transfer* 39:2599–2608.

Peng, X.F., G.P. Peterson, and B.X. Wang. 1994a. Frictional flow characteristics of water flowing through microchannels. *Experimental Heat Transfer* 7:249–264.

Peng, X.F., G.P. Peterson, and B.X. Wang. 1994b. Heat transfer characteristics of water flowing through microchannels. *Experimental Heat Transfer* 7:265–283.

Peng, X.F., B.X. Wang, G.P. Peterson, and H.B. Ma. 1995. Experimental investigation of heat transfer in flat plates with rectangular microchannels. *International Journal of Heat and Mass Transfer* 38:127–137.

Peng, X.F., G.P. Peterson, and B.X. Wang. 1996. Flow boiling of binary mixtures in microchannel plates. *International Journal of Heat and Mass Transfer* 39:1257–1264.

Peng, X.F., H.Y. Hu, and B.X. Wang. 1998. Flow boiling through V-shape microchannels. *Experimental Heat Transfer* 11:87–90.

Peng, X.F., D. Liu, D.J. Lee, Y. Yan, and B.X. Wang. 2000. Cluster dynamics and fictitious boiling in microchannels. *International Journal of Heat and Mass Transfer* 23: 4259–4265.

Peng, X.F., D. Liu, and D.J. Lee. 2001a. Dynamic characteristics of microscale boiling. *Heat and Mass Transfer/Waerme-und Stoffuebertragung* 37:81–86.

Peng, X.F., Y. Tien, and D.J. Lee. 2001b. Bubble nucleation in microchannels: Statistical mechanics approach. *International Journal of Heat and Mass Transfer* 44:2957–2964.

Peng, X.F., Y. Tian, and D.J. Lee. 2002. Arguments on microscale boiling dynamics. *Microscale Thermophysical Engineering* 6:75–83.

Perret, C., C. Schaeffer, A. Bricard, and J. Boussey. 1998. Microchannel integrated heat sinks in silicon technology. *Proceedings of IEEE Industry Applications Society Annual Meet* 2:384–389.

Peterson, G.P. 1987a. Analysis of a heat pipe thermal switch. *Proceedings of the 6th International Heat Pipe Conference,* Grenoble, France, 1:177–183.

Peterson, G.P. 1987b. Heat removal key to shrinking avionics. *Aerospace America* 8:20–22.

Peterson, G.P. 1988a. Investigation of miniature heat pipes. Final Report, Wright Patterson AFB, Contract No. F33615 86 C 2733, Task 9.

Peterson, G.P. 1988b. Heat pipes in the thermal control of electronic components. *Proceedings of the 3rd International Heat Pipe Symposium,* Tsukuba, Japan, 2–12.

Peterson, G.P. 1990. Analytical and experimental investigation of micro heat pipes. *Proceedings of the 7th International Heat Pipe Conference*, Minsk, USSR, Paper No. A-4.

Peterson, G.P. and Y.X. Wang. 2003. Flat heat pipe cooling devices for mobile computers. *IMECE 2003,* Washington DC.

Peterson, G.P. 1992. An overview of micro heat pipe research. *Applied Mechanics Review* 45:175–189.

Peterson, G.P. 1994. *An Introduction to Heat Pipes: Modeling, Testing and Applications.* New York: John Wiley & Sons.

Peterson, G.P. 1996. Modeling, fabrication and testing of micro heat pipes: an update. *Applied Mechanics Review* 49:175–183.

Peterson, G.P. and J.M. Ha. 1998. Capillary performance of evaporating flow in micro grooves: An approximate analytical approach and experimental investigation. *Journal of Heat Transfer* 120:743–751.

Peterson, G.P. and H.B. Ma. 1996a. Analysis of countercurrent liquid–vapor interactions and the effect on the liquid friction factor. *Experimental Thermal and Fluid Sciences* 12:13–24.

Peterson, G.P. and H.B. Ma. 1996b. Theoretical analysis of the maximum heat transport in triangular grooves: A study of idealized micro heat pipes. *Journal of Heat Transfer* 118:731–739.

Peterson, G.P. and H.B. Ma. 1996c. Theoretical analysis of the maximum heat transport in triangular grooves: A study of idealized micro heat pipes. *Journal of Heat Transfer* 118:731–739.

Peterson, G.P. and H.B. Ma. 1999. Temperature response of heat transport in micro heat pipes. *Journal of Heat Transfer* 121:438–445.

Peterson, G.P. and A.K. Mallik. 1995. Transient response characteristics of vapor deposited micro heat pipe arrays. *ASME Journal of Electronic Packaging* 117:82–87.

Peterson, G.P. and A. Ortega. 1990. Thermal control of electronic equipment and devices. *Advances in Heat Transfer* 20:181–314.

Peterson, G.P. and C.B. Sobhan. 2006. Micro heat pipes and micro heat spreaders. In Mohammed Gad-el Hak (Ed.), *The MEMS Handbook. 2nd ed. Applications* (pp. 11:1–11:37). Boca Raton: Taylor & Francis.

Peterson, G.P., A.B. Duncan, A.K. Ahmed, A.K. Mallik, and M.H. Weichold. 1991. Experimental investigation of micro heat pipes in silicon devices. *ASME Winter Annual Meeting*. Atlanta, Georgia. ASME DSC-32:341–348.

Peterson, G.P., A.B. Duncan, and M.H. Weichold. 1993. Experimental investigation of micro heat pipes fabricated in silicon wafers. *Journal of Heat Transfer* 115: 751–756.

Peterson, G.P., L.W. Swanson, and F.M. Gerner. 1998. Micro heat pipes, in *Microscale Energy Transport*, Tien, C.L., Majumdar, A., and Gerner, F.M., Eds., Washington D.C: Taylor & Francis.

Pfahler, J., J. Harley, H.H. Bau, and J. Zemel. 1991. Gas and liquid flow in small channels. *Micromechanical Sensors, Actuators and Systems. ASME DSC*-32:49–60.

Pfahler, J., J. Harley, H.H. Bau, and J. Zemel. 1990. Liquid transport in micron and submicron channels. *Journal of Sensors Actuators A*, 21–23:431–434.

Phillips, R.J., L.R. Glicksman, and R. Larson. 1987. Forced-convection, liquid cooled microchannel heat sinks for high power density microelectronics. *Proceedings of the International Symposium on Cooling Technology for Electronic Equipment*, Honolulu, Hawaii, 295–316.

Piasecka, M. and M.E. Poniewski. 2004. Hysteresis phenomena at the onset of subcooled nucleate flow boiling in microchannels. *Heat Transfer Engineering* 25:44–51.

Qu, W. and I. Mudawar. 2002. Analysis of three-dimensional heat transfer in micro-channel heat sinks. *International Journal of Heat and Mass Transfer* 45:3973–3985.

Qu, W., G.M. Mala, and D. Li. 2000a. Pressure-driven water flows in trapezoidal silicon microchannels. *International Journal of Heat and Mass Transfer* 43:353–364.

Qu, W., G.M. Mala, and D. Li. 2000b. Heat transfer for water flow in trapezoidal silicon microchannels. *International Journal of Heat and Mass Transfer* 43:3925–3936.

Rahman, M.M. and F. Gui. 1993. Experimental measurements of fluid flow and heat transfer in microchannel cooling passages in a chip substrate. *Advances in Electronic Packaging. ASME EEP* 4-2:685–692.

Raju, R. and S. Roy. 2005. Hydrodynamic study of high-speed flow and heat transfer through a microchannel. *AIAA Journal of Thermophysics and Heat Transfer* 19:106–113.

Ramadas, P. 1993. *Silicon Micromachined Heat Pipe Fabrication Technology for Electronic and Space Applications*. MS Thesis, University of Cincinnati, Cincinnati, Ohio.

Ramadas, P., B. Badran, F.M. Gerner, T.H. Henderson, and K.W. Baker. 1993. Liquid metal micro heat pipes incorporated in waste-heat radiator panels. *Proceedings of the 10th Symposium on Space Power and Propulsion*, Albuquerque, New Mexico, Jan. 10–14.

Ravigururajan, T.S. 1998. Impact of channel geometry on two-phase flow heat transfer characteristics of refrigerants in microchannel heat exchangers. *Journal of Heat Transfer* 120:485–491.

Reddy, S.V. and C.B. Sobhan. 2006. Computational analysis of wire-sandwiched micro heat pipes. *Proceedings of the International Conference on Renewable Energy Scale-up Development (ICRESD)*, Southeast University, Nanjing, China.

Ren, L., W. Qu, and D. Li. 2001. Interfacial electrokinetic effects on liquid flow in microchannels. *International Journal of Heat and Mass Transfer* 44:3125–3134.

Richardson, D.H., D.P. Sekulic, and A. Campo. 2000. Reynolds number flow inside straight microchannels with irregular cross-sections. *Heat and Mass Transfer/Waerme-und Stoffuebertragung* 36:187–193.

Richter, M., P. Woias, and D. Wei. 1997. Microchannels for applications in liquid dosing and flow rate measurement. *Sensors and Actuators A* 62:480–483.

Roach, G.M. Jr., S.I. Abdel-Khalik, S.M. Ghiaasiaan, M.F. Dowling, and S.M. Jeter. 1999. Low-flow critical heat flux in heated microchannels. *Nuclear Science and Engineering* 131:411–425.

Rostami, A.A., A.S. Mujumdar, and N. Saniei. 2002. Flow and heat transfer for gas flowing in microchannels: A review. *Heat and Mass Transfer/Waerme-und Stoffuebertragung* 38:359–367.

Roy, S.K. and B.L. Avanic. 1996. Very high heat flux microchannel heat exchanger for cooling of semiconductor laser diode arrays. *IEEE Transactions on Components, Packaging, Manufacturing Technology Part B: Advanced Packaging* 19:444–451.

Ryu, J.H., D.H. Choi, and S.J. Kim. 2002. Numerical optimization of the thermal performance of a microchannel heat sink. *International Journal of Heat and Mass Transfer* 45:2823–2827.

Ryu, J.H., D.H. Choi, and S.J. Kim. 2003. Three-dimensional numerical optimization of a manifold microchannel heat sink. *International Journal of Heat and Mass Transfer* 46:1553–1562.

Sane, N.K. and S.P. Sukhatme. 1974. Natural convection heat transfer from horizontal rectangular fin arrays. *Proceedings of the 5th International Heat Transfer Conference*, Tokyo, Japan, 3:114–118.

Sartre, V., M.C. Zaghdoudi, and M. Lallemand. 2000. Experimental study on silicon micro heat pipe arrays. *International Journal of Thermal Science* 39:498–504.

Sobhan, C.B. and G.P. Peterson. 2004. Modeling of the flow and heat transfer in micro heat pipes. *Proceedings of the 2nd International Conference on Microchannels and Mini Channels*, Rochester, New York, 883–890.

Sobhan, C.B. and S.V. Garimella. 2001. A comparative analysis of investigations on heat transfer and fluid flow in microchannels. *Microscale Thermophysical Engineering* 5:293–311.

Sobhan, C.B., H. Xiaoyang, and C.Y. Liu. 2000. Investigations on transient and steady-state performance of a micro heat pipe. *AIAA Journal of Thermophysics and Heat Transfer* 14:161–169.

Son, S.Y. and J.S. Allen. 2004. Visualization of wettability effects on microchannel two-phase flow resistance. *Journal of Heat Transfer* 126:498–499.

Sparrow, E.M., T.S. Chen, and V.K. Jonsson. 1964. Laminar flow and pressure drop in internally finned annular ducts. *International Journal of Heat and Mass Transfer* 7:583–585.

Steinke, M.E. and S.G. Kandlikar. 2004a. An experimental investigation of flow boiling characteristics of water in parallel microchannels. *Journal of Heat Transfer* 126:518–526.

Steinke, M.E. and S.G. Kandlikar. 2004b. Control and effect of dissolved air in water during flow boiling in microchannels. *International Journal of Heat and Mass Transfer* 47:1925–1935.

Sturgis, J.C. and I. Mudawar. 1999a. Critical heat flux in a long, rectangular channel subjected to one-sided heating—I. Flow visualization. *International Journal of Heat and Mass Transfer* 42:1835–1847.

Sturgis, J.C. and I. Mudawar. 1999b. Critical heat flux in a long, rectangular channel subjected to one-sided heating—II. Analysis of critical heat flux data. *International Journal of Heat and Mass Transfer* 42:1849–1862.

Suh, J.S. and Y.S. Park. 2003. Analysis of thermal performance in a micro heat pipe with axially trapezoidal groove. *Tamkang Journal of Science and Engineering* 6:201–206.

Suman, B. and N. Hoda. 2005. Effect of variations in thermophysical properties and design parameters on the performance of a V-shaped micro grooved heat pipe. *International Journal of Heat and Mass Transfer* 48:2090–2101.

Suman, B. and P. Kumar. 2005. An analytical model for fluid flow and heat transfer in a micro heat pipe of polygonal shape. *International Journal of Heat and Mass Transfer* 48:4498–4509.

Suman, B., S. De, and S.D. Gupta. 2005. Transient modeling of micro grooved heat pipe. *International Journal of Heat and Mass Transfer* 48:1633–1646.

Sun, H. and M. Faghri. 2003. Effect of surface roughness on nitrogen flow in a microchannel using the direct simulation Monte Carlo method. *Numerical Heat Transfer A* 43:1–8.

Sunil, S., J.R.N. Reddy, and C.B. Sobhan. 1996. Natural convection heat transfer from a thin rectangular fin with a line source at the base—a finite difference solution. *Heat and Mass Transfer/Waerme-und Stoffuebertragung* 31:127–135.

Takamatsu, K., N. Fujimoto, Y.F. Rao, and K. Fukuda. 1997. Numerical study of flow and heat transfer of superfluid helium in capillary channels. *Cryogenics* 37:829–835.

Tardu, S. 2004. Analysis of the electric double layer effect on microchannel flow stability. *Microscale Thermophysical Engineering* 8:383–401.

Tchikanda, S.W., R.H. Nilson, and S.K. Griffiths. 2004. Modeling of pressure and shear-driven flows in open rectangular microchannels. *International Journal of Heat and Mass Transfer* 47:527–538.

Thome, J.R. 2004. Boiling in microchannels: A review of experiment and theory. *International Journal of Heat and Fluid Flow* 25:128–139.

Thome, J.R., V. Dupont, and A.M. Jacobi. 2004. Heat transfer model for evaporation in microchannels. Part I: Presentation of the model. *International Journal of Heat and Mass Transfer* 47:3375–3385.

Toh, K.C., X.Y. Chen, and J.C. Chai. 2002. Numerical computation of fluid flow and heat transfer in microchannels. *International Journal of Heat and Mass Transfer* 45:5133–5141.

Triplett, K.A., S.M. Ghiaasiaan, S.I. Abdel-Khalik, and D.L. Sadowski. 1999a. Gas–liquid two–phase flow in microchannels. Part I: Two-phase flow patterns. *International Journal of Multiphase Flow* 25:377–394.

Triplett, K.A., S.M. Ghiaasiaan, S.I. Abdel-Khalik, A. LeMouel, and B.N. McCord. 1999b. Gas–liquid two-phase flow in microchannels. Part II: Void fraction and pressure drop. *International Journal of Multiphase Flow* 25:395–410.

Tso, C.P. and S.P. Mahulikar. 1998. Use of the Brinkman number for single phase forced convective heat transfer in microchannels. *International Journal of Heat and Mass Transfer* 41:1759–1769.

Tso, C.P. and S.P. Mahulikar. 1999. Role of the Brinkman number in analyzing flow transitions in microchannels. *International Journal of Heat and Mass Transfer* 42:1813–1833.

Tso, C.P. and S.P. Mahulikar. 2000a. Combined evaporating meniscus-driven convection and radiation in annular microchannels for electronics cooling application. *International Journal of Heat and Mass Transfer* 43:1007–1023.

Tso, C.P. and S.P. Mahulikar. 2000b. Experimental verification of the role of brinkman number in microchannels using local parameters. *International Journal of Heat and Mass Transfer* 43:1837–1849.

Tuckerman, D.B. and R.F.W. Pease. 1981. High-performance heat sinking for VLSI. *IEEE Electron Device Letters* EDL-2:126–129.

Tuckerman, D.B. and R.F.W. Pease. 1982. Ultrahigh thermal conductance microstructures for cooling integrated circuits. *Proceedings of the 32nd Electronic Components Conference, IEEE, EIA, CHMT*, 145–149.

Tunc, G. and Y. Bayazitoglu. 2002. Heat transfer in rectangular microchannels. *International Journal of Heat and Mass Transfer* 45:765–773.

Turner, S.E., L.C. Lam, M. Faghri, and O.J. Gregory. 2004. Experimental investigation of gas flow in microchannels. *Journal of Heat Transfer* 126:753–763.

Van Male, P., M. de Croon, R.M. Tiggelaar, A.Van den Berg, and J.C. Schouten. 2004. Heat and mass transfer in a square microchannel with asymmetric heating. *International Journal of Heat and Mass Transfer* 47:87–99.

Wang, B.X. and X.F. Peng. 1994. Experimental investigation on liquid forced convection heat transfer through microchannels. *International Journal of Heat and Mass Transfer* 37:73–82.

Wang, E.N., S. Devasenathipathy, J.G. Santiago, K.E. Goodson, and T.W. Kenny. 2004. Nucleation and growth of vapor bubbles in a heated silicon microchannel. *Journal of Heat Transfer* 126:497–498.

Wang, Y., H.B. Ma, and G.P. Peterson. 2001. Investigation of the temperature distributions on radiator fins with micro heat pipes. *AIAA Journal of Thermophysics and Heat Transfer* 15:42–49.

Wang, Y.X. and G.P. Peterson. 2002a. Analysis of wire-bonded micro heat pipe arrays. *AIAA Journal of Thermophysics and Heat Transfer* 16:346–355.

Wang, Y.X. and G.P. Peterson. 2002b. Optimization of micro heat pipe radiators in a radiation environment. *AIAA Journal of Thermophysics and Heat Transfer* 16: 537–546.

Wang, Y.X. and G.P. Peterson. 2002c. Capillary evaporation in microchanneled polymer films. *Proceedings of the 8th AMSE/AIAA Joint Thermophysics and Heat Transfer Conference,* St. Louis, Missourri, AIAA-2002: 2767.

Wayner, Jr., P.C., Y.K. Kao, and L.V. LaCroix. 1976. The interline heat-transfer coefficient of an evaporating wetting film. *International Journal of Heat and Mass Transfer* 19:487–492.

Weichold, M.H., G.P. Peterson, and A. Mallik. 1993. Vapor deposited micro heat pipes, U.S. Patent No. 5,179,043.

Weisberg, A., H.H. Bau, and J. Zemel. 1992. Analysis of microchannels for integrated cooling. *International Journal of Heat and Mass Transfer* 35:2465–2474.

Welty, J.R., C.E. Wicks, R.E. Wilson, and G.L. Rorrer. 2000. *Fundamentals of Momentum, Heat and Mass Transfer*. New York: Wiley.

White, F.M. 2003. *Fluid Mechanics*. New York: McGraw Hill.

Wu, D. and G.P. Peterson. 1991. Investigation of the transient characteristics of a micro heat pipe. *AIAA Journal of Thermophysics and Heat Transfer* 5:129–134.

Wu, D., G.P. Peterson, and W.S. Chang. 1991. Transient experimental investigation of micro heat pipes. *AIAA Journal of Thermophyics and Heat Transfer* 5:539–544.

Wu, H.Y. and P. Cheng. 2003. Friction factors in smooth trapezoidal silicon microchannels with different aspect ratios. *International Journal of Heat and Mass Transfer* 46:2519–2525.

Wu, H.Y. and P. Cheng. 2004. Boiling instability in parallel silicon microchannels at different heat flux. *International Journal of Heat and Mass Transfer* 47:3631–3641.

Wu, H.Y. and P. Cheng. 2005. Condensation flow patterns in silicon microchannels. *International Journal of Heat and Mass Transfer* 48:2186–2197.

Wu, P.Y. and W.A. Little. 1983. Measurement of friction factor for the flow of gases in very fine channels used for micro miniature Joule Thompson refrigerators. *Cryogenics* 23:273–277.

Wu, P.Y. and W.A. Little. 1984. Measurement of heat transfer chracteristic of gas flow in fine channel heat exchangers for micro miniature refrigerators. *Cryogenics* 24:415–420.

Xu, J.L., Y.H. Gan, D.C. Zhang, and X.H. Li. 2005. Microscale heat transfer enhancement using thermal boundary layer redeveloping concept. *International Journal of Heat and Mass Transfer* 48:1662–1674.

Yang, C., D. Li, and J.H. Masliyah. 1998. Modeling forced liquid convection in rectangular microchannels with electrokinetic effects. *International Journal of Heat and Mass Transfer* 41:4229–4249.

Yin, J.M., C.W. Bullard, and P.S. Hrnjak. 2002. Single-phase pressure drop measurements in a microchannel heat exchanger. *Heat Transfer Engineering* 23:3–12.

Yu, D., R. Warrington, R. Barron, and T. Ameel. 1995. An experimental and theoretical investigation of fluid flow and heat transfer in microtubes. *Proceedings of the ASME/JSME Thermal Engineering Conference* 1:523–530.

Yu, S. and T.A. Ameel. 2001. Slip-flow heat transfer in rectangular microchannels. *International Journal of Heat and Mass Transfer* 44:4225–4234.

Yu, S. and T.A. Ameel. 2002. Slip flow convection in isoflux rectangular microchannels. *Journal of Heat Transfer* 124:346–355.

Yu, S., T. Ameel, and M. Xin. 1999. An air-cooled microchannel heat sink with high heat flux and low pressure drop. *Proceedings of the 33rd National Heat Transfer Conference*, Albuquerque, New Mexico, NHTC 99–162:1–7.

Zhang, L., E.N. Wang, K.E. Goodson, and T.W. Kenny. 2005. Phase change phenomena in silicon microchannels. *International Journal of Heat and Mass Transfer* 48:1572–1582.

Zhao, C.Y. and T.J. Lu. 2002. Analysis of microchannel heat sinks for electronics cooling. *International Journal of Heat and Mass Transfer* 45:4857–4869.

Zhuang, Y., C.F. Ma, and M. Qin. 1997. Experimental study of local heat transfer with liquid impingement flow in two dimensional micro-channels. *International Journal of Heat and Mass Transfer* 40:4055–4059.

Zohar, Y. 2006. Microchannel heat sinks. In Mohammed Gad-el Hak (Ed.), *The MEMS Handbook. 2nd ed. Applications* (pp.12:1–12:29). Boca Raton: Taylor & Francis.

5 Microscale Radiative Heat Transfer

5.1 MACROSCOPIC APPROACH

Radiative heat transfer is the mode of energy transport in which electromagnetic waves of the thermal range carry energy and convert it to heat while they interact with surfaces or media. Any body above a temperature of absolute zero emits radiation, and the distribution of the emitted power of an ideal emitter is described theoretically by Planck's distribution function (Siegel and Howell 1992):

$$E_{b\lambda}(\lambda, T) = \frac{C_1}{\lambda^5 [\exp{(C_2/\lambda T)} - 1]} \tag{5.1}$$

where
$E_{b\lambda}$ denotes the emitted power at a wavelength (monochromatic emissive power)
T is the absolute temperature

C_1 and C_2 are given by

$$C_1 = 2hc^2 = 11,909 \text{ W } \mu m^4/cm^2 \text{ and}$$
$$C_2 = hc/k_B = 14,388 \text{ } \mu m \text{ K}$$

with
h denoting the Planck's constant
k_B the Boltzmann constant
c the light velocity

The maximum emissive power of the emitting body corresponds to the peak of the Planck's distribution function, and its relationship to the corresponding wavelength is given by the Wien's displacement law (Siegel and Howell 1992):

$$\lambda_{max} T \approx 2898 \text{ } \mu m \text{ K} \tag{5.2}$$

An ideal emitter is one that emits the maximum possible radiation at a given temperature and is called a blackbody. The emissivity of a body is defined as the ratio of the actual emissive power of a body to that of a blackbody, which has a value unity. Thus for any body, the emissivity is less than 1. Idealized bodies, which emit less than a blackbody at the same temperature, with an emissivity independent of the wavelength, are termed gray bodies. Most of the classical radiation calculations utilize a gray body assumption for the emitter. Natural bodies emit radiation at different wavelengths, including the visible spectrum and are called colored bodies, where the emissivity is a function of the wavelength of the emitted radiation.

The total emissive power of a blackbody, defined as the emitted energy per unit area of the surface per unit time, can be obtained theoretically by integrating the wavelength-dependent emissive power (monochromatic emissive power) in the Planck's distribution, over the range of wavelengths between zero and infinity, and utilizing the principles of statistical thermodynamics. The result is often given by the following equation for macroscopic radiative systems, which incorporates the Stefan–Boltzmann constant to establish a relationship between the emissive power and the fourth power of the absolute temperature of the body. This governing equation is called the Stefan–Boltzmann law. For gray bodies, the calculation also involves the introduction of the emissivity into the equation. Thus,

$$E = \sigma\varepsilon\, T^4 \tag{5.3}$$

where
  E is the total emissive power
  σ is the Stefan–Boltzmann constant
  ε is the emissivity

Three surface properties are defined for radiation analysis—the absorptivity, the reflectivity, and the transmissivity (sometimes also termed absorptance, reflectance, and transmittance) of a surface. These represent the fractions of the impinging radiation that is absorbed, reflected, and transmitted by the surface, respectively, where the sum of these three quantities must always equal 1. At equilibrium conditions, it can be shown that the emissivity of a surface is equal to its absorptivity. This is called Kichhoff's law in radiative heat transfer.

In calculating the heat exchange by radiation between surfaces, the shape factor or configuration factor between the two surfaces is used. The shape factor determines what fraction of the radiation emitted from a surface falls on an interacting surface. On the basis of the definition of the intensity of radiation, it is possible to derive expressions for the shape factor in terms of the mutual geometric configuration of the interacting surfaces and their surface geometries. For combinations of standard surfaces, shape factors are available in data books. The calculation of the radiative exchange between two gray surfaces, incorporating the shape factors is given as follows:

$$E = \varepsilon\, \sigma\, A_1\, F_{12}\left(T_1^4 - T_2^4\right) \tag{5.4}$$

where F_{12} is the shape factor that denotes the fraction of energy leaving surface 1, which reaches surface 2 (also, $A_1 F_{12} = A_2 F_{21}$, called the reciprocity relationship). Here, A_1 and A_2 denote the surface areas.

For multiple surfaces, the interaction can be written using a radiation network model, with resistances representing the surface emissivities (surface resistances) and the mutual configuration of the surfaces (shape resistances). Calculation of radiation exchange using the Stefan–Boltzmann equation, emissivity values of surfaces, and shape factor algebra is the essence of most of the radiation analysis applied to macroscopic engineering systems. This can be found in all fundamental engineering heat transfer books (Siegel and Howel 1992, Incropera and DeWitt 2001). Further, in the presence of participating media, the basic approach is modified to incorporate the effect of the medium due to the absorption and emission of radiation, while radiative heat transfer takes place between surfaces.

5.2 MICROSCOPIC APPROACH

The microscopic approach to obtain the radiative heat transfer involves the use of the concept of photons, which carry quanta of energy, and the summation (or integration) of all such quantized energy transport over the number of photons. Owing to the observation that accurate prediction of experimental data requires modifications on the wave assumption of radiation (such as standing transverse waves), the average number of photons with any angular frequency has been derived as follows (Kittel 1996, Kumar and Mitra 1999) through application of the quantum theory to modify the wave hypothesis:

$$n = \frac{1}{\exp\left(h\omega / 2\pi k_B T\right) - 1} \tag{5.5}$$

This essentially modifies the kinetic theory approach, which averages the energy over a large number of particles as $k_B T / 2$ (the equipartition theorem), by the application of quantum theory and produces a spectrum for the number of phonons, over the angular frequency.

The total energy at any angular frequency is given by (Kumar and Mitra 1999)

$$\Xi = \left(n + \frac{1}{2}\right) \frac{h}{2\pi} \omega \tag{5.6}$$

This considers also the zero point (datum) energy, given by $0.5\,(h\omega / 2\pi)$.

With the modification described above, the energy is calculated for the standing transverse wave describing radiation in a cavity. This approach thus leads to an integration of the energy over the frequency-dependent range of photons, considering also the two directions of polarization of the transverse waves. Such an integration also leads to the Stefan–Boltzmann law for the radiative transfer resulting from the macroscopic approach.

The wavelength-dependent energy density and emissive power for a blackbody obtained through the microscopic approach are (Kumar and Mitra 1999)

$$\rho_{\lambda b} = \frac{8\pi hc}{\lambda^5} \frac{1}{\exp\left(\frac{h\omega_{em}}{2\pi k_B T}\right) - 1} \tag{5.7}$$

$$E_{\lambda b} = \frac{2\pi hc^2}{\lambda^5} \frac{1}{\exp\left(\frac{h\omega_{em}}{2\pi k_B T}\right) - 1} \tag{5.8}$$

where
  c is the light velocity
  h is the Planck's constant
  ρ is the energy density
  E is the emissive power
  λ is the wavelength
  ω is the angular frequency of the radiation

The macroscopic Stefan–Boltzmann law can be obtained by integrating the spectral emissive power over the wavelength range from zero to infinity.

The photon theory of radiative heat transfer explains the phenomenon due to the absorption of incident photons by phonons and electrons. The optical phonons produced on the absorption of photons conserve momentum and energy. The absorption by electrons produces a scattering process, and relaxation in electrons.

Of the models used to analyze the phenomenon, the Lorentz model is a basic model, which uses an analogy of the simple spring-mass system to describe the photon absorption. The analysis essentially gives an expression for the complex refractive index of the medium in the following form (Bohren and Huffman 1983, Kumar and Mitra 1999):

$$\bar{n}^2 = (n + i\kappa)^2 = 1 + \frac{\omega_p^2}{\omega_o^2 - \omega^2 - i\gamma\omega} \tag{5.9}$$

where
  ω_p is a constant called the plasma frequency
  γ is the damping constant
  ω is the incident frequency
  ω_o is the characteristic resonant frequency

Similarly, to describe the absorption of photons by electrons, considering the relaxation effects by means of the value zero assigned to the spring constant, a model has been developed. This is called the Drude model. This results in the following expression for the complex refractive index (Kumar and Mitra 1999):

$$\bar{n}^2 = (n + i\kappa)^2 = 1 - \frac{\omega_p^2}{\omega^2 + i\gamma\omega} \tag{5.10}$$

The plasma frequency ω_p found in the above expressions is given by

$$\omega_p^2 = \frac{\bar{n}e^2}{m\varepsilon_0} \tag{5.11}$$

where
   m is the electron mass
   e the electron charge
   ε_0 the dielectric constant for free space

   In a semiconductor, the absorption of a photon can excite an electron from the valance band to the conduction band. In materials with permanent dipoles, a mechanism of photon absorption is postulated to be the Debye relaxation, which involves a change in the alignment of dipoles. Detailed discussions on the various photon absorption mechanisms, including these, can be found in Kumar and Mitra (1999).

5.3 MICROSCALES IN RADIATIVE TRANSFER

The spatial and temporal microscales identified for radiative energy transport are discussed in this section. Thermal radiations and laser radiations are the two mechanisms of radiative heat transfer encountered in common engineering applications. As the applications of short pulse lasers have become increasingly popular, microscale radiation regimes have been identified where the laser pulse duration, in comparison with characteristic microscales, becomes the major criterion to determine the method of analysis. Longtin and Tien (1998) give an extensive discussion on the spatial and temporal length scales associated with radiative heat transfer, a brief review of which is presented below. Some cases of modeling of microscale radiation phenomena under various microscale regimes, presented by Kumar and Mitra (1999), are also summarized later in this chapter.

5.3.1 SPATIAL MICROSCALES FOR RADIATION

The most important of the characteristic length scales associated with radiative heat transfer is the wavelength of the radiation (λ); a more useful value being the peak wavelength. At any temperature, the peak wavelength is determined by Wien's law. Another characteristic length scale is the coherence length. Two photons emitted from the same source can interfere, if they are within the distance equal to the coherence length. The coherence length for blackbody thermal radiation in a vacuum can be obtained by the following formula (Tien and Chen 1994, Longtin and Tien 1998):

$$L_c T = 0.15 ch/k_B \approx 2160 \ \mu\text{m K} \tag{5.12}$$

To give an idea of the order of magnitude of the coherence length, the value for blackbody thermal radiation at 25°C is 7.3 μm, as reported by Longtin and Tien (1998). Figure 5.1 gives the peak wavelength and coherence length for blackbody as a function of temperature, also reproduced from this paper.

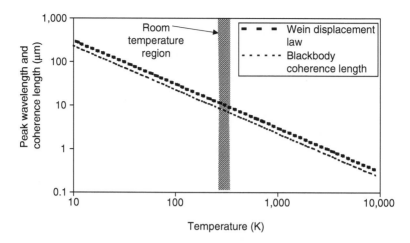

FIGURE 5.1 Variation of the peak wavelength (according to Wien's displacement law) and the coherence length for a blackbody as function of the absolute temperature. (Adapted from Longtin, J.P. and Tien, C.L., *Microscale Energy Transport*, C.L. Tien, A. Majumdar, and F.M. Gerner (Eds.), Taylor & Francis, New York, 1998. With permission.)

Another characteristic length scale called the penetration depth is also used. The concept is explained here. The intensity of radiation gets attenuated, as it moves through a medium. This attenuation, according to classical theory, is given by the Beer's law (Brewster 1992, Siegel and Howell 1992):

$$I(x) = I_0 \exp(-\mu_a x) \tag{5.13}$$

where

I_0 is the intensity of the incident radiation at the surface

x is the distance of radiation penetration into the medium

μ_a is called the absorption coefficient (m^{-1})

The term penetration depth represents the reciprocal of the absorption coefficient, and is indicated by δ. It can be shown that the intensity reduces to $1/e$ times the value at the surface of the medium, at the penetration depth. As the reciprocal of the absorption coefficient, the electromagnetic theory gives the expression for the penetration depth as follows:

$$\delta = \frac{\lambda}{4\pi\kappa} \tag{5.14}$$

where

κ is the extinction coefficient of the medium

λ is the radiation wavelength

The penetration depth depends on the material of the medium. Typical values for metals are in the range 20–40 nm, and for dielectrics like water and glass are in the range 1–10 m (Longtin and Tien 1998).

The above discussions point to the fact that the penetration depth is a natural characteristic length scale, which has its origin in the radiation–medium interaction. However, in certain cases, characteristic length scales dependent on the physical extent of the medium can also be utilized, such as the length dimensions of the medium itself (for instance, the thickness of a film).

Three microscale regimes have been defined by Tien and Chen (1994) based on the characteristic spatial microscales. Extensive discussions on these can be found in Longtin and Tien (1998). However, the concepts are summarized below.

The first spatial microscale regime is defined by

$$L/l_c \leq O(1) \tag{5.15}$$

Comparing the Wien's Law (Equation 5.2) and the definition of the coherence length (Equation 5.12), it can be seen that the wavelength for peak intensity and the coherence length are of the same order of magnitude (it follows that $l_c \approx 0.75\lambda_{max}$) for classical radiative transfer. So, it can be assumed that the microscale effects become important when the characteristic dimension of the material becomes comparable to the coherence length. In a general sense, the above is the correct criterion to be used while considering radiative heat transfer where wave-interference is important. In other words, in this regime, modeling based on Maxwell's equations for the electric field in the material would be appropriate, due to the wave nature of radiation. Typical examples demanding such an analysis could be multilayered thin films, media with particles, and micromachined structures (Longtin and Tien 1998). An example of analysis of particulate systems as reviewed by Kumar and Mitra (1999) belongs to this microscale regime and is discussed later in this chapter.

The second spatial microscale regime is defined for a case where $L/\lambda_c \geq O(1)$ as follows (Longtin and Tien 1998):

$$L/\Lambda \leq O(1) \tag{5.16}$$

$$\delta/\Lambda \leq O(1) \tag{5.17}$$

where Λ is the carrier mean free path. In the first case, when the characteristic dimension is smaller than the mean free path, size effect is introduced due to the dominance of collisions of the carriers with the boundaries. In the second case, a comparison of the penetration depth with the mean free path decides the criterion. As the electric field varies substantially within the penetration depth, the condition given in Equation 5.17 subjects the carriers near the surface to nonuniform conditions between collisions. This means that the classical models assuming a uniform electric field should be modified in this microscale regime. A suggested method is to assume an effective mean free path, to take care of both the effects mentioned above.

The third spatial microscale regime is given by

$$L/\lambda_c \leq O(1) \tag{5.18}$$

In this case, when the characteristic material dimension is less than the energy carrier wavelength, the boundary interaction makes classical equations (wave equations for instance) incapable of describing the physical problem. Quantum mechanics should be used for the analysis of such cases (Longtin and Tien 1998).

Longtin and Tien (1998) also describe other microscales, which become significant while dealing with microstructures. Two length-scale criteria have been established to decide whether special treatment (to modify the electromagnetic theory) is required for such cases. In defining these, the characteristic length dimension considered is the distance (l_s) between the adjacent elements in the (say, periodic) microstructure. These criteria are

$$l_s \leq \lambda \tag{5.19}$$

$$l_s \leq l_c \tag{5.20}$$

In the first case, neighboring elements disturb the electromagnetic field, and in the second, the interference of radiation from adjacent elements will have to be considered. An example of an analysis of dependent scattering, that is, interaction between the scattered electromagnetic radiation from particles in a medium (Kumar and Mitra 1999), is explained later in this chapter. This pertains to the microscale regime mentioned above.

5.3.2 Temporal Microscales

Various temporal microscales have been defined, with reference to short pulse lasers propagating through media (Longtin and Tien 1998). The laser pulse duration itself is a useful temporal microscale. This is the duration for which the laser pulse intensity exceeds half of the maximum value (as the laser pulse has a time-varying Gaussian profile). The thermal diffusion time is another temporal microscale, which is based on the diffusion of heat through a conducting medium. This becomes useful in dealing with radiation–conduction interaction systems, and is given by

$$\tau_d \approx \frac{L^2}{\alpha} \tag{5.21}$$

where
  α is the thermal diffusivity of the material
  L is the characteristic length through which thermal energy diffuses in the medium

The radiation propagation time is defined as the time required for the electromagnetic radiation to propagate through a medium of characteristic length L, and is given by

$$\tau_c \approx \frac{Ln}{c_0} \tag{5.22}$$

where
  n is the refractive index of the medium
  c_0 is the velocity of light in vacuum

Two intrinsic timescales of a material are the thermalization and relaxation times. Thermalization time is the time taken by an electron to impart the energy to the phonon system. The relaxation time is the time taken by an atom or molecule to get back to the original state, from an excited state attained by the absorption of a photon.

Temporal microscale regimes have been delineated based on the duration of the laser pulse in comparison with the characteristic temporal microscales discussed above (Longtin and Tien 1998). On the basis of thermal diffusion time, a regime is defined by

$$\tau_p \leq \tau_d \tag{5.23}$$

which can also be written as

$$\tau_p \leq L^2/\alpha \tag{5.24}$$

Equation 5.23 means that the rate of laser pulse is such that the energy cannot diffuse through the medium at the rate at which it is incident. As indicated by the square of the length in Equation 5.24, this timescale is highly size dependent.

Another temporal microscale regime is defined as

$$\tau_p \leq \tau_c \tag{5.25}$$

Here, the fastness of the laser pulse, being comparable or shorter than the light propagation time, results in an intensity variation in the medium. This will necessitate modification of the wave propagation equation for the intensity, in a nonscattering, nonabsorbing medium, which is normally written as

$$\frac{1}{c}\frac{\partial I}{\partial t} + \frac{\partial I}{\partial x} = 0 \tag{5.26}$$

by including suitable terms in the right-hand side (Longtin and Tien 1996).

Another temporal microscale regime becomes important when analyzing the phenomena that arise in durations that are comparable with the relaxation times of the carriers. Examples of such phenomena are photon absorption by free electrons in metals, absorption in dielectrics, and interactions of fast lasers with semiconductors (Longtin and Tien 1998).

5.4 INVESTIGATIONS OF MICROSCALE RADIATION

Modern trends in microminiaturization have made the analysis of radiative phenomena in the microscale very important. When the characteristic lengths of structures and systems become comparable with wavelengths, such considerations become extremely important. For quite some time, exploration into size-affected surface properties and studies on fabrication-associated radiative interaction on miniature systems, structures and solid films have been significant topics of research. Although very extensive studies were not performed regarding radiation, to the level available in other heat transfer modes, there have been some important contributions over the early years of the development of microscale heat transfer as a topic of research. The following section discusses mainly such early investigations as reviewed in the literature by Duncan and Peterson (1994). More modern trends in microscale radiation and its application are reviewed later in this chapter.

5.4.1 RADIATION INTERACTION WITH MICROSTRUCTURES AND MATERIALS

Focused research on microscale radiation phenomena has been reported since the 1980s. Hesketh and Zemel (1988) measured the normal, polarized spectral (3–14 μm) emittances of highly doped, micromachined, periodic structures on heavily phosphorus-doped silicon for pattern repeat scales of 10, 14, 18, and 22 μm and depths ranging from 0.7 to 45 μm. The s-polarization vector in these measurements was perpendicular to the grating vector. The data demonstrated that the emittance from the deep structures was dominated by standing electromagnetic waves in a direction normal to the surface similar to those in an organ pipe. It was concluded that these measurements, particularly the s-polarized emission from deep gratings, could not be explained by calculations of electromagnetic singularities on lamellar gratings.

Hesketh et al. (1988) also reported measurements of both the spectral and specular thermal radiation emission characteristics of regularly microgrooved surfaces in a silicon substrate at 300°C and 400°C. Samples having groove depths of 0–42 μm and groove widths of 12.6–14 μm were evaluated. The geometry repeat distance for all grooves was 22 μm, corresponding to 455 grooves per cm. The grooves correspond directly in size to the band of principle emission wavelengths that arise at these temperature levels. The measurements showed strong spectral effects for normal emissions, including highly favored frequencies in certain cases which suggested a cavity "organ pipe" mode of emission. Similar, though modified, effects were found in directional emission, away from the normal. There were also strong polarization effects, with the cross-groove polarization mode dominant. The spectral and specular measurements were compared with classical predictions, which assume that wavelengths are much less than the groove characteristic dimension. Comparison of the data and calculations showed a strong departure from simple geometric behavior in both the normal and directional spectral emittance of highly doped silicon microgrooves.

Timans (1992) determined the total emissivity of specimens of GaAs by using a technique that combined isothermal electron beam heating with a temperature measurement method that exploited the temperature dependence of the band gap of GaAs. Emission spectra from the GaAs specimens were recorded for a range of heating power densities. These spectra displayed a maximum near the semiconductor absorption edge, due to the fact that the blackbody radiation increases with increasing wavelength while the spectral emissivity decreases rapidly once the photon energy falls below the band gap. The temperature was determined by fitting the Planck radiation function to the high energy side of the maximum. This allowed a self-consistent determination of the temperature dependence of the position of the absorption edge. The results were used to calibrate a second set of experiments in which a corresponding set of reflection spectra were recorded. The reflection spectra exhibited a large change in reflectivity at the absorption edge due to the fact that light was being reflected from the back surface of the wafer as well as from the front when it became transparent. The reflection spectra provided a convenient method of temperature measurement because the intensity of the reflected light could be much greater than that of the emitted light. These high signal levels permitted temperature measurements to be made with good spatial and temporal resolution, even for samples at low temperatures. The temperature measurements could be performed with a precision of about 2°C, and in this study the accuracy was $+/-5$°C. As a result of these measurements, the total hemispherical emissivity of the GaAs specimens was determined over a temperature range from 350 to 630°C. The values for the total emissivity of GaAs reported in this paper were believed to be the first direct, experimental measurements of this quantity. It was found that the total hemispherical emissivity of 485 μm thick specimens increased from 0.08 to 0.28 over the measured temperature range.

Timusk and Tanner (1989) presented the results of a study of the temperature dependence of the infrared properties over a wide range of frequencies in samples of $YBa_2Cu_3O_{7-d}$. It was found that the mid-infrared region was dominated by a strong, nearly temperature-independent absorption band, possessing around 80% of the infrared oscillator strength.

Phelan et al. (1991) also presented reflectance predictions based on two of the major theories of the superconducting state, the classical Drude–London theory and the quantum mechanical Mattis–Bardeen (MB) theory, and compared them with available experimental data to determine the best method for predicting the radiative properties of thin film Y–Ba–Cu–O. It was observed that the Mattis–Bardeen theory was more successful than the Drude–London theory in predicting the reflectance. Consequently, approximate models for the Mattis–Bardeen theory were developed, thus enhancing the theory's usefulness for engineering calculations. The theoretical results for the MB theory were within 12% of the experimental data and the agreement improved as the wavelength increased.

Cohn and Buckius (1993) investigated scattering from microcontoured surfaces having length scales of the order of the incident radiation. This required a complete consideration of electromagnetic scattering theory. The bidirectional reflection function and directional spectral emissivity for periodic rectangular surfaces were presented. The results focused on general scattering phenomena and on cases comparable with the experimental observations of Hesketh et al. (1988) and Hesketh and Zemel (1988). The predictions showed excellent agreement with this existing data.

5.4.1.1 Radiation Scattering by Microstructures

Raad and Kumar (1992) considered radiation scattering by periodic microgrooved surfaces of conducting materials. A theoretical model for the scattering properties of microgrooves was constructed via a rigorous solution of the electromagnetic wave theory, which implicitly accounts for nonclassical effects of guided waves in the cavity and interference of scattered waves. Exact solutions for perfectly conducting materials were presented and the reflectivities were determined for normal and oblique incidence. The effects of the microgroove size, periodicity, and wavelength of the incident radiation on the scattering characteristics were examined. The analytical results were shown to asymptotically approach the smooth surface solutions in the limit of large width or small height.

5.4.2 STUDIES ON SILICON FILMS

Grigoropoulos et al. (1986) studied a simplified model of the fundamental mechanisms governing the radiatively induced recrystallization of a thin polysilicon layer on a heat sink structure. It had been observed that instabilities (the penetration of solid lamellae into the molten zone) in laser-induced crystal growth on thin polysilicon layers occurred frequently. To study this phenomenon, a linear stability analysis was used. The reflectivity difference between the liquid and the solid phases was shown to be a fundamental source of thermal instability. A qualitative understanding of the effect of the reflectivity change on the stability of laser-driven recrystallization of thin films was developed. The shape of the external heating intensity distribution, as well as the total power input, was found to determine the stability of the solidification interface.

During thermal processing of multilayer thin film structures (e.g., silicon oxide film–silicon film–silicon oxide film–silicon substrate) with a radiant heat source, the layer thicknesses are on the same order of magnitude as the wavelengths of the incident radiation (0.1–12 μm). Wong and Miaoulis (1990) presented a technique that allowed the coupling of these radiation microscale heat transfer effects with the conductive heat transfer in the film structure. A parametric study was conducted to observe the effects of varying the thickness of the different layers on the maximum temperature attained during processing. Results indicated that the variation of the thickness of either silicon oxide layer caused significant fluctuations on the reflectivity and the temperature of the film. Increasing the thickness of the thin silicon layer resulted in a nearly linear increase in temperature. Layer thickness ranges where small variations in thickness may result in large changes in reflectivity were identified.

Further investigation into this phenomenon (Wong et al. 1991) was conducted for the case of an oxide-capped SOI structure where the capping film and insulating film are of the same material. Evidence suggested that the sensitivity of reflectance upon thickness was due to the peaking of the cross correlation of the reflectance of the two films when they were identical. In fact, it was determined that when any two films are optically identical, the average wavelength of the radiation spectrum has to be of the order of, or less than, the thickness of the films for the correlation effects to occur. Continuing this effort, the stability of the moving solidification interface of thin silicon films undergoing zone-melting recrystallization with a graphite strip heater was examined numerically (Yoon and Miaoulis 1992). Unstable growth of the interface occurred for scanning speeds greater than 250 μm/s for typical processing conditions. The growth rate of the perturbed interface varied with scanning speed, showing a distinct increase at a velocity of 400 μm/s. At this processing speed, a steeper temperature gradient in the liquid was induced by a decrease in the radiative cooling of the melt.

Xu and Grigoropoulos (1993) presented an experimental procedure for measuring radiative properties of thin polysilicon films in a temperature range from room temperature to approximately 400 K at the He–Ne laser wavelength (0.633 μm). The complex refractive index of polysilicon films was determined by combined ellipsometric and normal incidence reflectivity measurements in an inert gas environment. Transient reflectivity measurements were performed during pulsed excimer laser heating of thin polysilicon films of nanosecond duration. Numerical heat transfer and optical response analysis based on the measured radiative properties were conducted. The calculated results were compared with the experimental data to reveal the transient temperature field.

5.4.3 SUPERCONDUCTING MATERIALS AND FILMS

Rao et al. (1990) studied the optical reflectance of thin superconducting $Bi_2Sr_2Ca-Cu_2O_x$ films in the frequency range of 80–48,000 cm^{-1} and the temperature range of 5 K $< T <$ 300 K. The dielectric function of the superconductor was calculated taking into account the effect of the quasi-particle lifetime. Structure in the reflectivity spectra was identified with Bi 2:2:1:2 phonons, $SrTiO_3$ substrate phonons, and evidence for a superconducting gap was observed. The reflectance was measured using two spectrometers, a prism-based instrument, and a Fourier transform spectrometer equipped with a triglycine sulfate detector and an Si bolometer detector.

It was predicted that one of the first applications of thin-film high-temperature superconductors would be in the construction of liquid nitrogen temperature superconducting bolometers (Phelan et al. 1990). Phelan et al. (1990) developed a predictive model employing the Drude-free-electron theory applied to films of the order of 1 μm thick. The only measured parameter required by the model was the direct current electrical resistivity. On the basis of the comparisons with available data, the authors reached the conclusion that a simple Drude-free-electron model is capable of predicting the normal state for infrared radiative properties of 1 μm or thicker thin film $YBa_2Cu_3O_7$ for normal incidence. This model was successful for the temperature range of 100–300 K. The temperature-dependent electron number density was approximated from the electrical resistivity.

For the preparation of high quality films of high T_c superconductors on crystalline substrates, it is necessary to control the substrate temperature accurately during deposition. Flik et al. (1992) conducted a study that indicated that thermal radiation heat transfer in the deposition chamber governs the substrate temperature. The application of thin film optics yielded the emittance of the substrate holder substrate film composite as a function of the thickness of the growing film. In a single target off-axis sputtering system, the substrate temperature was measured during film deposition using a novel method for the attachment of a thermocouple to the substrate front surface. For constant heater power, the measurements indicated a decrease of the substrate temperature, in agreement with the theoretical prediction. On the basis of the substrate emittance variation determined in this work, a pyrometric in situ temperature measurement technique could be developed. The basic mechanisms for substrate temperature change during high T_c superconducting film deposition were presented. The film emittance was found to change during film growth due to both interference and absorption effects. In the single target sputtering system investigated, the decrease of the substrate temperature was also caused by heat flow redistribution between the substrate and the remaining substrate holder area.

Choi et al. (1992) also investigated the room temperature radiative properties of YBaCu$_3$O$_{7-d}$ superconducting films with varying oxygen content ($d = 0–1$) on a LaAlO$_3$ substrate. The fitting of reflectance spectra using thin film optics and the Drude–Lorentz model function with multiple oscillators was used to determine the optical constants. The film with high oxygen content ($d = 0$) exhibited a metallic optical behavior. A decrease in the oxygen content caused a removal of free carriers, and the film with low oxygen content ($d = 1$) acted like a dielectric with a very weak free electron contribution.

Zhang et al. (1991) investigated the refractive index of thin YBa$_2$Cu$_3$O$_7$ super-conducting films. The primary intent was to determine whether thin film optics with a constant refractive index could be applied to the thinnest high T_c superconducting films. The reflectance and transmittance of thin YBa$_2$Cu$_3$O$_7$ films on LaAlO$_3$ substrates were measured by a Fourier transform infrared spectrometer at wavelengths from 1 to 100 μm at room temperature. The reflectance of these samples at 10 K in the wavelength region from 2.5 to 25 μm was measured using a cryogenic reflectance accessory. The film thickness varied from 10 to 200 nm. By applying dispersion relations for the normal and superconducting states and electromagnetic theory, the film complex refractive index was obtained with a fitting technique. It was found that a thickness-independent refractive index could be applied even to 25 nm films. Average values of the spectral refractive index for film thicknesses between 25 and 200 nm were recommended for engineering applications. For film thicknesses from 25 to 200 nm, a systematic size effect on the refractive index was not observed. For the 10 nm film, since the mean free path was of the order of film thickness, boundary scattering could play a role in the electron scattering process. The deviation between the refractive index of the 10 nm film and those of the other films could also have been caused by the different microstructure of the 10 nm film (Zhang et al. 1991).

5.5 MODELING OF MICROSCALE RADIATION

In the classical radiation analysis, the energy flux and the intensity are given by Equations 5.27 and 5.28, respectively:

$$S = E \times H = \frac{1}{\mu} E \times B \qquad (5.27)$$

where
   E is the electric field
   H and B are the magnetic field and the magnetic field density, respectively
   μ is the permeability of the surrounding medium

$$I = |\langle S \rangle_T|, \tag{5.28}$$

which represents the magnitude of the energy flux vector (Poynting vector), averaged over the wave period.

The classical approach using the intensity becomes inadequate when the sizes decrease, because the interaction of waves becomes significant and the above averaging becomes more or less meaningless. Fundamental models incorporating wave interference effects, for various physical situations, have been presented by Kumar and Mitra (1999) in their treatise on microscale aspects of thermal radiation. Earlier investigations on the analysis of thin films using such models have also been reviewed in this paper.

Various cases of microscale radiation have been dealt with, using theoretical models by Kumar and Mitra (1999). Analysis incorporating electromagnetic wave interference, scattering and absorption in particulate systems and radiative transfer in microgrooved surfaces are considered. Mathematical modeling of thin metallic films, which leads to prediction of size-affected surface properties, has been discussed. Short-pulse radiation transport has been modeled to study the interactions in scattering absorbing media, metals, liquids, and organic materials. While the interested reader is encouraged to go through the extensive discussions presented by Kumar and Mitra (1999), a brief overview will be provided here.

5.5.1 PARTICULATE SYSTEMS: MODELING WITH ELECTRICAL FIELD

The effects of the interference of waves are neglected in conventional modeling of radiation propagation through particulate systems, assuming independent absorption and scattering of radiation by the particles. This implies that the overall effect can be obtained by summing up the effects of the individual particles (Bohren and Huffman 1983). In densely packed systems, this assumption leads to significant errors in predictions. Deviations also result from impingement of scattered radiation from particles on adjacent particles. These situations lead to what is termed dependent scattering.

The deviations from predictions in densely packed systems have been experimentally observed as well as considered in analysis by various researchers, as reviewed by Kumar and Mitra (1999). Tien and Drolen (1987) have produced a map, as shown in Figure 5.2, which shows regions corresponding to various engineering applications of particulate systems, delineating independent and dependent scattering of radiation in particulate systems.

For a randomly packed particulate system, the intensity in a system with N particles with dependent scattering is shown to be related to the values with an independent scattering assumption and can be related as follows (Kumar and Mitra 1999):

$$I_{sN}(\theta) = NI_{sI}F(\theta)|\gamma^2| \tag{5.29}$$

where the term NI_{sI} on the right-hand side is the intensity with the independent scattering assumption, which is N times the effect of a single particle. The term γ is given by

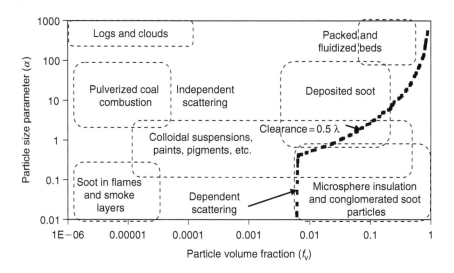

FIGURE 5.2 Map of engineering applications and the effect of radiation scattering on them, given by Tien and Drolen (1987). (Adapted from Kumar, S. and Mitra, K., *Adv. Heat Transfer*, 33, 187, 1999. With permission.)

$$\gamma = \left[1 - \frac{\varepsilon - 1}{\varepsilon + 1} f_v (1 + 1.6\alpha^2) \right] \qquad (5.30)$$

where
   ε is the square of the electron mass
   f_v is the particle volume fraction
   α is the size parameter given by $\frac{\pi D}{\lambda}$, where D is the particle diameter

The expressions for $F(\theta)$ for spherical particles and randomly positioned fibers are derived and given by Kumar and Mitra (1999). A typical result showing the effect of the solid volume fraction on the ratio of the absorption efficiencies comparing dependent and independent absorption by particles, as presented, is given in Figure 5.3.

Theoretical models have been used to analyze the scattering properties of microgrooves also, based on the electromagnetic theory. This approach does not use the classical intensity formulation, but accounts for the interference of the waves, considering guided, evanescent, and propagating waves inside and above the microgrooves. The periodicity of the grooves is also incorporated in the analysis. Perturbation solutions are obtained, and the analytical results are shown to approach those for smooth surfaces in the limiting cases of large width or small height of the grooves (Kumar and Mitra 1999).

5.5.2 THIN METALLIC FILMS: BOLTZMANN AND MAXWELL EQUATIONS

Theoretical modeling of the radiative interaction in thin metallic films has been presented by Kumar and Mitra (1999). Simultaneous solution of the two governing

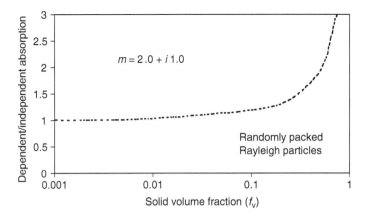

FIGURE 5.3 Effect of solid volume fraction on the ratio of absorption efficiencies, assuming dependent and independent absorption in random particulate media. (Adapted from Kumar, S. and Mitra, K., *Adv. Heat Transfer*, 33, 187, 1999. With permission.)

equations namely the Boltzmann equation (discrete carrier interaction) and the Maxwell equation (continuum field) has been utilized to obtain the reflectivity and transmissivity of the film. These surface properties are introduced as variables into the formulation of the thin film medium, at the boundaries. The Boltzmann equation is simplified with a relaxation time approximation. A detailed description of many further simplifications applied to the Boltzmann equation has been given in the paper. For the normal incidence of a monochromatic wave on a thin film, the forms of the simplified Boltzmann equation and Maxwell Equation solved are given as follows (Kumar and Mitra 1999):

$$\frac{\partial g_1(v,y,t)}{\partial t} - \frac{eE_x(y,t)}{m}\frac{\partial g_0(v,y,t)}{\partial v_x} + v_y\frac{\partial g_1(v,y,t)}{\partial y} = -\frac{g_1(v,y,t)}{\tau(v)} \quad (5.31)$$

$$\frac{\partial^2 \hat{E}_x(y,t)}{\partial y^2} + \omega^2\mu\varepsilon\hat{E}_x(y,t) = i\omega\mu j_x(y,t)E_x = \text{Re}[\hat{E}_x] \quad (5.32)$$

In the above equation, g_0 and g_1 represent the equilibrium distribution and the deviation from the equilibrium distribution of the number of free electrons per unit volume. As obvious, these are functions of the electron velocity v, position y for normal incidence, and time t. The real part of the complex electrical field E_x appears in Equation 5.32. The electron charge is given by e and ω is the angular frequency of the thermal radiation electric field. The electrical permittivity and the magnetic permeability are given by ε and μ, respectively.

The electric current density j_x in Equation 5.32 is given by

$$j_x(y,t) = -e\int\int\int v_x g_1(v,y,t)\mathrm{d}v_x\mathrm{d}v_y\mathrm{d}v_z \quad (5.33)$$

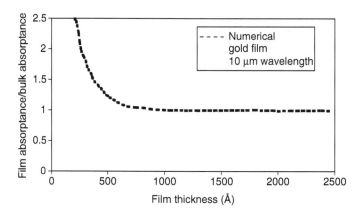

FIGURE 5.4 Effect of film thickness on normal absorptance in thin gold films. (Adapted from Kumar, S. and Mitra, K., *Adv. Heat Transfer*, 33, 187, 1999. With permission.)

Numerical computations have been performed to solve the above governing equations, and the influence of the film thickness on the film absorptance has been brought out as shown in Figure 5.4.

5.5.3 SHORT-PULSE LASER INTERACTIONS: DEVIATIONS FROM CLASSICAL MODELS

To analyze the physical problem of the transport of a pulsed radiation incident on a surface of a scattering and absorbing medium, the transient radiative heat transfer equation in the following form has been utilized by Kumar and Mitra (1999).

$$\frac{1}{c}\frac{\partial I(\vec{r},\hat{s},t)}{\partial t} + \hat{s}\cdot\vec{\nabla}I(\vec{r},\hat{s},t) = -\sigma_e I(\vec{r},\hat{s},t) + \frac{\sigma_s}{4\pi}\int I(\vec{r},\hat{s}',t)p(\hat{s}'\longrightarrow\hat{s})\,d\Omega'$$

$$+ S(\vec{r},\hat{s},t) \tag{5.34}$$

Here
  I is the intensity
  c is the velocity of light in the medium
  t is the time in seconds
  $\hat{s}$ represents the unit vector in the direction of intensity, around which the solid angle is given by Ω'
  r is the spatial location vector
  S is the incident radiation
  p is a scattering phase function

The effects of the extinction and scattering are captured through the inclusion of a radiative coefficient given as σ in the above equation where subscripts e and s represent the extinction and scattering effects, respectively. To solve the equation, the integral on the right-hand side can be reduced to a simpler form using various assumptions, in terms of a known radiative coefficient due to absorption (σ_a).

The solution with various approximations for the integral has been described in detail by Kumar and Mitra (1999). Transport of short-pulsed laser has been treated for two specific cases of different scales, propagation through biological tissues, and through ocean water containing schools of fish. Reflectivity and transmissivity are obtained as a function of nondimensional time, from the application of various approximations of the governing equation, and parametric studies on the influence of optical properties of the medium on the solution are also presented. A typical result on the reflectivity of a medium with various approximations represented by P1, P2, P3, two-flux, and discrete ordinates (Kumar and Mitra 1999) is shown in Figure 5.5. (Note here that the optical depth τ is defined by the relation, $\frac{I}{I_0} = e^{-\tau}$ where I and I_0 are the intensities at the source and at a given distance, and is a measure of the transparency of a scattering and absorbing medium).

Various instances of radiation analysis pertaining to short pulse lasers have been presented by Kumar and Mitra (1999) in their extensive publication. These are briefly reviewed below, highlighting the salient features of the analysis.

The interaction between a laser and a conducting metal involves energy exchanges in which photons, electrons, and phonons take part. The initial phenomenon is an exchange of energy between photons and electrons, producing extremely hot electrons for a short period, subsequently redistributing to reach the equilibrium Fermi–Dirac distribution (and producing a corresponding equilibrium temperature). Further, the electrons lose energy by emitting optical phonons; these optical phonons emit acoustic phonons, producing lattice heat conduction in the material. Taking into account all of these phenomena, a fundamental analysis can be done by solving the Boltzmann transport equations of the photons, electrons, and phonons. The conservation equations, with a simplifying assumption of thermodynamic equilibrium between optical and acoustic phonons, and substitution of energy densities of

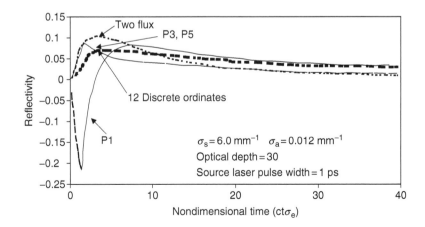

FIGURE 5.5 Variation of reflectivity of the medium of optical depth 30 as a function of nondimensional time, for the case of propagating source pulse, that is, pulsed laser with Gaussian temporal distribution. (Adapted from Kumar, S. and Mitra, K., *Adv. Heat Transfer*, 33, 187, 1999. With permission.)

electrons and phonons in terms of temperatures have been rewritten as follows (Kumar and Mitra 1999):

$$C_e(T_e)\frac{\partial T_e}{\partial t} = -\nabla . q - G(T_e - T_1) + S \qquad (5.35)$$

$$C_1(T_1)\frac{\partial T_1}{\partial t} = G(T_e - T_1) \qquad (5.36)$$

$$q + \tau\frac{\partial q}{\partial t} = -\kappa\nabla(T_e) \qquad (5.37)$$

where
G is a factor for coupling the electron and lattice terms
τ is the electron relaxation time
q is the heat flux vector
T_e and T_1 are the electron and lattice temperatures, respectively
C_e and C_1 are the corresponding specific heats

Suitable models have been used for the heat flux–temperature gradient relation. The simplest model is the Fourier law, but this is quite inadequate when the characteristic heating time is small. In such a situation, hyperbolic models can be used to relate between the flux and the temperature gradient. For instance, in a two-dimensional case, the flux models have been written as follows (for a cylindrical coordinate system):

$$q_r + \tau\frac{\partial q_r}{\partial t} = -\kappa_r\frac{\partial T_e}{\partial r} \qquad (5.38)$$

$$q_z + \tau\frac{\partial q_z}{\partial t} = -\kappa_z\frac{\partial T_e}{\partial z} \qquad (5.39)$$

where q_r and q_z represent the components of the heat flux vector in the radial and axial directions, respectively.

The source term in Equation 5.35 can be obtained from the intensity, which is in turn obtained by solving appropriate radiative transfer equation in terms of intensity. In the two-dimensional case mentioned above, the intensity equation is (Kumar and Mitra 1999)

$$\frac{1}{c}\frac{\partial I(r,z,t)}{\partial t} + \frac{\partial I(r,z,t)}{\partial z} = -\alpha I(r,z,t) \qquad (5.40)$$

where
I is the intensity
c is the light velocity
α is the absorption coefficient

The intensity is related to the source term in Equation 5.35 according to the following equation:

$$S(r,z,t) = -\frac{\partial I(r,z,t)}{\partial z} \qquad (5.41)$$

Another method of obtaining the source term is by solving the time-dependent Maxwell's electromagnetic wave equations to obtain the intensity in terms of the complex refractive index (Kumar and Mitra 1999). The final result for intensity is given by the following expression:

$$I = \frac{1}{2}\mathrm{Re}\left[\frac{m}{\mu c_0}\right]|E_x|^2 \qquad (5.42)$$

where m is the complex refractive index.

The approach has been demonstrated by obtaining a solution for simplified cases of one- and two-dimensional temperature distributions, with hyperbolic heat conduction model (one-dimensional case) and with a combination of exponentially decaying source in the axial direction and hyperbolic heat conduction in the radial direction (the two-dimensional case). Also a parabolic model, which is a fairly good assumption when the characteristic heating time is much more than the relaxation time, and a simple Fourier conduction model for large characteristic heating time have been utilized in obtaining solutions. A typical result with hyperbolic and parabolic models and comparisons with experimental results for normalized electron temperature changes with respect to time is given in Figure 5.6.

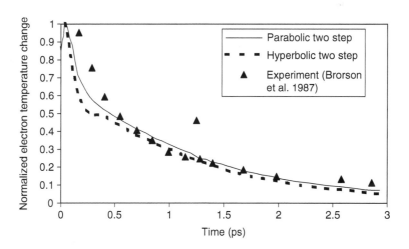

FIGURE 5.6 Normalized electron temperature changes from two models and experiments, for short pulse radiation interaction with metallic film. The data are for the front surface of the metallic film. (Adapted from Kumar, S. and Mitra, K., *Adv. Heat Transfer*, 33, 187, 1999. With permission.)

The case of radiation transport in liquids and organic materials has also been dealt with by Kumar and Mitra (1999). Here, the photon absorption can be modeled as transitions between energy levels of molecules. Depending on the phenomenon, the process can be taken as due to a single-photon process (a molecule goes from the ground to the first energy state by absorbing a photon), or two-photon process where it is taken to a subsequent second energy state by absorption of one more photon. The problem of radiation absorption and heating in liquids has been analyzed as a one-photon process, and ablation of polymers with high intensity laser has been treated as a two- or multiple-photon process (Kumar and Mitra 1999).

The analysis of the saturable absorption phenomenon has been discussed by Kumar and Mitra (1999). A brief review of the analysis is attempted here to introduce the phenomenon and the analysis method.

The classical Beer's law used for obtaining the intensity attenuation in a liquid is given as (Siegel and Howell 1992)

$$I_\lambda(s) = I_\lambda(0) \exp\left[-\int_0^s \kappa_\lambda(s^*)\,ds^*\right] \tag{5.43}$$

where
  $I_\lambda(0)$ is the incident intensity
  s represents the distance into the medium
  s^* is a dummy variable of integration

This assumes an absorption coefficient (κ_λ) independent of incident intensity, and gives an exponentially decreasing intensity in the liquid. This breaks down at high intensities of the incident radiation, as the absorption becomes intensity dependent in this case. This phenomenon is called saturable absorption (Kumar and Mitra 1999). The equations for single-photon absorption have been solved to obtain the intensity variation into the fluid, and to calculate the intensity-dependent temperature of the liquid. On the basis of these calculations, a characteristic intensity, called the saturation intensity, has been defined such that, for greater values of incident intensity, the calculated values using the photon absorption model deviates considerably from the classical Beer's law model. The variation of the intensity-dependent temperature with respect to the distance into the fluid, for the case of a commercial organic dye is given in Figure 5.7, as presented in the literature (Longtin and Tien 1996) and cited by Kumar and Mitra (1999).

As mentioned earlier, a multiple absorption model is required for the analysis of processes such as high intensity laser ablation of organic polymers. Computations based on a two-photon absorption model have been presented by Kumar and Mitra (1999). The etching depths predicted in a polymide, using the single-photon model, two-photon model, and classical Beer's law have been compared with experimental data by Pettit and Sauerbrey (1993) and reproduced as shown in Figure 5.8.

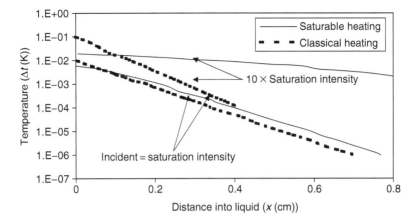

FIGURE 5.7 Variation of intensity-dependent laser absorption temperature in an organic liquid. (Adapted from Kumar, S. and Mitra, K., *Adv. Heat Transfer*, 33, 187, 1999. With permission.)

5.6 RADIATION PROPERTIES IN THE MICROSCALE REGIME

Various attempts have been reported in computing and measuring radiation properties in the microscale regime. The cases studied include thin films of semiconductor and other materials, and the problems are associated with fabrication as well as applications of these. Quite a large amount of work has been reported by Professor Z.M. Zhang and his group at the Georgia Institute of Technology. This section provides some discussions on the major radiation properties, and reviews some of the investigations reported in the literature for obtaining size-affected radiation properties.

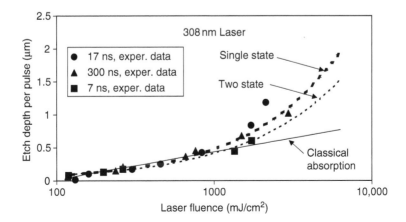

FIGURE 5.8 Predicted etching depth in a polymide using various models, compared with experimental results. (Adapted from Kumar, S. and Mitra, K., *Adv. Heat Transfer*, 33, 187, 1999. With permission.)

Transmittance is the fraction of the incident radiation that passes through a medium or specimen. This fraction of incident radiation transmitted through a film is affected by the reflection at the two interfaces and the absorption by the medium. So, the externally measured transmittance and the actual internal transmittance are different, which are related through the reflectance of the medium (the fraction of the incident radiation that is reflected) by Zhang et al. (1997) as follows:

$$T = (1 - \rho)^2 \tau \qquad (5.44)$$

where
    T is the external transmittance
    ρ is the reflectance
    τ is the internal transmittance

The internal transmittance of a film for the incidence of radiation on its surface at an arbitrary direction (and a corresponding angle of refraction θ in the medium) is given by Zhang (1999).

$$\tau = \exp\left(-\frac{4\pi\kappa}{\lambda} d/\cos\theta\right) \qquad (5.45)$$

In the above equation
    d is the thickness of the film
    κ is the extinction coefficient of the film
    λ is the wavelength of the radiation

The schematic of a general oblique incidence is shown in Figure 5.9. Note that d and θ in Equation 5.53 correspond to d_2 and θ_2 in Figure 5.9. The angles θ_1, θ_2, and θ_3 in Figure 5.9 are related through Snell's law as follows:

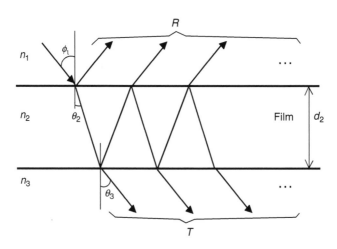

FIGURE 5.9 Schematic of oblique incidence and multiple reflections in a film. (Adapted from Zhang, Z.M., *Appl. Optics*, 38, 205, 1999. With permission.)

$$n_1 \sin \theta_1 = n_2 \sin \theta_2 = n_3 \sin \theta_3 \tag{5.46}$$

Here n_1, n_2, and n_3 are refractive indices of the media as shown in Figure 5.9.

Some mention should be made again of the mathematical nature of the complex refractive index of a medium at this stage. The complex refractive index, resulting from electromagnetic theory, has a real part and an imaginary part, and can be written as

$$\bar{n} = n + i\kappa \tag{5.47}$$

n and κ, the real and imaginary parts (called the refractive index and the extinction coefficient, respectively), are wavelength dependent. The absorption coefficient is given by $a = \frac{4\pi\kappa}{\lambda}$.

For normal incidence, $\cos \theta$ in Equation 5.45 is equal to 1. Thus, for normal transmission, the internal transmittance τ is e^{-ad} where d is the thickness of the film.

For normal incidence, the reflectance can be shown to be

$$\rho = \frac{\left[(n-1)^2 + \kappa^2\right]}{\left[(n+1)^2 + \kappa^2\right]} \tag{5.48a}$$

If the extinction coefficient κ is much smaller than $(n-1)$, this becomes

$$\rho = \frac{(n-1)^2}{(n+1)^2} \tag{5.48b}$$

The optical density can be defined in terms of the external transmittance T (Zhang et al. 1997).

$$OD = -\log_{10} T \tag{5.49}$$

The absorption coefficient, the thickness of the film, and the optical density can be related utilizing Equation 5.44. For normal incidence, this becomes

$$a = \frac{OD + 2\log_{10}(1-\rho)}{d \log_{10} e} \tag{5.50}$$

Zhang (1999) have presented simple equations for calculating the transmittance and reflectance of a slightly absorbing film when the incident radiation is oblique. This is obtained by considering multiple reflections in the medium, tracing the reflected and transmitted waves, as shown in Figure 5.9. The final results presented by Zhang (1999) are as follows:

$$T = \frac{n_3 \cos \theta_3}{n_1 \cos \theta_1} \times \frac{(1+r_1)^2(1+r_2)^2 \tau}{1 + r_1^2 r_2^2 \tau^2 + 2r_1 r_2 \tau \cos \phi} \tag{5.51}$$

$$R = \frac{r_1^2 + r_2^2 \tau^2 + 2r_1 r_2 \tau \cos \phi}{1 + r_1^2 r_2^2 \tau^2 + 2r_1 r_2 \tau \cos \phi} \tag{5.52}$$

where

$$\phi = \frac{4\pi n_2}{\lambda} d_2 \cos \theta_2 \qquad (5.53)$$

Here, r_1 and r_2 are the Fresnel reflection coefficients at the interfaces between medium 1 and medium 2 (the film), and medium 2 and medium 3, respectively. The Fresnel coefficients depend on the angle of incidence and polarization, and the expressions for these are available in text books (Siegel and Howell 1992).

Experimental investigations have been reported in the literature, for measurement of the transmittance of optical filters (Zhang et al. 1997). From measured transmittance values, the optical density is calculated, which is in turn used for evaluating the absorption coefficients of filters with various optical densities. Measurements were reported on optical filters with optical densities in the range of 1–10 in this work. The general experimental method adopted by Zhang et al. (1997) is described briefly here.

The schematic of the setup (Zhang et al. 1997) to measure the transmittance is shown in Figure 5.10. The principal components of the experimental setup shown in Figure 5.10 are an Nd:YAG laser with a power stabilizer and linear photodiode detectors. The laser source used had an output power of 3 W and a wavelength

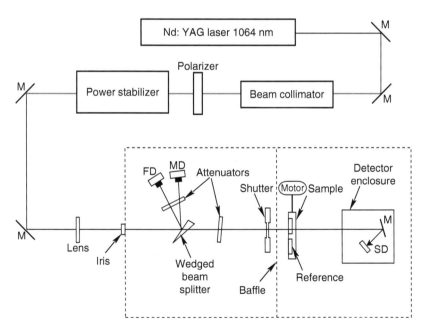

FIGURE 5.10 Optical setup for transmittance measurements at 1064 nm wavelength. M, mirror; FD, feedback detector; MD, monitor detector; SD, signal detector. (Adapted from Zhang, Z.M., Gentile, T.R., Migdall, A.L., and Datla, R.U., *Appl. Optics*, 36, 8889, 1997. With permission.)

(infrared) of 1064 nm. A quartz wedge beam splitter is used, which reflects a small portion (approximately 4%) of the laser power to a feedback detector that controls the power stabilizer, and another portion to a monitor to normalize the laser power that reduces the effect of residual power fluctuation, these two portions being reflected from the two surfaces of the wedge. The transmitted beam passes through the sample, or the reference filter (based on the vertical position of the motor-controlled filter holder), on to a signal detector, which is coupled with a built-in amplifier and a digital voltmeter, to obtain a voltage signal corresponding to the detected radiation. The relative transmittance (relative to the reference filter) is obtained from six voltage signals as given below.

$$T_{relative} = \frac{[(V_{s1} - V_{s0})/V_m]_{sample}}{[(V_{s1} - V_{s0})/V_m]_{reference}} \tag{5.54}$$

The transmittance of the sample filter is calculated as

$$T_{sample} = T_{relative} \times T_{reference} \tag{5.55}$$

where T_{ref} is the transmittance of the reference filter.

In Equation 5.54, V is the output voltage of the detector; subscripts s and m indicate the signal detector and the monitor detector, respectively, and 0 and 1 correspond to the closed or open conditions of the shutter. The signal obtained while the shutter is closed is subtracted from the signal when the shutter is open, to (see Equation 5.54) ensure that the background signals are eliminated in the calculation. Detailed description of the experimental setup and data analysis are given by Zhang et al. (1997). Typical variations of the measured transmittance values and the optical density calculated from them are shown in Figure 5.11a and 5.11b.

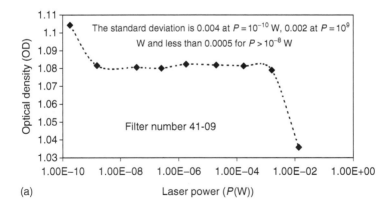

(a)

FIGURE 5.11 (a) Optical density OD (nominal OD = 1) of filter measured at different laser powers. (Adapted from Zhang, Z.M., Gentile, T.R., Migdall, A.L., and Datla, R.U., *Appl. Optics*, 36, 8889, 1997. With permission.)

(*continued*)

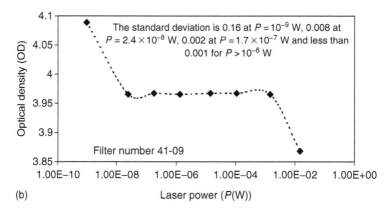

(b)

FIGURE 5.11 (continued) (b) Optical density OD (nominal OD = 4) of filter measured at different laser powers. (Adapted from Zhang, Z.M., Gentile, T.R., Migdall, A.L., and Datla, R.U., *Appl. Optics*, 36, 8889, 1997. With permission.)

Transmittance measurements have been used to obtain the refractive index and extinction coefficients of films. Measurements have been reported on polyimide films (Zhang et al. 1998a) (which find extensive applications in optoelectronics), made by spin coating, followed by a thermal curing process. The films were in the range of 0.1–4 μm thickness. In these measurements, a Fourier transform infrared (FT-IR) spectrometer has been used to obtain the transmittance spectrum (transmittance variation with respect to the wave number), a typical plot of which is shown in Figure 5.12.

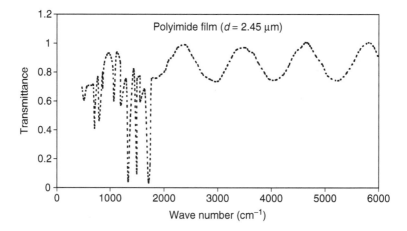

FIGURE 5.12 Transmittance spectrum of a 2.45 μm thick polyimide film. (Adapted from Zhang, Z.M., Button, G.L., and Powell, F.R., *Int. J. Thermophys.*, 19, 905, 1998. With permission.)

In the high wave number (low-absorbing) region, for normal incidence, the transmittance can be shown to be related to the wave number $\left(\nu = \frac{1}{\lambda}\right)$ as follows (Zhang et al. 1998a).

$$T = \frac{(1-\rho)^2}{1 + \rho^2 - 2\rho \cos{(4\pi n \nu d)}} \tag{5.56}$$

According to this expression, in the high wave number region, the maxima and minima correspond to wave numbers given by

$$\nu_{\text{max},m} = \frac{m}{2nd} \tag{5.57}$$

and

$$\nu_{\text{min},m} = \frac{\left(m + \frac{1}{2}\right)}{2nd} \tag{5.58}$$

where $m = 0, 1, 2, \ldots$. These relations can be used to determine the refractive index, for known reflectivity (material) and thickness of the sample. In the work reported by Zhang et al. (1998a), the transmittance minima are used to obtain the refractive index of the polyimide films. As mentioned earlier, for normal incidence, and for negligible absorption (where $\kappa = 0$), the reflectance is given by Equation 5.48b. This, along with Equation 5.55, gives the minimum transmittance:

$$T_{\text{min}} = \left[\frac{2n}{(1+n^2)}\right]^2 \tag{5.59}$$

Thus, from the measured minimum transmittance, the refractive index can be calculated in the low absorption region. It is also to be noted that the maximum transmittance in this region is close to 1. The difference in the wave numbers between two adjacent maxima is $\Delta\nu = \frac{1}{2nd}$. This gives a method to obtain the thickness of the film also from the measurement. Another method mentioned by Zhang et al. (1998a) to obtain the refractive index and the thickness is by a curve fitting of the transmittance to Equation 5.55.

In the high absorptance (low wave number) region of Figure 5.12, the extinction coefficient is also significant. In this case, the general transmittance equation, also considering the extinction coefficient, is given by

$$T = \left[\frac{(1 - r)(1 + r)e^{i\delta}}{1 - r^2 e^{i2\delta}}\right]^2 \tag{5.60}$$

where δ represents the optical phase change inside the film. For normal incidence, the Fresnel coefficient r is given by

$$r = \frac{(1 - \bar{n})}{(1 + \bar{n})} \tag{5.61}$$

and

$$\delta = 2\pi \bar{n}\nu d. \tag{5.62}$$

The wavelength-dependent refractive index and extinction coefficient in the above case have been determined using experimental data of the spectral transmittance of two films of the same material, having different thicknesses. Here, a multiple parameter, least square fitting program, was employed to fit the transmittance equation given above, where the method employs minimizing the difference between the measured and calculated values of transmittance. Comparisons with theoretical results obtained using a Lorentzian oscillator model for the transmittance is also presented in the literature (Zhang et al. 1998a). The theoretical approach models the dielectric function, which is the square of the complex refractive index, by a superposition of many Lorentzian oscillators, and obtains the results through a best fit parameter, which provides a good match between the experimental and predicted values of the transmittance. Extensive discussions on the calculation methodology can be found have been presented by Zhang et al. (1998a), and a detailed discussion of this is not attempted here.

Temperature- and thickness-dependent radiation properties of thin YBCO superconducting films ($YBa_2Cu_3O_{7-\delta}$) have been measured at a temperature range of 10–300 K (Zhang et al. 1998b, Kumar et al. 1998). The film side and back side (substrate side) reflectances of the films deposited on silicon substrates by laser ablation were measured using a Fourier transform spectrometer. The radiation used was far-infrared. Transmittance spectra were obtained using a Michelson interferometer, and the absorption spectra were deduced from these two measurements. The interference fringes associated with the measurements were also used to analyze the interaction of radiation with the film–substrate interface. A significant difference was obtained between the absorptance for radiation incident on the film side and the substrate side.

A comparison of the measured data on absorptance with those calculated using a simple Drude model is also reported by Kumar et al. (1998). The Drude model assumes that the dielectric function of the film in the far-infrared region is the sum of two components as given below:

$$\varepsilon(\nu) = \varepsilon_{\text{Drude}} + \varepsilon_{\text{h}} \tag{5.63}$$

$$\varepsilon_{\text{Drude}} = -\frac{\nu_{\text{p}}^2}{\nu^2 + i\nu\gamma} \tag{5.64}$$

Here, the Drude term accounts for the temperature-dependent free carrier absorption, and ε_{h} is a constant term which accounts for the mid-infrared and high frequency contributions to the dielectric function. The dielectric function is related to the complex refractive index as follows:

$$\overline{n_{\text{f}}} = n_{\text{f}} + i\kappa_{\text{f}} = \sqrt{\varepsilon} \tag{5.65}$$

In the above equations

ν is the frequency

ν_p is termed the plasma frequency (which is independent of temperature)

γ is the scattering rate that is dependent on temperature

The values of ν_p and γ have been obtained by fitting calculated transmittance values to measured spectra, and ε_h taken as 100 based on assumptions for the far-infrared region considered, that phonons and mid-infrared bands do not have significant effects below a value of 100 (Kumar et al. 1998). The theoretical calculations have shown a good agreement with the measured absorptance. The effect of the film thickness was also studied theoretically in the range 5–300 nm, while the experimental study was conducted for a thickness of 35 nm (Kumar et al. 1998).

Application of a three-axis automated scatterometer (TAAS) for measurement of the bidirectional distributions (with respect to the zenith and azimuthal angles) of reflectance and transmittance of semiconductor wafers has been reported. The method has been utilized for measurements on silicon wafers with rough surfaces (Shen et al. 2003), silicon wafer surfaces with periodic microstructures (Chen et al. 2004), and silicon surfaces coated with silicon dioxide (Lee and Zhang 2006). The major components of the scatterometer are a diode laser optical source, a goniometric table with three rotary stages and the detection and data acquisition system. The schematic of an arrangement for measurement of the bidirectional reflectance is shown in Figure 5.13 as reproduced from Lee and Zhang (2006). A beam splitter is used along with two detectors as shown in the diagram, one of which (A) monitors the reflected radiation, and the other (B) the incident radiation simultaneously, so that fluctuations in the laser output can be eliminated from the measurement. Measurements are done by positioning the sample at required orientations with the goniometric table. Extensive discussions of the setup, the measurement method, and the results can be found in the literature (Shen et al. 2003, Chen et al. 2004, Lee and Zhang 2006). The result of a measurement on the polished side of a silicon wafer (a smooth surface), conducted as a test for validation of the experimental setup, is shown in Figure 5.14, which demonstrates the accuracy of the experimental method by comparing it with the theoretical results.

Reflectance measurements on thin films of Ta_2O_5 deposited on Si and SiO_2 substrates were conducted using a Fourier transform infrared spectrometer equipped with a multiple angle reflectometer, before and after exposure to high-temperature heat treatment (Chandrasekharan et al. 2007). The films were 0.85 and 3 μm thick. The reflectance data were used to compute the optical constants of the deposited film, crystalline Ta_2O_5, and the interfacial layer (formed by the diffusion of silicon from the substrate to the film, and having a thickness in the range 1.6–10 μm), using a least square optimization technique. The interfacial layer was found to be more absorptive than the crystalline Ta_2O_5.

The radiative properties of semitransparent silicon wafers with rough surfaces have been predicted using the Monte Carlo simulation model. The numerically obtained bidirectional reflectance distributions showed a similar trend as the experimental data. Discussions on the simulation results, as well as the merits and limitations of the model, have been presented by Zho and Zhang (2003).

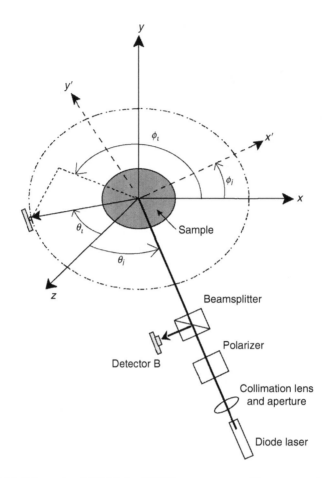

FIGURE 5.13 Schematic of TAAS for BRDF measurements. Here, a sample is vertically mounted and the laser beam is parallel to the optical table, which is in the x–z plane. Rotating the sample around the y axis changes θ_i, and rotating detector A in the x–z plane changes θ_i. φ_i can be changed by rotating the sample around the z axis. To measure the in-plane BRDF, the detector is kept within the horizontal plane. (Adapted from Lee, H.J. and Zhang, Z.M., *Int. J. Thermophys.*, 27, 820, 2006. With permission.)

A comprehensive review and discussions on several critical issues associated with microscale radiation in thermo-photovoltaic devices are presented by Basu et al. (2007). The application of microscale radiation principles as a means of increasing the conversion efficiencies of thermo-photovoltaic systems is also discussed in this paper.

5.7 RECENT DEVELOPMENTS IN THEORETICAL MODELING

A number of theoretical studies of microscale radiative heat transfer have been reported recently. Some of these focused on specific applications of microscale radiation phenomena related to manufacturing processes, while others were aimed

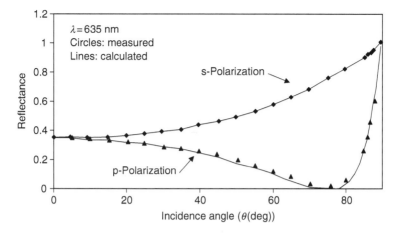

FIGURE 5.14 Reflectance of the polished side of a silicon wafer at 15,635 nm, compared with that calculated using Fresnel's equations. (Adapted from Shen, Y.J., Zhu, Q.Z. and Zhang, Z.M., *Rev. Sci. Instrum.*, 34, 4885, 2003. With permission.)

at a better understanding of the processes for modifications of end use applications. A brief review of such publications is presented in this section.

Knowledge of the topography-dependent radiation properties is of extreme importance in controlling the heating processes and monitoring surface temperatures (using optical pyrometry) during manufacturing processes. Fue and Hsu (2007) presented a theoretical model for the calculation of the radiative properties of surfaces with random roughnesses at the microscale. In this work, the Maxwell equations for the electromagnetic wave reflection from the surface were solved using a finite difference scheme, assuming a Gaussian probability density distribution function for the surface height. The computation was done for perfect electric conductors and for silicon. The bidirectional reflectivity predicted by the model was found to be in good agreement with ray-tracing calculations and integral solutions.

Theoretical analysis of periodic structures such as multilayer structures and patterned wafer surfaces has gained attention. Lee and Zhang (2007) investigated coherent emission from a multilayer structure with an SiC film coated on top of a dielectric photonic crystal, the coherent emission being produced by exciting surface waves at the interface. The calculations were based on the fundamental and governing equations for electromagnetic waves. The calculated emissivity showed sharp peaks in a narrow wavelength band and at a well-defined direction, which is the characteristic property of coherent sources (Lee and Zhang 2007). Chen and Zhang (2007) conducted a parametric study of the radiative properties of patterned wafers with the smallest dimension as low as 30 nm. The intention of the investigation was to investigate the effects of temperature, wavelength, polarization and angle of incidence of radiation. The motivation of the investigation was the problem associated with the nonuniform absorption of thermal radiation by patterned wafers, and resulting temperature nonuniformity, which introduce thermal stresses. Numerical solutions were obtained for Maxwell's equations using the rigorous coupled wave analysis (RCWA) computer program (Lee and Zhang 2007). Effective medium formulations were also developed and compared with the results of this analysis.

Theoretical calculations to compute the total hemispherical emissivity of doped silicon lamellar gratings, where the thermal emission is anisotropic and polarized, have been presented by Marquier et al. (2007). The calculation strategy involves modification of the radiative properties of the sources while performing analysis in a plane, which is not perpendicular to the grooves of the gratings, to accommodate for the directional effects. Guo and Quan (2007) presented an analysis of micro/ nanoscale radiation energy transfer in optical microcavity and waveguide coupling structures which utilize optical resonance for energy storage. The problem was formulated with the Helmholtz equations (simplified Maxwell's equations for time harmonic waves) for energy transfer of the electromagnetic waves, which were solved using the finite element method. It was found that the submicron/nanoscale gap is crucial for efficient energy transfer and storage.

Photon tunneling phenomenon is due to the interaction of evanescent waves in gaps smaller than the wavelength of the emitting source. Evanescent waves are exponentially decaying electromagnetic waves in space. Tunneling has found widespread use in applications such as optical microscopy and microscale thermo-photovoltaic devices. Fu et al. (2007) theoretically analyzed the energy transmission by photon tunneling in multilayer structures, including negative refractive index materials. (Negative refractive index materials were first proposed as hypothetical materials (Veselago 1968). These have peculiar properties such as the focusing of light by a flat slab. The analysis presented in the paper predicted transmittance spectra in the multilayer structures, and examined the influence of the number of layers on the transmittance.

Ablation of dielectrics using femtosecond laser pulses was analyzed using an electron density model by Heltzel et al. (2007). The material removal was found to be strongly dependent on the optical response of the dielectric during the formation of the plasma. The model also predicted the response of borosilicate glass to an enhanced electromagnetic field, due to the presence of microspheres on the substrate surface. Experimental studies using AFM and SEM were also conducted to characterize the surface and verify the predicted effects due to the surface modifications.

Other important theoretical analyses presented recently on microscale radiative transport include an electromagnetic field solution using a Green function approach in microscale spherical regions bounded by silicon carbide (Hammonds 2007), analysis based on a two-photon absorption model for high energy laser heating of volume-absorbing glass to obtain its temperature history (Zhang and Chen 2007), and theoretical prediction of the scattering due to gold nanoparticles and two-dimensional agglomerates of metallic particles on surfaces (Venkata et al. 2007).

5.8 CONCLUDING REMARKS

Understanding of the phenomenon of thermal radiation, with a microscopic viewpoint, is essential in all advances leading to applications of radiative heat transfer for microscale systems. The classical radiation theory would be insufficient to describe the radiation interaction of microscale structures. Analysis based on the particle (photon) theory and on the electromagnetic field theory, and more generally, utilizing the particle-wave duality becomes necessary in most of the microscale radiative transport problems.

The objective of this chapter was to develop the basics required in the microscale approach to radiative transport and to cite occasions where analysis different from the classical methods has been utilized and reported in the literature. A general review of the state of the art in microscale radiative heat transfer was also attempted in this chapter. A few problems described in detail in the literature (Kumar and Mitra 1999) have been presented in the summarized form to provide a preliminary understanding of the mathematical modeling. Some experimental and theoretical methods utilized in investigations for obtaining radiative properties, which find applications in microscale radiation systems, were also reviewed in this chapter.

REFERENCES

Basu, S., Y.B. Chen, and Z.M. Zhang. 2007. Microscale radiation in thermophotovoltaic devices–A review. *International Journal of Energy Research* 31: 689–716.

Bohren, C.F. and Huffman, D.R. 1983. *Absorption and Scattering of Light by Small Particles*. New York: Wiley.

Brewster, M.Q. 1992. *Thermal Radiative Heat Transfer and Properties*. New York: Wiley.

Brorson, S.D., J.G. Fujimoto, and E.P. Ippen. 1987. Femtosecond electronic heat transfer dynamics in thin gold film. *Physical Review Letters* 59: 1962–1965.

Chandrasekharan, R., S. Prakash, and M.A. Shannon. 2007. Change in radiative optical properties of Ta_2O_5 thin films due to high-temperature heat treatment. *Journal of Heat Transfer* 129: 27–36.

Chen, Y.B. and Zhang, Z.M. 2007. Radiative properties of patterned wafers with nanoscale line width. *Journal of Heat Transfer* 129: 79–90.

Chen, Y.B., Q.Z. Zhu, T.L. Wright, W.P. King, and Z.M. Zhang. 2004. Bidirectional reflection measurements of periodically microstructured silicon surfaces. *International Journal of Thermophysics* 25: 1235–1252.

Choi, B.I., Z.M. Zhang, M.I. Flik, and T. Siegrist. 1992. Radiative properties of Y-Ba-Cu-O films with variable oxygen content. *Journal of Heat Transfer* 114: 958–964.

Cohn, D.W. and R.O. Buckius. 1993. Scattering from microcontoured surfaces. Proceedings of Japan-U.S. Seminar on Molecular and Microscale Transport Phenomena, Kanazawa, Japan. Paper A-9.

Duncan, A.B. and G.P. Peterson. 1994. Review of microscale heat transfer. *Applied Mechanics Review* 47: 397–428.

Flik, M.I., B.I. Choi, A.C. Anderson, and A.C. Westerheim. 1992. Thermal analysis and control for sputtering deposition of high-T_c super-conducting films. *Journal of Heat Transfer* 114: 255–263.

Fu, C.J., Z.M. Zhang, and D.B. Tanner. 2007. Energy transmission by photon tunneling in multilayer structures including negative index materials. *Journal of Heat Transfer* 129: 1046–1052.

Fue, K. and P. Hsu. 2007. Modeling the radiative properties of microscale random roughness surfaces. *Journal of Heat Transfer* 129: 71–78.

Grigoropoulos, C.P., R.H. Buckholz, and G.A. Domoto. 1986. The role of reflectivity change in optically induced recrystallization of thin silicon films. *Journal of Applied Physics* 59: 454–458.

Guo, Z. and H. Quan. 2007. Energy transfer to optical microcavities with waveguides. *Journal of Heat Transfer* 129: 44–51.

Hammonds, Jr, J.S. 2007. Thermal radiative transport enhancement via electromagnetic surface modes in microscale spherical regions bounded by silicon carbide. *Journal of Heat Transfer* 129: 94–97.

Heltzel, A., A. Battula, J.R. Howell, and S. Chen. 2007. Nanostructuring borosilicate glass with near-field enhanced energy using a femtosecond laser pulse. *Journal of Heat Transfer* 129: 53–59.

Hesketh, P.J., B. Gebhart, and J.N. Zemel. 1988. Measurements of spectral and directional emission from microgrooved silicon surfaces. *Journal of Heat Transfer* 110: 680–686.

Hesketh, P.J. and J.N. Zemel. 1988. Polarized spectral emittance from periodic micromachined surfaces. *Physics Review B* 37: 10795–10813.

Incropera, F.P. and D.P. DeWitt. 2006. *Fundamentals of Heat and Mass Transfer* 5th ed. New York: Wiley.

Kittel, C. 1996. *Introduction to Solid State Physics*. 6th ed. New York: Wiley

Kumar, S. and K. Mitra. 1999. Microscale aspects of thermal radiation transport and laser applications. *Advances in Heat Transfer* 33: 187–294.

Kumar, S. and C.L. Tien. 1989. Effective diameter of agglomerates for radiative extinction and scattering. *Combustion Science and Technology* 66: 199–216.

Kumar, A.R., Z.M. Zhang, V.A. Boychev, and D.B. Tanner. 1998. Temperature dependent far-infrared absorptance of thin $YBa_2Cu_3O_{7-\delta}$ films in the normal state. *Journal of Microscale Thermophysical Engineering* 3: 5–15.

Lee, H.J. and Z.M. Zhang. 2006. Measurement and modeling of the bidirectional reflectance of SiO_2 coated Si surfaces. *International Journal of Thermophysics* 27: 820–838.

Lee, B.J. and Z.M. Zhang. 2007. Coherent thermal emission from modified periodic multilayer structures. *Journal of Heat Transfer* 129: 17–26.

Longtin, J.P. and C.L. Tien. 1996. Saturable absorption during high-intensity laser heating of liquids. *Journal of Heat Transfer* 118: 924–930.

Longtin, J.P. and C.L. Tien. 1998. Microscale radiation phenomena. In *Microscale Energy Transport*, C.L. Tien, A. Majumdar, and F.M. Gerner (Eds.), New York: Taylor & Francis, pp. 119–147.

Marquier, F., M. Laroche, R. Carminati, and J.J. Greffet. 2007. Anisotropic polarized emission of doped silicon lamellar grating. *Journal of Heat Transfer* 129: 11–15.

Pettit, G.H. and R. Sauerbrey. 1993. Pulsed ultraviolet laser ablation. *Applied Physics A* 56: 51–63.

Phelan, P.E., M.I. Flik, and C.L. Tien. 1990. Normal state radiative properties of thin-film high temperature superconductors. *Advances in Cryogenic Engineering* 36: 479–486.

Phelan, P.E., M.I. Flik, and C.L. Tien. 1991. Radiative properties of superconducting Y-Ba-Cu-O thin films. *Journal of Heat Transfer* 113: 487–493.

Raad, N. and S. Kumar. 1992. Radiation scattering by periodic microgrooved surfaces of conducting materials. *Proceedings of ASME Winter Annual Meeting: Micromechanical Systems DSC* 40: 243–258.

Rao, A.M., P.C. Eklund, G.W. Lehman et al. 1990. Far-infrared optical properties of superconducting Bi_2Sr_2Ca Cu_2O films. *Physical Review B* 42: 193–201.

Shen, Y.J., Q.Z. Zhu, and Z.M. Zhang. 2003. A scatterometer for measuring the bidirectional reflectance and transmittance of semiconductor wafers with rough surfaces. *Review of Scientific Instruments* 34: 4885–4893.

Siegel, R. and J.R. Howel 1992. *Thermal Radiation Heat Transfer*. 3rd ed. New York: Hemisphere Publ. Co.

Tien, C.L. and G. Chen. 1994. Challenges in microscale conductive and radiative heat transfer. *Journal of Heat Transfer* 116: 799–807.

Tien, C.L. and B.L. Drolen. 1987. Thermal radiation in particulate media with dependent and independent scattering. *Annual Review of Numerical Fluid Mechanics and Heat Transfer* 1: 349–417.

Timans, P.J. 1992. The experimental determination of the temperature dependence of the total emissivity of gas using a new temperature measurement technique. *Journal of Applied Physics* 72: 660–670.

Timusk, T. and D.B. Tanner. 1989. Infrared studies of a-b plane oriented oxide superconductors. *Physical Review B* 38: 6683–6688.

Venkata, P.G., M.M. Aslan, and M.P. Menguc. 2007. Surface plasmon scattering by gold nanoparticles and two-dimensional agglomerates. *Journal of Heat Transfer* 129: 60–70.

Veselago, V.G. 1968. The electrodynamics of substances with simultaneously negative values of ε and μ. *Soviet Physics Uspeki* 10: 509–514.

Wong, P.Y. and I.N. Miaoulis. 1990. Microscale heat transfer phenomena in multilayer thin film processing with a radiant heat source. *American Society of Mechanical Engineers Dynamic Systems and Control* 19: 159–170.

Wong, P.Y., L.M. Trefethen, and I.N. Miaulis. 1991. Cross-correlation thermal radiation phenomena in multilayer thin-film processing. *American Society of Mechanical Engineers Dynamic Systems and Control* 32: 349–359.

Xu, X. and C.P. Grigoropoulos. 1993. Measurements of high temperature radiative properties of this polysilicon films at a wavelength of 0.633 micrometers and its application to the development of a fast temperature probe. *ASME HTD* 253: 1–12.

Yoon, S.M. and I.N. Miaoulis. 1992. Effect of scanning speed on the stability of the solidification interface during zone-melting recrystallization of thin silicon films. *Journal of Applied Physics* 72: 316–318.

Zhang, Z.M. 1999. Optical properties of a slightly absorbing film for oblique incidence. *Applied Optics* 38: 205–207.

Zhang, Z.M., G.L. Button, and F.R. Powell. 1998a. Infrared refractive index and extinction coefficient of polyimide films. *International Journal of Thermophysics* 19: 905–915.

Zhang, Z.M., A.R. Kumar, V.A. Boychev, D.B. Tanner, L.R. Vale, and D.A. Rudman. 1998b. Back-side reflectance of high T_c superconducting thin films in the far infrared. *Applied Physics Letters* 73: 1907–1910.

Zhang, Z.M. and D.H. Chen. 2007. Thermal analysis of multiphoton absorption at 193 nm in volume-absorbing glass. *Journal of Heat Transfer* 129: 391–395.

Zhang, Z.M., B.I. Choi, T.A. Le, M.I. Flik, M.P. Siegal, and J.M. Phillips. 1991. Infrared refractive index of thin $YBa_2Cu_3O_7$ superconducting films. *1991 ASME Winter Annual Meeting*, Atlanta, GA, Dec. 1–6.

Zhang, Z.M., C.J. Fu, and Q.Z. Zhu. 2003. Optical and thermal radiative properties of semiconductors related to micro/nanotechnology. *Advances in Heat Transfer* 37: 179–296.

Zhang, Z.M., T.R. Gentile, A.L. Migdall, and R.U. Datla. 1997. Transmittance measurements for filters of optical density between one and ten. *Applied Optics* 36: 8889–8895.

Zho, Y.H. and Z.M. Zhang. 2003. Radiative properties of semitransparent silicon wafers with rough surfaces. *Journal of Heat Transfer* 125: 462–470.

6 Nanoscale Thermal Phenomena

6.1 INTRODUCTION

Principles of heat transfer in size-affected domains have been discussed extensively in the previous chapters, referring to the microscale regime. The basic differences in the analysis of heat transfer in the microscale originate from the difference of the approach required, as the governing laws based on macroscopic continuum considerations break down in the size-affected domains. In the case of the microscopic approach, the first task is to identify whether the analysis should include size effects at all, that is, the primary concern is whether a macroscopic continuum approach is sufficient, or whether the analysis should deal with discrete, particle-based phenomena to describe the transport of thermal energy in the system.

Nanoscale heat transfer, on the other hand, invariably deals with physical phenomena in domains where the sizes are so small that they affect the definitions of thermophysical properties based on macroscopic considerations. So, the possibility of a continuum analysis does not exist here, and the approach should be microscopic. The interaction of the microscale physical domain and the microscopic phenomenon of energy transport is found to produce several anomalous effects compared to what would be expected in the macroscopic domain. Whether these are due to the influence of the physical boundaries of the domain on the thermophysical behavior of materials, or due to the molecular or submolecular level interactions in the media, the topic of nanoscale heat transfer puts forward an interesting and challenging area of study and research.

Heat transfer in many of the nanoscale systems and structures can be analyzed at the microscopic level, more or less along the same lines as has been described in the previous chapters, but the analysis pertains to much smaller dimensions. For example, when analyzing nanoscale processes, the physical size of the domain places the calculations in the free molecular flow regime, and there is little scope for continuum or modified continuum modeling in this case. The applications of nanoscale heat transfer analysis are concerned with the thermal performance of nanoscale systems, as well as the thermal phenomena associated with the manufacture of nanoscale systems.

Another area of importance is the thermal behavior and properties of nanomaterials and nanocomposites. Materials with special properties can be produced by the introduction of nanometer-sized particles (nanoparticles) in solid matrices or fluids. It has been noticed that the inclusion of nanoparticles in solid matrices or fluids can produce anomalously enhanced thermophysical properties. Determination and characterization of these thermophysical properties is an important and challenging area of research.

Some of the fundamental concepts required in the analysis and understanding of nanoscale thermal phenomena will be briefly introduced at the beginning of this chapter. A few applications of nanoscale heat transfer will also be discussed. Most of the contents of this chapter are devoted to discussions on a category of nanomaterials—nanofluids or suspensions of nanoparticles in liquid media. As experimental and theoretical analysis of nanofluids are some of the focus areas of ongoing research at the authors' laboratories, extensive discussions on nanofluids and studies on the anomalous thermal behaviors observed in them are presented in the subsequent sections of this chapter.

Comprehensive and extensive discussions on nanoscale heat transfer, especially from a fundamental perspective can be found in the literature (Chen et al. 2004). While an elaborate treatment of the conceptual and theoretical aspects is not attempted here, a brief discussion is presented, to provide an introduction to the basics of nanoscale heat transfer.

6.1.1 Length Scales for Nanoscale Heat Transfer

In nanoscale physical dimensions, analysis of phenomena often considers the energy carriers to have both wave and particle characteristics (wave–particle duality). In order to understand when the wave nature has to be considered, the characteristic length dimension has to be compared with certain characteristic length scales, as in the case of microscale energy transport processes discussed previously. Various length scales have been suggested, based on physical reasoning, and considering these, a nanoscale regime map has been introduced (Chen et al. 2004). The mean free path, the average distance of travel of energy carriers between successive collisions, is one of the characteristic length scales. Knowing the thermal conductivity, the mean free paths can be estimated from kinetic theory based expressions as follows:

$$k = \frac{1}{3}Cv^2\tau = \frac{1}{3}Cv\Lambda \text{ (gases)} \tag{6.1}$$

$$k = \frac{\pi^2 n k_B^2 T}{m v_F}\Lambda \text{ (electrons in metals)} \tag{6.2}$$

$$k = \frac{1}{3}\int_0^{\omega_{max}} C_\omega v_\omega^2 \tau_\omega \, d\omega = \frac{\Lambda}{3}\int_0^{\omega_{max}} C_\omega v_\omega \, d\omega \text{ (phonons)} \tag{6.3}$$

where
  k is the thermal conductivity
  Λ is the mean free path
  C is the volumetric specific heat
  τ is the relaxation time
  v is the velocity of carriers
  m is the mass of electron mass

n is the electron number density at the Fermi surface
C_ω, v_w, and τ_ω are the volumetric specific heat, the velocity, and the relaxation
   time at each frequency, respectively
k_B is the Boltzmann constant

Here the relaxation time τ, represents a characteristic timescale.

It is a necessary condition that the mean free path should be comparable to the characteristic length of the structure, which is the most relevant physical length in a particular problem, for inclusion of wave effects.

Another length scale is the average wavelength of the energy carriers (thermal wavelength) given by

$$\lambda_t = \frac{h}{\sqrt{3mk_B T}} \text{ (for electrons or molecules)} \tag{6.4}$$

$$\lambda_t = \frac{2h\nu}{k_B T} \text{ (for photons)} \tag{6.5}$$

where
h is the Planck's constant
λ_t is the wavelength of energy carriers
v is the velocity of the energy carriers
m is the mass of energy carriers
k_B is the Boltzmann constant

A third characteristic length scale considers the spread in energy and wavelength of the carriers. This is called the coherence length in optics, which is the width of the wave packets formed by the superposition of waves. The coherence length is given by

$$l_c \approx \frac{c}{\Delta\nu} \tag{6.6}$$

where
c is the speed of light
$\Delta\nu$ is the bandwidth of radiation

Comparing the structure characteristic length with the length scales mentioned above, the various nanoscale size-effect regimes have been obtained. The nanoscale size-effect regime map is shown in Figure 6.1, which is used to determine the approach to be followed in analyzing nanoscale heat transport phenomena. Extensive discussions on the size-effect regimes can be found in the literature (Chen et al. 2004), which is suggested as reading material to gain an in-depth knowledge of the subject matter.

6.1.2 HEAT TRANSFER MODES IN NANOSCALE SIZE-AFFECTED DOMAINS

Analysis of heat transfer processes in nanoscale size-affected domains requires treatment different from that of macroscopic phenomena in large scale systems.

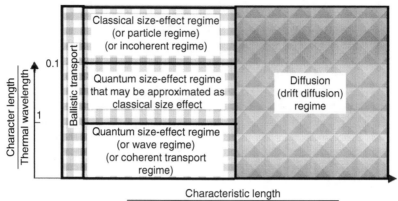

FIGURE 6.1 Different nanoscale heat transfer regimes. (Adapted from Chen, G., Borca-Tasciuc, D., and Yang, R.G., *Encyclopedia of Nanoscience and Nanotechnology*, American Scientific Publishers, California, 2004. With permission.)

The physical size of the system itself imposes restrictions to the continuum analysis, as the characteristic dimensions are comparable with the length scales such as those mentioned above. The size effect on the thermophysical properties, which are defined based on classical macroscopic considerations, will make the basis of the continuum governing equations itself wrong. In the conduction heat transfer problem, as the differential equation describing the macroscale process is based on Fourier's law which connects the temperature gradient and the heat flux through the thermal conductivity in a domain where the thermal conductivity cannot be considered as a size-independent quantity, the formulation will not have any validity. Nanoscale heat conduction problems may involve heat conduction at interfaces, inside and outside thin films, and heat conduction within nanostructures in between different heat carriers. The heat conduction problem is visualized as one of the electron and phonon scattering on phonons and interfaces.

The Knudsen number considerations introduced in Chapter 3 very often brings flow and convective heat transport problems to the regime of free molecular flow in nanoscale flow domains. Here also, continuum approximations and the application of the Navier–Stokes equations fail, because of the size effects on properties such as the thermal conductivity and viscosity, and because the size of the domain itself may demand the physical model to have only a few molecules or particles, with nothing in between, thus breaking down the whole concept of differential extrapolation in the continuum (such as the Taylor series expansion of field variables). Thus the continuum hypothesis is valid only for macroscale domain, or at the most, for cases where modifications such as slip flow boundary conditions would be valid. In the nanoscale domain, the method of analysis should be based on discrete computations, for instance, concentrating on each of the particles under consideration.

Radiative heat transfer in nanoscale structures impose special problems because of the subwavelength physical dimensions associated with the structures.

Special size effects have been observed in nanoscale films, such as those due to the wave phenomena like interference and diffraction due to the wave nature of the energy carriers, which may create altogether new energy states within the structure. The heat transport problem, in this case, cannot be approached using the classical radiation theory, but should be based on photon transport and subsequent energy transmission to electrons and dissipation through phonons, as discussed in Chapter 5.

Measurement and characterization of heat transport in the nanoscale is a challenging area, due to the extremely small sizes of the physical systems and structures. Standard nonintrusive measurement using optical techniques fail in many situations, because of the subwavelength dimensions of the physical systems. Probes suitable for measurement should essentially be manufactured with micro-electro-mechanical system (MEMS) and nano-electro-mechanical systems (NEMS), utilizing advances in nanotechnology and molecular level manufacturing.

An interesting class of heat transfer problems that require nanoscale consideration in analysis, is that of thermal phenomena in nanomaterials including nanofluids (nanoparticle suspensions), obtained by introducing nanometer-sized solid particles in solid matrices or liquids. The analysis of thermal phenomena in nanofluids will be dealt with extensively in later sections of this chapter.

6.1.3 Application Areas of Nanoscale Heat Transfer

A large number of application areas have been identified for nanoscale heat transfer. These involve areas where heat transfer analysis can be used to develop optimal control strategies for end-user applications, and where heat transfer analysis can be used to effectively produce and optimize manufacturing processes for systems and structures. Many of the developments in nanoscale heat transfer are aimed at micro/nanoelectromechanical systems. To be more specific, nanoscale heat transfer finds immense use in diverse areas such as biotechnology, nanotechnology (nanofabrication, nanomaterial synthesis), energy conversion and power generation systems such as heat exchange systems, photovoltaic conversion and thermoelectric energy conversion systems, thermal control of electronics including optimal thermal control of computer processors, optoelectronics, semiconductor manufacture and, in general, every application that involves miniaturization as an objective.

Nanomaterial synthesis and nanofabrication technologies are futuristic technologies, and all manufacturing processes of this type require precise control of temperature at the nanoscale. Molecular manufacturing is projected to be one of the key areas in materials technology. A few examples where nanoscale heat transfer can be used are applications involving the engineering of nanocrystals, multilayer structures, implantable materials, control and synthesis of biological matter such as DNA, and the immensely attractive projected applications of nanofluids, nanowires, and carbon nanotubes (Chen et al. 2004).

6.2 NANOPARTICLES AND NANOFLUIDS

Since the beginning of the twentieth century, a number of methods and techniques have been developed, capable of manufacturing a wide variety of nanometer-sized particles for engineering applications and scientific study. The most common method of production has been the "inert gas evaporation" method, in which a metal material evaporates in a crucible or a resistance filament in the presence of an inert gas and deposits as nanoparticles, on a cooled surface. The distribution of the particle diameters obtained through this process is determined by the evaporation rate, the inert gas pressure, the atomic mass of the gas, and the evaporation temperature. Gleiter (1969) and Granqvist and Buhrman (1976) produced nanometer-sized particles by evaporation in a temperature-regulated oven containing a reduced atmosphere of an inert gas with nanometer-sized crystal particles having a diameter less than 20 nm and a spherical shape. Larger nanometer-sized crystal particles typically assumed a crystalline structure that distorted this spherical shape. To date, a variety of nanometer-sized particles have been made from common metal or metal oxide materials, as well as from compounds such as ZnO, ZnS, CdS, CdSe, Cd_3P_2, Zn_3P_2, PbI_2, HgI_2, BiI_3, and GaN. The diameters of the particles range from several nanometers to several hundred nanometers and can be produced in a highly reliable process that can accurately control the composition, shape, and size of the particles.

Determination of the mean diameter and the diameter distribution can be made using a variety of techniques. The electron microscopy method utilizes a microscopic grid at the center of the cooling plate on which the nanoparticles deposit, or into a colloid made of nanoparticles. After the grid has been dried, the nanoparticles can be readily observed using an electron microscope. Other methods include the use of a differential mobility analyzer (DMA) to determine the size distribution of nanoparticles (Hummes et al. 1996) and to measure this distribution using an atomic force microscope and transmission electron microscopy (TEM).

The size effect of the nanoparticles has resulted in a number of unique properties. Examples are thermodynamic fluctuations on the superconducting transition of aluminum particles (in a radius range of 12.5–400 nm), under different temperatures and magnetic fields (Buhrman and Halperin 1973), and peculiarities in the static magnetic moment of a collection of small metallic spherical indium particles with a mean diameter of approximately 5 nm which indicated that the previous models of electronic quantum size effects are not alone sufficient to explain the experimental results (Meier and Wyder 1973). Another case reported is the anomalously large specific heat discontinuity observed at the superconducting transition temperature of indium particles of diameter less than 2.2 nm (Novotny and Meincke 1973). All of these phenomena have been assumed to be the result of the size effect.

Tanner and Sievers (1975) measured the far infrared adsorption of small metallic particles of Cu, Al, Sn, and Pb, with diameters ranging from 6.5 to 35 nm, and concluded that the adsorption ability of the particles had a linear relationship with the frequency in the low frequency region and in a temperature range from 1.2 through 20 K. This investigation also indicated that the assumption of a constant internal field would fail at a diameter of 20 nm, when calculating the absorption. Yee and Knight (1975) conducted nuclear magnetic resonance line shift experiments of copper particles having diameters ranging from 2.5 to 45 nm, which indicated that quantum size effects exist for nanoscale particles, and suggested modifications in theoretical predictions.

Kamat and Dimitrijevic (1990a,b) studied the size quantization effects, nonlinear optical and enhanced photoredox properties of nanometer size particles as semiconductors. Ball and Garwin (1992) investigated the optical effects of nanometer-sized particles and found that the electronic structure does not necessarily acquire the true band-like characteristics of the bulk solid and that the molecular orbitals might remain valid. The influence of size on the dielectric behavior of nanoparticles has also been studied. Experiments conducted by Xu et al. (2001) indicated that GaN coarse-grain powders have a lower relative dielectric constant when compared to nanometer size particles, whose effect was attributed to the existence of a larger amount of interfaces in nano-sized powders. Nalwa (2004) found that at nanometer length scales, the electrons and phonons by which heat is conducted will be significantly influenced by the conduction dimensions and the interfaces between the different regions due to scattering at the interface. This phenomenon is a classic example of the size effect, in that, if the size of the particles is less than the thermal wavelength, the quantum effects will be an important factor which will reduce the thermal conductivity significantly. The size effects are found to be especially strong at low temperatures and have been studied in detail by Guczi and Horvath (2000) who measured the morphology, electron structure, and catalytic activity of Au, FeO_x, SiO_2, and Si nanoparticle samples in CO oxidation with x-ray photoelectron spectroscopy, UV photoelectron spectroscopy, and TEM.

Nanofluids, which are fluid suspensions of nanometer-sized solid particles, have been found to improve the performance of heat transfer liquids (Choi 1995a,b). Recently, nanofluids have attracted great interest in various heat transfer applications because of their enhanced thermal properties. Experiments conducted by Xuan and Li (2000a) on nanofluids have brought out significant increases in thermal conductivity compared to the base fluid. In some of the experiments, the extent of enhancement of thermal conductivity was found to exceed the predictions by well established theories (Choi 1995a,b, Mapa and Sana 2005). Eastman et al. (1997) observed that the addition of nanoparticles of less than 100 nm diameter provides an effective way of improving the heat transfer characteristics of fluids. Experimental results showed that an increase in thermal conductivity of approximately 60% can be obtained in nanofluids consisting of water and 5% volume fraction of copper oxide nanoparticles.

Forced convection measurements by Xuan and Li (2003) indicated that the heat transfer coefficient of water containing 0.9% by volume of copper oxide nanoparticles was enhanced by more than 15% compared to the base liquid. Choi (1998) conducted experiments on the boiling characteristics of nanofluids. You et al. (2003) measured the critical heat flux in pool boiling of aluminum oxide, water nanofluid systems, and reported a threefold increase in the critical heat flux over that of pure water.

Mapa and Sana (2005) conducted experiments on a mini heat exchanger and observed an improvement in the heat flow due to the addition of nanoparticles, even at very low concentrations. The experimental results showed that the heat transfer rate was increased by 5.5% at a mass flow rate of 0.005 kg/s. and Yulang Kongsheng and Yulong (2004) investigated the convective heat transfer of nanofluids in laminar conditions and reported an increase in the heat transfer coefficient with respect to the percentage volume of the nanoparticles. Lee and Choi (1996) experimentally observed that the presence of nanoparticles in the base fluid significantly alters the cooling characteristics of advanced cooling systems. The experimental results showed that the efficiency of the cooling system increased by 7.5% even at very low concentrations of nanoparticles. Vadasz and Govender (2005) reported that even very small quantities, less than 1% by volume, of copper nanoparticles can improve the thermal conductivity of the base fluid by 40%.

6.2.1 PREPARATION OF NANOFLUIDS

Nanoparticle suspensions, which are termed "nanofluids," are produced by mixing nanoparticles with a fluid, such that the nanoparticles remain in suspension for a long time. There are two principal methods of producing these nanoparticle suspensions: The one-step method in which nanoparticles are produced directly in the base fluid to obtain the suspension and the two-step method where the nanoparticles are produced independently and then mixed with a base fluid to obtain the suspension. In a common single-step approach, the nanoparticles are produced in a vessel by evaporation and fall into the base fluid to form the suspension in a single process. This technique is called the VEROS (Vacuum Evaporation on Running Oil Substrate) technique.

Because of the ease with which the concentration and size distribution can be controlled, a large number of the experimental investigations have utilized the two-step method, and so are most of the original experimental studies reported in this book, which have used an ultrasonic agitation technique to produce the nanofluid from nanoparticles.

Changing the pH value of the suspension, adding surfactants or a suitable surface activator, or using ultrasonic or microwave vibration, are all techniques that have been used with the two-step method to better disperse and more evenly distribute the nanoparticles in the base fluid and maintain the stability of the suspension. When the nanoparticles are small, the weight to volume ratio is suitable, and the dispersion method is applied correctly, the nanoparticles will be very well dispersed and the suspension will be stable for several days. However, in a majority of the investigations studied, the literature reports that the suspension samples can typically be maintained in a homogeneous stable state for no more than 24 h.

Like nanoparticles, because of the extremely small size of the particles dispersed in a base fluid, nanoparticle suspensions have a number of interesting and seemingly unusual properties. A number of recent investigations using well-prepared suspensions have been conducted to investigate the different physical aspects of nanoparticle suspensions, such as viscosity, effective thermal conductivity, flow behavior, and heat transfer characteristics. Among the most interesting property variations, is the effect that the addition of nanoparticles to a base fluid can have on the effective thermal conductivity of the nanoparticle suspension. Several experimental investigations have demonstrated that nanoparticle suspensions have remarkably high effective thermal conductivity compared with traditional heat transfer fluid media, like water, engineering oil and organic fluids and that the increase in the effective thermal conductivity is greater than what might be predicted by the conventional expressions developed for larger particles suspended in liquids. To better utilize the novel thermal properties of nanofluids, it is necessary to more fully understand the fundamental mechanisms that govern the behavior of nanoparticles.

6.2.2 STATE OF THE ART IN EXPERIMENTAL INVESTIGATIONS

As mentioned previously, a large number of experimental studies reported in literature have brought out the characteristic changes in the thermophysical properties of fluids due to the addition of nanoparticles, and differences in thermal phenomena in nanofluids. Measurement of thermal conductivity has been the focus of a majority of the investigators, though there have been reports on studies on various transport phenomena in nanofluids as well. In this section, an overall review of the important experimental techniques and investigations will be given, with discussions on the salient features of the observations made.

6.2.2.1 Determination of Effective Thermal Conductivity

There are two principal experimental approaches that have been used to measure the effective thermal conductivity of nanoparticle suspensions: the transient approach, which typically utilizes a hot-wire method; and steady-state methods, which utilize a guarded hot plate, a temperature oscillation technique, or a cut-bar apparatus.

The most commonly used apparatus in the measurement of the effective thermal conductivity of fluids and nanoparticle suspensions is the transient hot-wire system. Nagasaka and Nagashima (1981) first utilized this method to measure the thermophysical properties of electrically conducting liquids. In this approach, a coated platinum hot-wire is suspended symmetrically in a liquid contained within a vertical cylindrical container. This hot-wire serves as both a heating element through electrical resistance heating and as a thermometer by measuring the electrical resistivity of the fluid. The thermal conductivity can be calculated from the following equation derived using the relationship between the electrical and thermal conductivity as

$$T(t) - T_{\text{ref}} = \frac{q}{4\pi k} \ln\left(\frac{4K}{a^2 C}t\right) \tag{6.7}$$

where

$T(t)$ is the temperature of the platinum hot-wire in the fluid at time t
T_{ref} is the temperature of the test cell
q is the applied electric power applied to the hot-wire
k is the thermal conductivity
K is the thermal diffusivity of the test fluid
a is the radius of the platinum hot-wire
C is a constant

$\ln C = g$, where g is Euler's constant. This relationship between δT and $\ln(t)$ is linear and the data of δT is valid over a range of $\ln(t)$, namely between time t_1 and time t_2, the thermal conductivity of the fluid can be calculated as

$$k = \frac{q}{4\pi(T_2 - T_1)} \ln\left(\frac{t_2}{t_1}\right) \qquad (6.8)$$

where $T_2 - T_1$ is the temperature difference of the platinum hot-wire between times t_1 and t_2.

Since this system has a wire coated with a layer of electrically insulating material, a number of problems that may arise have been analyzed. These problems may include (1) the effects of the insulation layer on the temperature rise of the metallic wire, (2) the effects of the insulation layer on the reference temperature, (3) the effects of the thermal contact resistance between the metallic wire and the insulation layer, and (4) the effects of the finite length of the wire. Experimental calibrations using NaCl solutions verified that the whole system could be used with an overall accuracy of $\pm 0.5\%$.

A number of investigators have applied the transient hot-wire method to measure the effective thermal conductivity of different nanoparticle suspensions. These investigations included those of Choi (1995), Eastman et al. (1997, 2001), Lee et al. (1999), Xuan and Li (2000), and Xie et al. (2002a,b). The accuracy of this method has been found to be excellent.

The second method for measuring effective thermal conductivity, the steady-state method, utilizes a number of different approaches. The first of these was used by Wang et al. (1999) and designed by Challoner and Powell (1956). As illustrated in Figure 6.2, this one-dimensional parallel-plate method utilizes a guarded hot plate that surrounds two round copper plates positioned parallel to each other and set compactly in an aluminum cell. The nanoparticle suspension is placed between the

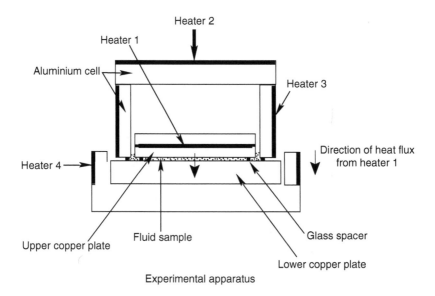

FIGURE 6.2 The steady-state one-dimensional method apparatus. (Adapted from Wang X., Xu, X., and Choi, S.U.S., *AIAA J. Thermophys. Heat Transfer,* 13, 474, 1999. With permission.)

two plates and by measuring the temperature of the plates and the cell, the overall thermal conductivity of the two copper plates and the sample fluid can be calculated using the one-dimensional heat conduction equation relating the heat flux, q, the temperature difference ΔT, and the geometry of the liquid cell.

$$k = (qL_{\mathrm{g}})/(S\Delta T) \qquad (6.9)$$

$$k_{\mathrm{e}} = \frac{kS - k_{\mathrm{g}}S_{\mathrm{g}}}{S - S_{\mathrm{g}}} \qquad (6.10)$$

where
   L_{g} is the thickness of the glass spacer between the two copper plates
   S is the cross-sectional area of the top copper plate
   k_{g} and S_{g} are the thermal conductivity and the total cross-sectional area of the
      glass spacers, respectively

The accuracy of this type of system has been reported to be greater than 0.02°C for the temperature gradient, which yields an overall experimental uncertainty of less than ±3%.

   The temperature oscillation technique was first reported by Czarnetzki and Roetzel (1995) and is illustrated in Figure 6.3. This approach has been applied to measure the effective thermal conductivity of a variety of materials (Das et al. 2003b) and utilizes the temperature oscillation in a test section through a number

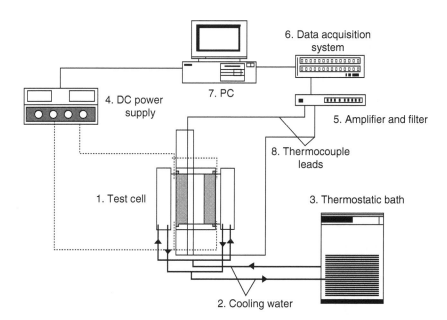

FIGURE 6.3 The temperature oscillation system. (Adapted from Das, S.K., Putra, N., Thiesen, P., and Roetzel. W., *J. Heat Transfer,* 125, 567, 2003. With permission.)

of thermocouples to determine the thermal diffusivity of the nanoparticle suspension. Then, by comparison with a reference layer, the thermal conductivity can be calculated. The thermal diffusivity and thermal conductivity are related using the one-dimensional form of the energy equation

$$\frac{\partial^2 T}{\partial \xi^2} = \frac{\partial T}{\partial \tau} \tag{6.11}$$

where

$\xi = x\sqrt{\frac{\omega}{\alpha}}$ and represents the dimensionless x-space coordinate
α is the thermal diffusivity
ω is the constant angular frequency
τ is a dimensionless time

Using the assumption that the periodic temperature oscillations are generated with the same frequency in the boundary regions, but different amplitude and phases than at the two surfaces, the thermal diffusivity of the nanoparticle suspensions can be obtained by measuring the phase shift and amplitude of the temperature oscillation at these two surfaces and the center point of the fluid sample. The limitation of this method is that the accuracy of the thermal conductivity measurement of the nano-particle suspensions depends directly upon the accuracy of the measured thermal diffusivity of the reference layer. The average deviation of the measured thermal diffusivity from the standard handbook values has been shown to be 2.7% in the temperature range of 20°C to 50°C and less than 2.11% in the range of 20°C to 30°C (Das et al. 2003b).

Another steady-state measurement system is the cut-bar apparatus used exten-sively by Peterson and Fletcher (1989) to measure the thermal conductivity of both saturated and unsaturated dispersed ceramics and the contact resistance of various interfacial materials (Duncan et al. 1989). This experimental method first described by Miller and Fletcher (1974) has been used for measuring the thermal contact resistance. This has also been applied with good success to the measurement of the effective thermal conductivity of materials and saturated porous materials and has been shown to have an accuracy of nearly 0.02°C and an overall experimental uncertainty of less than ±3%. A variation of the original facility has been used for the measurement of the effective thermal conductivity of nanofluids, which will be described in detail later in this chapter.

6.2.2.2 Studies on Thermal Conductivity

Over the past 45 years, there have been a number of reports documenting the results of experimental investigations conducted to determine the effective thermal conductivity of different types of particle suspensions, with particles ranging in size from milli-meters to nanometers. Meredith and Tobias (1961) investigated the thermal conduct-ivity of water and propylene carbonate emulsions with volume fractions ranging from 0.0% to 0.5%. As observed in this investigation, when water was the discontinuous phase and the propylene carbonate was the continuous phase, the effective thermal conductivity of the mixture was very sensitive to the volume ratio between the two phases. However, when water was the continuous phase, the effective thermal conductivity was relatively insensitive to the volume ratio of the two phases.

In 1962, Hamilton and Crosser measured the effective thermal conductivity of mixtures containing millimeter-sized aluminum and balsa particles with silastic rubber at a mean temperature of approximately 95°F using an apparatus that consisted of an electrically heated sphere surrounded by a spherical shell of the mixture. The particles used in the experiments were different in shape, such as spheres, cylinders, disks, and cubes. The thermal conductivity enhancement of the suspensions with different shapes of particles varied considerably. However, the thermal conductivity of the mixture containing nonsphere-shaped balsa wood particles showed no shape effect.

Eastman et al. (1997) measured the effective thermal conductivities of CuO/water nanoparticle suspensions with a mean diameter of the CuO nanoparticles of 36 nm and Al_2O_3/water nanoparticle suspensions with a mean diameter of the Al_2O_3 particles of 33 nm. The results indicated that for volume ratios of 0, 0.01, 0.02, 0.03, 0.04, and 0.05, the CuO/water nanoparticle suspensions demonstrated an effective thermal conductivity enhancement of as much as 60% and as much as 30% for the Al_2O_3/water nanoparticle suspension. The effective thermal conductivity enhancement was found to have a linear relationship to the volume ratio of the nanoparticle to water. Further experiments indicated that both Cu/Duo-Seal oil nanoparticle suspensions and Cu/HE-200 oil nanoparticle suspensions, exhibited even higher thermal conductivity enhancements than those observed in the Al_2O_3/water or the CuO/water nanoparticle suspensions.

Lee et al. (1999) measured the thermal conductivity enhancement of Al_2O_3/ethylene glycol and CuO/ethylene glycol nanoparticle suspensions. It was concluded that the particle shape and size were both dominant in the thermal conductivity enhancement of nanoparticle suspensions of this type, a result that is consistent with that of Hamilton and Crosser (1962). The effect of agglomeration of nanoparticles on the thermal conductivity enhancement was also discussed.

Nanoparticle suspensions comprised of CuO and Al_2O_3 nanoparticles with different sizes (23 and 28 nm, respectively) and differing base fluids (pump fluid, engine oil, ethylene glycol, and water) were tested by Wang et al. (1999). Three different methods were utilized to produce the nanoparticle suspensions—dispersion of nanoparticles into the base fluid with an ultrasonic bath; addition of polymer, coatings (styrene-maleic anhydride, 5000 mol. wt., 2.0% by weight) keeping the pH value at 8.5 to 9.0; and filtering and removing particles larger than 1 μm in diameter. Nanoparticle suspensions produced using the third method resulted in the greatest enhancement in the thermal conductivity. The results were also compared with a number of different theoretical models. The effect of the Brownian motion of the nanoparticles was considered, as were the effects of electric double layer and the van der Waals force. Micromotion caused by the electric double layer, van der Waals force, and the chain structure of the nanoparticles were considered to be the principal reasons and provided the main explanation for the enhancement in this particular case.

Xuan and Li (2000) used the hot-wire method to evaluate copper nanoparticles in transformer oil and copper nanoparticles in water, in the presence of surface activators. The suspensions were prepared using ultrasonic agitation. The diameter of the copper nanoparticles ranged from 30 to 100 nm and the volume ratio between the nanoparticle and the base fluid varied from 2.5% to 7.5%. In a different investigation, the thermal conductivity of several nanoparticle suspensions using different stabilizing agents and Cu/ethylene glycol combinations were evaluated experimentally. The diameter of the Cu nanoparticles was less than 10 nm (Eastman et al. 2001). The nanoparticle suspensions containing thioglycolic acid as a stabilizing agent showed improved behavior when compared to nonacid samples containing the nanoparticle suspensions, while fluids containing thioglycolic acid, but no particles, showed no improvement in thermal conductivity. Fresh nanoparticle suspensions tested within two days of preparation exhibited slightly higher conductivities than fluids that were stored up to two months prior to measurement. The measured effective thermal conductivities of metallic and oxide nanoparticle suspensions were compared, indicating that nanoparticle suspensions with metallic nanoparticles have higher thermal conductivities than nanoparticle suspensions with oxide nanoparticles, which implies that the higher intrinsic thermal conductivity of the metallic nanoparticles was the principal cause of this increase over the oxide nanoparticle suspensions. However, it is important to note that the metallic nanoparticles used in this experiment had a considerably smaller diameter than the oxide nanoparticles evaluated (Eastman et al. 2001).

Experimental data from Al_2O_3 nanoparticle suspensions were reported by Xie et al. (2002), in which the Al_2O_3 nanoparticles had a specific surface area ranging from 5 to 124 m^2/g and a particle size ranging from 30.2 to 12.2 nm with different crystalline phases. It was found that with the same type of nanoparticles, the thermal conductivity enhancement reduced with the increasing value of thermal conductivity of the base fluid. When the same base fluid was used, however, the enhancement was shown to be dependent on the pH value of the suspension and the specific surface area of the nanoparticles. Nanoparticle suspensions of Al_2O_3/water and CuO/water were measured by Das et al. (2003b) at various nanoparticle volume ratios and different temperatures, using a temperature oscillation method. Figure 6.4 shows the

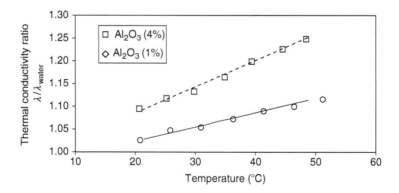

FIGURE 6.4 Temperature dependence of thermal conductivity enhancement for Al_2O_3/water nanoparticle suspensions. (Adapted from Das, S.K., Putra, N., Thiesen, P., and Roetzel. W., *J. Heat Transfer*, 125, 567, 2003. With permission.)

enhancement of thermal conductivity of Al_2O_3/water nanoparticle suspensions as a function of the temperature. The enhancement ratio relationship was also studied and the experimental data were compared with the prediction of the model by Hamilton and Crosser (1962). These experimental data showed a dependence on the volume fraction and temperature and when coupled with other previous investigations, indicated that in addition to the relative volume fraction, thermal conductivity and particle size, a stochastic motion of nanoparticles may be a probable explanation of the difference between the prediction of the Hamilton and Crosser theory and the experimental data.

Patel et al. (2003) found that the enhancement of the effective thermal conductivity did not increase linearly with the temperature, for the same volume fraction of nanoparticle suspensions composed of Au-thiolate nanoparticles with 10–20 nm diameters and a base fluid of water and toluene. The enhancement ratio in this investigation was found to be 5% to 21% in the temperature range of 30°C–60°C at a volume ratio of 0.00026 in toluene, and 7% to 14% for Au particles stabilized with a monolayer of octadecanethiol, even for a loading of 0.001% in water. This nonlinear variation is quite similar to findings by Wang et al. (2003), who utilized SiO_2 nanoparticle suspensions. In both these investigations, the transient hot-wire method was used to measure the effective thermal conductivity.

Li and Peterson (2006a) have applied the steady-state method to measure the effective thermal conductivities of Al_2O_3/water and CuO/water nanofluids at various temperatures and volume fractions. These experimental investigations will be explained in greater detail later.

The unusual enhancement of thermal conductivity observed in carbon nanotube suspensions has been the focus of attention of various investigations in recent times. A number of researchers have conducted theoretical studies of the thermal conductivity of carbon nanotubes and found that nanotubes have very high thermal conductivity, as mentioned by Li and Peterson (2005). Che et al. (2000) conducted a molecular dynamic numerical simulation of the thermal conductivity of carbon nanotubes and found that the thermal conductivity of the individual nanotubes was dependent on the structure, defects, and vacancies, and that the anisotropic character of the highly thermally conductive nanotubes is very similar to diamond crystal and in-plane graphite sheets. Driven by both a desire for a more complete understanding of the fundamental behavior of these nanoparticle suspensions and also of carbon nanotubes themselves, experimental investigations on the application of carbon nanotubes in nanoparticle suspensions has begun to attract the attention of a number of researchers. Hone et al. (1999) found that bulk samples of single-walled nanotube bundles had very large thermal conductivities and that the thermal conductivity increased linearly with the temperature. Kim et al. (2001) and Shi et al. (2003) measured the thermal conduction in 148 and 10 nm diameter single-wall carbon nanotube bundles. These materials showed a 1.5 power relation between the thermal conductance and the temperature, in the range of 20–100 K, which is quite different from the quadratic relationship between thermal conductance and temperature for individual multiwall carbon nanotubes (MWCNs). The effective thermal conductivity of carbon nanotube suspension has been shown to have an anomalously larger value than for other more classical nanoparticle suspensions and has demonstrated a nonlinear relationship with the nanotube volume fraction (Xie et al. 2003).

As demonstrated in various experimental investigations, there appears to be unusual phenomena occurring in nanoparticle suspensions that have particular relevance in the application of these materials to both scientific study and engineering applications. However, no single method has yet been developed that is capable of accurately accounting for or predicting the effect of the size, volumetric ratio, and thermophysical effects on the effective thermal conductivity of nanoparticle suspensions. The equations currently in use are only capable of predicting the effective thermal conductivities for the limited type of suspensions from which they have been developed. The experimental data found in the literature mentioned above indicate that new and generally satisfactory correlations are necessary to completely understand and better predict the heat transfer in nanofluids.

6.2.2.3 Transport Phenomena in Nanoparticle Suspensions

Although determination of the effective thermal conductivity has been the focus of most of the investigations of the thermophysical properties of nanoparticle suspensions, a number of researchers have recently begun to examine other parameters, such as the heat transfer coefficient in nanoparticle suspensions. Lee and Choi (1996) observed nanoparticle suspensions using an x-ray beam mirror system that was cooled using a microchannel system with water and liquid nitrogen as coolants. In this investigation, however, little information has been provided regarding the kinds of nanoparticles and fluids used to make the suspensions, other than the inference that the effective thermal conductivities were two or three times greater than that of water. The experimental results demonstrated that the nanoparticle suspension with an effective thermal conductivity three times greater than that of water had a performance curve similar to that of liquid nitrogen.

Xuan and Roetzel (2000) described two different methods used to study the convective heat transfer occurring in nanoparticle suspensions. The first was the traditional method in which the nanoparticle suspension was treated as a single-phase fluid and the second included the features of the dispersed nanoparticles using a multiphase fluid model. Using this latter technique, Xuan and Li (2003) developed the experimental test facility illustrated in Figure 6.5 to determine the heat transfer coefficient of nanoparticle suspensions consisting of Cu particles which have diameters below 100 nm and volume fractions of 0.3%, 0.5%, 0.8%, 1.0%, 1.2%, 1.5%, and 2.0%. The system uncertainty of this test facility was less than 3%. The results were presented in terms of the heat transfer coefficient as a function of velocity and are shown in Figure 6.6. As shown, the results clearly indicate that the convective heat transfer coefficient of the suspension increased with the flow velocity as well as with the volume fraction.

On the basis of this experiment, an equation to calculate the Nusselt number of the Cu nanoparticle suspension was developed. The resulting expression,

$$\text{Nu} = \left(0.005991.0 + 7.6286\phi^{0.6886}\text{Pe}_{\text{particle}}^{0.001}\right)\text{Re}^{0.9238}\text{Pr}^{0.4} \qquad (6.12)$$

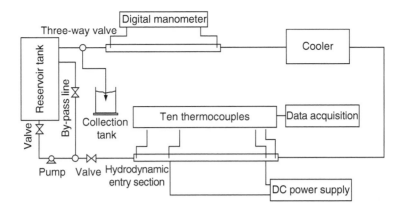

FIGURE 6.5 The experimental system for study of the convective heat transfer in nanoparticle suspensions. (Adapted from Xuan, Y., and Li, Q., *J Heat Transfer*, 125, 151, 2003. With permission.)

was able to predict the Nusselt number of different volume fractions of Cu nanoparticle suspensions to within 8% when compared with the experimental data. In this expression, ϕ represents the volume fraction and $Pe_{particle}$ is the Peclet number ($Pe = Re\ Pr$), based on the diameter of the Cu nanoparticle. The increase in the enhancement of the convective heat transfer coefficient for the nanoparticle suspensions was attributed to the increase in the effective thermal conductivity of the nanoparticle suspension and the chaotic movement of the nanoparticle, which accelerated the energy exchange process in the fluid.

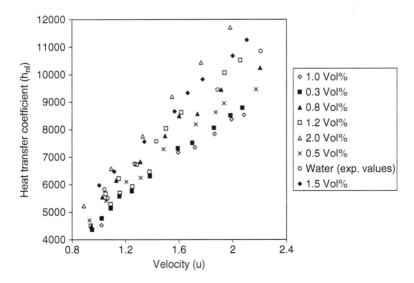

FIGURE 6.6 The convective heat transfer coefficient versus velocity. (Adapted from Xuan, Y., and Li, Q., *J Heat Transfer*, 125, 151, 2003. With permission.)

In addition to the study of the forced convection effects of nanoparticle suspensions, a number of researchers have studied the natural convection occurring in these suspensions. Okada and Suzuki (1997) investigated the natural convection occurring in two types of suspensions with micro soda glass beads in water with an average diameter of 4.75 and 6.51 μm, respectively. The weight fractions of both suspensions were less than 5%. The suspensions were heated in a rectangular cell at the center of the horizontal bottom wall. During the experiments, it was clear that several layers were formed in the test cell along the direction of the gravitational vector, and that each layer had several circular flows that were constrained to that cell and did not move or extend into the neighboring layers. Kang et al. (2001) conducted a similar experiment on SiO_2 particles having a mean diameter of 2.97 μm with a 0.03 μm standard deviation, in water. The difference between these two experiments was the method of heating. In the first, the heat was added to the bottom and in the other it was added from one of the vertical walls, opposite the cold vertical wall. In the latter experiment, it was found that the natural convection of the suspension could be classified into one of five distinguishable patterns, and that the critical wall temperature difference that would determine the natural convection pattern, decreased with a decrease in the particle concentration in the suspension.

Putra et al. (2003) conducted an experimental study of the steady-state natural convection of two Al_2O_3 and CuO nanoparticle suspensions, one with a mean diameter of 131.2 nm and the other with a mean diameter of 87.3 nm in distilled water, under different conditions. The results indicated a systematic degradation of natural convective heat transfer, i.e., that the Nusselt number decreased with the increase in the nanoparticle volume fraction in the suspension. A correlation between the Nusselt number and the Rayleigh number was given as $Nu = CRa^n$, where n was weakly dependent, and the constant, C, was strongly dependent upon the volume fraction.

Another aspect of the heat transfer performance of nanoparticle suspensions is the effect of these suspensions on boiling heat transfer. You and Kim (2003) experimentally investigated the enhancement of the critical heat flux in pool boiling for Al_2O_3 nanoparticle suspensions. The results indicted that in the nucleate boiling regime, the boiling heat transfer coefficient was not influenced by the addition of the Al_2O_3 nanoparticles. However, there was a dramatic increase in the critical heat flux (CHF), for different concentrations of the nanoparticle suspension that ranged as high as 300% of that measured for pure water. This increase in the CHF in Al_2O_3 nanoparticle suspensions varied with the concentration of nanoparticles.

In the experimental investigation of Das et al. (2003a), the boiling performance of the water was found to deteriorate with the addition of the nanoparticles, and a higher wall superheat was found to exist for the same heat flux. With the increase of the nanoparticle concentration in the suspension, the deterioration was more apparent. It was observed that the nanoparticles with a diameter of 20 to 50 nm filled the cavities of an uneven heating surface and changed the surface characteristics by effectively making it smoother. With the increase of nanoparticle concentration, a layer of sedimentation of the nanoparticles developed, which deteriorated the boiling performance even further.

Vassallo et al. (2004) studied the boiling characteristic of silica nanoparticle and microparticle suspensions with diameters of 15, 50, and 3000 nm in boiling situations, ranging from nucleate boiling to film boiling under atmospheric pressure. All of the suspensions expressed no difference in the boiling performance of water, below the CHF point, but increased the CHF point dramatically. Silica coating on the wire was also found after the test and was determined to be part of the reason for the increase in the CHF.

6.3 MEASUREMENTS IN NANOFLUIDS

Various attempts to observe thermal phenomena in nanofluids, and to measure and characterize associated properties and parameters have been made by both the authors' research groups, at their laboratories—The Two Phase Heat Transfer Laboratory at the University of Colorado at Boulder, USA, and the Nanotechnology Research Laboratory at the National Institute of Technology, Calicut, India. Attention has been devoted to the measurement of thermophysical properties such as the thermal conductivity, viscosity, and surface tension of nanofluids, and to study the enhancement and alteration produced by the addition of nanoparticles to the base fluid. Thermal phenomena such as convective heat transfer and phase change also have been studied. Discussions on the experimental methods and the important results from such experimentation are presented in some of the sections to follow.

6.3.1 THERMAL CONDUCTIVITY

The steady-state method, based on a modification of the "cut-bar apparatus," mentioned earlier has been used to measure the effective thermal conductivity of different nanoparticle suspensions. The use of the method in investigating the effect of various parameters, namely the temperature, volume fraction of the nanoparticles in the fluid, and the particle size, will be described here, with the significant results from the experimental study conducted at the Two Phase Heat Transfer Laboratory at the University of Colorado at Boulder, USA. Detailed descriptions of the experimental studies can be found in various publications (Peterson and Li 2006, Li and Peterson 2005, 2006a,b,c).

The schematic arrangement of the standard cut-bar apparatus used for the effective thermal conductivity of porous media and other materials (Peterson and Fletcher 1987, 1988, 1989) is shown in Figure 6.7. In the modified test setup for the measurement of the thermal conductivity of nanofluids, the test cell of the apparatus has been modified as shown in Figure 6.8. This consisted of a pair of 2.54 cm diameter copper rods, separated by an O-ring to form the test cell. The experimental test facility was placed in a vacuum chamber and all tests were conducted in a vacuum environment of less than 0.15 Torr. Neglecting the convective and radiation losses, the steady-state effective thermal conductivity of the fluid or nanoparticle suspensions can be modeled as shown in Figure 6.9.

The heat flux in a traditional cut-bar apparatus can be determined by measuring the temperature differences in the upper and lower copper bars as indicated in Equation 6.13 and then averaging the computed heat fluxes.

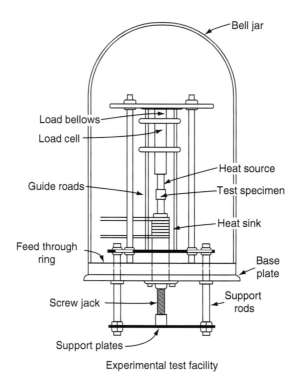

FIGURE 6.7 Schematic diagram of the standard cut-bar apparatus.

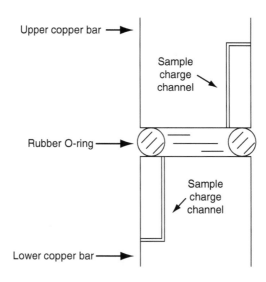

FIGURE 6.8 Diagram of test cell.

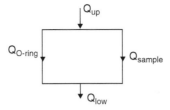

FIGURE 6.9 Heat flux diagram.

$$q = k_{copper} A_{bar} \frac{\Delta T_{bar}}{\Delta Z_{bar}} \qquad (6.13)$$

Using the measured heat flux from Equation 6.13 and the surface temperature, determined by averaging the response of six thermocouples located at the surface of the upper and lower bars, the effective thermal conductivity of the fluid or the nanoparticle suspension inside the test cell can be calculated as

$$k_{eff} = \left(q \frac{\Delta Z_{cell}}{\Delta T_{cell}} - k_{O\text{-}ring} A_{O\text{-}ring} \right) \Big/ A_{cell} \qquad (6.14)$$

Prior to initiating the tests, the two copper bars were carefully aligned vertically, and a load was applied using a load screw to ensure good contact and hence, an airtight seal for the test cell. The sample was charged through the lower sample change inlet until the suspension filled the test cell and exited the charge outlet, as shown in Figure 6.8.

To calibrate the test rig, the experimental results were compared with tabular data available in the literature for ethylene glycol and distilled water (base fluids, in the study undertaken), which indicated an overall experimental uncertainty in the measurement of the thermal conductivity that is well below ±2.5%.

In the investigation, nanoparticle suspensions were produced by dispersing nanoparticles into a base fluid of distilled water. A two-step method was utilized in which the nanoparticles were first produced and then dispersed into the base fluid. Two types of nanoparticles were used, copper oxide and aluminum oxide, both of which were prepared using a gas condensation technique and provided by Nanophase Technologies Corporation in Burr Ridge, Illinois. The purity of the aluminum oxide nanoparticles was greater than 99.5%, and had an average area weighted diameter of 36 nm. The density was 3.6 g/cc in the gamma crystal phase. The copper oxide nanoparticles had a purity of greater than 99.0%, an average diameter of 29 nm, a density of 6.5 g/cc, and the crystal phase was monoclinic. The nanoparticles were characterized using a scanning electron microscope (SEM), and the size distribution was found to be log-normal with the mean diameters mentioned above. Figure 6.10 illustrates an SEM image of the aluminium oxide nanoparticles.

To prepare the nanoparticle suspensions, the base fluid, distilled water, was boiled in a closed container for approximately 4 h, to eliminate any dissolved gases. The appropriate nanoparticle powder was then carefully weighed using an

FIGURE 6.10 SEM image of Al_2O_3 nanoparticles.

electronic balance, and the volume calculated. The powder was dispersed into a predetermined volume of the base fluid to obtain the desired volume fraction suspension. The powder and base fluid were blended by immersion in an ultrasonic bath for 3 h to break up any residual agglomerations and ensure that the suspension was well dispersed. Following this blending process, the suspensions were examined and found to be very stable, with essentially no sedimentation or stratification. All of the suspensions were made without changing the pH value.

6.3.1.1 Effect of Temperature and Volume Fraction on Thermal Conductivity

An experimental investigation was used to determine the effect of variations in the temperature and volume fraction on the steady-state effective thermal conductivity of the two nanoparticle suspensions. The experimental results were compared for different volume fractions and different mean sample temperatures. The results for aluminum oxide are shown in Figure 6.11, where the normalized thermal conductivity enhancement is an increasing function of temperature, for various volume fractions of Al_2O_3 in the suspension. It is important to note here that the thermal conductivity of Al_2O_3 as a bulk material decreases with increasing temperature, while the thermal conductivity of water increases with increasing temperature within the range of temperatures evaluated in the current experiment. As a result, it is clear that the effective thermal conductivity of the nanoparticle suspension is not a simple function of the combined thermal properties of the base fluid and the Al_2O_3 nanoparticles.

A number of trends are clearly apparent from these data. First, the rate of enhancement of the effective thermal conductivity at different volume fractions increases significantly with respect to increasing temperature, indicating a strong

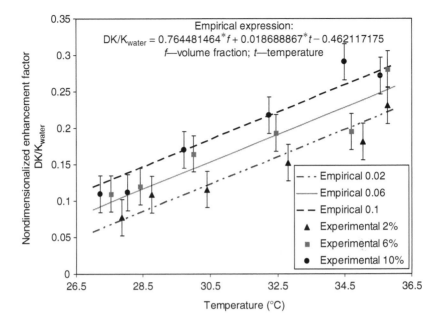

FIGURE 6.11 Comparison of the empirical expression for the nondimensionalized enhancement factor and the experimental data at different volume fractions as a function of temperature for Al_2O_3/water nanoparticle suspensions.

dependence on temperature. Additionally, at each volume fraction, the effective thermal conductivity increases with increasing temperature and the higher the volume fraction, the greater the increase in effective thermal conductivity.

The experimental data also indicate that the predictive model of Maxwell, shown here as Equation 6.15, is not valid for these types of nanoparticle suspensions.

$$\frac{k_{eff}}{k_f} = 1 + \frac{3(\alpha - 1)\phi}{(\alpha + 2) - (\alpha - 1)\phi} \tag{6.15}$$

where
    k_{eff} is the effective thermal conductivity of the suspension
    k_f is the thermal conductivity of the base fluid
    α is the ratio of the thermal conductivities of the nanoparticle and base fluid
    ϕ is the volume fraction of the nanoparticle suspension

Using the data, the rate of increase in the effective thermal conductivity for the Al_2O_3/water suspension can be expressed as a function of both the volume fraction and temperature. Here, a two-factor linear regression analysis was employed to obtain an empirical expression, shown as Equation 6.16 below, with an R-square value of 0.9171. A linear analysis was used here after studying the experimental data, which demonstrated a linear relationship between the nondimensionalized enhancement factor and both the temperature and the nanoparticle volume fraction.

Comparison of the two-factor linear regression analysis and the experimental data are in good agreement as also shown in Figure 6.11.

$$\frac{(k_{\text{eff}} - k_{\text{f}})}{k_{\text{f}}} = 0.7644\phi + 0.01869T - 0.46215 \qquad (6.16)$$

where the temperature T is expressed in degree Celsius.

It is quite interesting to note that the nanoparticle volume fraction has a coefficient which is almost 40 times greater than the temperature coefficient, indicating the relative importance of these two parameters.

The experiments were repeated for various volume fractions of CuO/water nanoparticle suspensions, and similar results were observed. Comparison of the two sets of data indicated that for the same volume fraction, the CuO/water nanoparticle suspension has a higher enhanced effective thermal conductivity as seen in Figure 6.12 (for a volume fraction of 2%). The reasons for these differences are not immediately clear, due to insufficient data on the bulk thermal conductivity of CuO and its variation with temperature, but observation did indicate that (1) the CuO nanoparticles used were much smaller than the Al_2O_3 nanoparticles (29 nm diameter vs. 36 nm diameter), (2) the CuO particles were nonspherical in shape unlike the predominantly spherical shape for the Al_2O_3 particles, and (3) for the same volume fraction, the CuO/water nanoparticle suspensions were much more viscous. Because of the interrelationship between these three factors, it is not clear which is most significant.

The two-factor linear regression analysis was also employed to develop an empirical relationship for the nondimensionalized enhancement factor and the

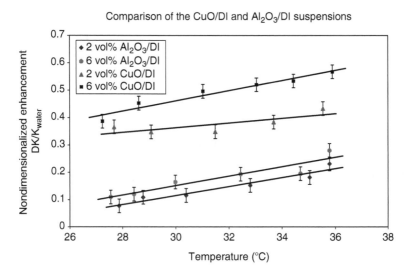

FIGURE 6.12 Comparison of the experimental data for 2% and 6% volume fraction Al_2O_3/water and CuO/water nanoparticle suspensions as a function of temperature.

experimental data at different volume fractions as a function of temperature for CuO/water nanoparticle suspensions. The result is shown as Equation 6.17 below and had an R-square value of 0.9078.

$$\frac{(k_{\text{eff}} - k_{\text{f}})}{k_{\text{f}}} = 3.761088\phi + 0.017924T - 0.30734 \qquad (6.17)$$

Figure 6.13 shows the comparisons between the experimental data on the Al_2O_3 nanoparticle suspension and the experimental data reported by Das et al. (2003b), which were obtained using the temperature oscillation method. Both of the experiments show that at a given volume fraction, the enhancement of the effective thermal conductivity increases with the increase of temperature, and the higher the volume fraction is, the sharper is this increasing trend. Similarly, at a given bulk temperature, the enhancement of the thermal conductivity increases with the increase of the volume fraction, and the enhancement rate becomes higher at larger volume fractions. However, these two sets of experimental data exhibit some differences in the magnitudes of the enhancement under various test conditions. The measured enhancement values of CuO nanoparticle suspensions also have shown differences on comparison with the data reported by Das et al. (2003b), where all of the experimental data on enhancement from the present experiments were found to be of higher magnitudes. The differences noticed in the experimental results are possibly due to the variations in particle size, shape, and the different experimental

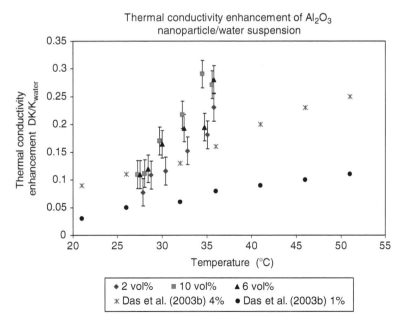

FIGURE 6.13 Comparison of the magnitude of thermal conductivity enhancement of Al_2O_3 nanoparticle suspension at different temperatures.

techniques utilized, which impose additional parameters which are not kept identical in the two experimental studies. However, all of the experimental results generally agree to the conclusion that the enhancement of thermal conductivity in nanofluids is a strong function of the volume fraction and the temperature level.

The results of the investigation clearly indicate that the effective thermal conductivity enhancement in nanoparticle suspensions is not simply a function of the thermal conductivity of the nanoparticles added to the base fluid, but is the result of one or more other mechanisms. The relative dependence on the volume fraction and temperature indicate that one possible explanation is the Brownian motion, which is closely related to both the volume fraction and increases with respect to temperature.

As indicated in the regression analysis, the linear relationship between the nondimensionalized enhancement factor is more strongly influenced by the volume fraction than the temperature. How exactly the volume fraction influences the enhancement is still unclear, but the mechanism of the effective thermal conductivity enhancement may be mainly due to a combination of the higher thermal conductivity of the nanoparticles, and the microconvection resulting from the motion of the particles in the base fluid. Jang and Choi (2004) and Prasher et al. (2005) have come up with two theoretical models which included the effects of Brownian motion of nanoparticles. In these two models, it has been considered that the Brownian motion of nanoparticles leads to a high heat transfer coefficient between them and the fluid. The Brownian motion of a large number of nanoparticles can have effects on the enhancement by stirring the base fluid particles around their mean positions, which in turn, enhances the heat transfer in the suspension. However, the effect of the Brownian motion has been considered insignificant by some investigators (Keblinski et al. 2002) stating that the thermal diffusion of the base fluid is much faster than the Brownian diffusion of the nanoparticles.

Another possible secondary order mechanism in the enhancement of the effective thermal conductivity of the nanofluid could be the higher thermal capacity of nanoparticle materials compared to that of the base fluid. In general, addition of nanoparticles to the base fluid is likely to produce two effects, one being the higher thermal properties of nanoparticle materials (higher thermal conductivity and higher heat capacity), and the other is the effect of the Brownian motion. This dual role of the nanoparticle in the base fluid requires additional study in order to more clearly understand the precise nature of the relationship of the temperature and volume fraction of nanoparticles on the effective thermal conductivity of suspensions.

6.3.1.2 Effect of Particle size on Thermal Conductivity

In order to investigate the effect of the particle size on the enhancement of the thermal conductivity in nanofluids, steady-state experiments were conducted using the cut-bar apparatus setup explained previously. Experimental studies were performed on suspensions of aluminum oxide nanoparticles with mean diameters of 36 and 47 nm. Tests were conducted over a temperature range of 27°C to 37°C for volume fractions ranging from 0.5% to 6.0%. The thermal conductivity enhancement of the two nanofluids demonstrated a nonlinear relationship with respect to temperature, volume fraction, and nanoparticle size, with increases in the volume fraction, temperature, and particle size, all resulting in an increase in the measured enhancement. The most significant finding was the effect that variations in particle size had on the effective thermal conductivity of the Al_2O_3/distilled water nanofluids. The largest enhancement difference observed, occurred at a temperature of approximately 32°C and at a volume fraction of between 2% and 4%. The experimental results exhibited a peak in the enhancement factor in this range of volume fractions for the temperature range evaluated, which implies that an optimal size exists for different nanoparticle and base fluid combinations. This phenomenon can be neither predicted nor explained using the theoretical models currently available in the literature.

It was noticed in Figures 6.14 and 6.15 that the 36 nm diameter Al_2O_3 nanofluids have a higher thermal conductivity enhancement than the 47 nm diameter Al_2O_3 nanofluids at every volume fraction and temperature, with the only exception being the enhancement at 28°C and a volume fraction of 4%. Figure 6.14 indicates that at a temperature of approximately 28.0°C, the enhancement for each of the nanoparticle sizes evaluated is relatively small for all volume fractions, while for increases in the bulk temperature to 30.5°C or 35.5°C, the 36 nm diameter Al_2O_3/DI nanofluids show increasingly higher thermal conductivity enhancements than the 47 nm diameter Al_2O_3/DI nanofluids for all volume fractions. Figure 6.15 shows that at 0.5% volume fraction, the differences between the thermal conductivity enhancement for the two sizes of nanofluids are very small at all temperatures. With changes in the volume fraction, the enhancement changes as a function of both the temperature and the particle size.

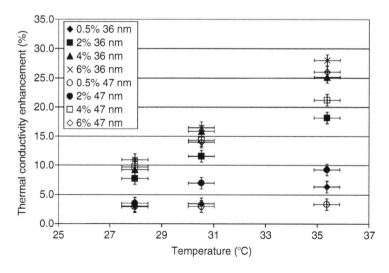

FIGURE 6.14 The thermal conductivity enhancement comparison between 36 and 47 nm diameter Al_2O_3/DI nanofluids versus temperature at different volume fractions.

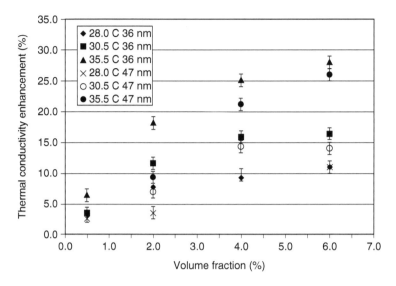

FIGURE 6.15 The thermal conductivity enhancement comparison between 36 and 47 nm diameter Al_2O_3/DI nanofluids versus volume fraction at different temperatures.

6.3.1.3 Transient and Steady-State Experimental Methods for Nanofluids

In order to examine whether the measured enhancement in the thermal conductivity of nanofluids is independent of the measurement technique, comparison has been made between measured effective thermal conductivity values obtained using a steady-state method (the cut-bar apparatus) and a transient method (hot-wire method). The relevant observations made at the Two Phase Heat Transfer Laboratory at the University of Colorado at Boulder are presented in this section (Li et al. 2008).

The transient hot-wire (THW) experimental facility employed here is the one described by Williams (2006). The THW is essentially made from a fine wire placed in a cell consisting mainly of a platinum/Isonel-coated wire ($\rho = 21.45$ g/cm^3, $k = 71.6$ W/m K, $c_p = 0.1325$ J/g K) with a bare diameter of 25 μm (28 μm with the insulating coating). The length of the wire used is variable between 25 and 40 mm. The supporting leads are two tantalum rods to keep the wire straight and to connect it to the electrical system; these rods are also electrically insulated. The electrical system is composed of a current source (Keithley-6221) and a nanovolt-meter (Keithley-2182A) to collect the voltage data: Using these instruments is a key factor in the development of this facility, because the variations in the resistance of the wire are usually quite small ($<$mV), making it extremely important to have a voltage meter with a very high degree of sensitivity. The vessel is made of stainless steel with an external diameter of approximately 2.5 cm and it is placed in a larger container that can be used to circulate water from a thermostatic bath in order to control the temperature at which the experiments are performed. The current source and the voltmeter provide the possibility of using a remote control via a GPIB port utilizing a MATLAB code to interface the instruments with a PC. One parameter of particular importance is the data acquisition rate, because it is important to have very rapid measurements to prevent the onset of convection. The range of acquisition rates for these instruments should go from 0.01 to 60 NPLC, where NPLC means "Number of Power Line Cycles" and 1 NPLC equal to 16.7 ms. In the measurements NPLC $= 0.1$ is used, which corresponds to an acquisition rate of about 17 ms. Each measurement takes around 20 s, in this partition. In the first 5 s the system acquires the voltage signal at a very low input current (usually 1 mA) in order to measure the resistance of the wire before the heating (R_0). In this way, using the resistance–temperature relationship, the temperature of the sample is known and compared to the thermostatic bath. In the last 15 s the input current switches to higher values (in the range of 50–100 mA) and the wire starts heating; at the end the current source is turned off and the collected data are transferred to the software. In order to calibrate the measurement system, thermal conductivity of water and ethylene glycol at 25°C were measured which showed very good agreement with the data available in the literature (Yaws, 1999).

The Al_2O_3/DI-water nanofluid test samples were produced by first blending the 47 nm diameter spherical Al_2O_3 nanoparticle powder (Nanophase Inc., Illinois) with degassed distilled water (DI water), and ultrasonically oscillating it for 90 min. A set of different volume fraction Al_2O_3/DI-water nanofluids are generated through 0.5%, 2%, 4%, to 6%, and their effective thermal conductivities were tested with both the methods described above.

The normalized enhancement of effective thermal conductivity of the nanofluids is compared to pure DI water in Figure 6.16, both at room temperature 27.5°C. The differences between the data measured by the cut-bar method and the hot-wire method are all within the uncertainty except for the 4% volume fraction. This difference, however, may be the result of other factors of the testing techniques. The average of repeated tests could eliminate this difference. The experiments also illustrated that there is almost no difference between the two methods regarding the time or technique used to measure the effective thermal conductivity of nanofluids.

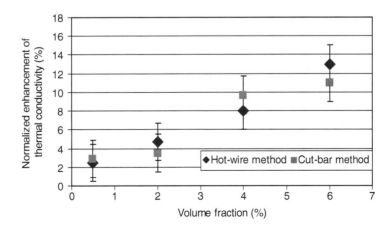

FIGURE 6.16 The comparison of the normalized enhancement measured by the transient method and steady-state method.

Because the samples were prepared identically, the effects of the ultrasonic vibration time, the size distribution of the nanoparticles, any possible multibody interactions, nanoparticle's shape, and the interfacial phenomena or other thermal property effects that might contribute to the effective thermal conductivity of the nanofluids, were minimized.

The comparison between the measured values of the sample set of $Al_2O_3/$ DI-water nanofluids samples shows good agreement between the transient hot-wire method and steady-state cut-bar method at room temperatures. The results confirm that the observed enhancement of the thermal conductivity of Al_2O_3/water nanofluids exists and is independent of the measurement technique.

6.3.2 Viscosity Measurements

The viscosity of a fluid, which accounts for the internal resistance to flow, is a manifestation of two phenomena—the intermolecular force of cohesion and the molecular momentum exchange. In the case of liquids, the intermolecular cohesion is more predominant even though the molecular momentum exchange also occurs. The viscosity of a fluid can be altered by the addition of various types of additives. It has been observed that nanoparticles of dialkyldithiophosphate-modified copper in liquid paraffin form a protective film that reduces the wear (Zhou et al. 1999). Wu et al. (2007) examined the tribological properties of lubricating oil with CuO, TiO_2, and nano-diamond nanoparticles as additives. With the addition of the copper oxide nanoparticles into the lubricating oil, the friction reduction and antiwear properties were enhanced. Ye et al. (2002) observed that $NiMoO_2S_2$ nanoparticles as additive possess good antifrictional performance with a smooth transition from fluid film lubrication at low temperature to solid film lubrication with elevated temperature, which could fit the needs of lubrication with large temperature ranges. Tribological studies conducted by Hu et al. (2002, 2000) using 10 nm-sized magnesium borates as additives in oil with sorbitol monostearate as the dispersing agent showed that by increasing the amount of nanoparticles, the nonseizure load increases up to a particular level. Friction coefficient also showed a decrease with a 0.65% by weight of magnesium borate in oil along with dispersing agent sorbitol monostearate. Sunqing et al. (1999) studied the tribological properties of spherical and cylindrical CeF_3 nanoparticles 25 nm in size dispersed in 500SN base oil along with dispersants T154, calcium, methyl-tricaprylamine chloride, etc. The dispersants were found to affect the load carrying capacity. Varying the nanoparticle weight, it was found that the load carrying capacity increases sharply when its weight is below 1%. The antiwear property was also affected by the dispersants with the trend similar as in nonseizure load test, but it was found to decrease with the increase in the dosage of CeF_3 nanoparticles. Tarasov et al. (2002) studied the effect of nano-sized copper additives in motor oil on friction, wear and load carrying capacity. It was reported that copper additives reduce friction and wear and increase load carrying capacity. The mechanism was explained as the nanoparticles replacing or covering the real wear particles with those of a more ductile metal. This effect is nullified at higher sliding speeds due to the subsurface layer of nanoparticle acting as a viscous liquid resulting in smoother surface and hence more contact, but at higher loads plastic deformation results in a rippled surface. Hu and Dong (1998a) found that wear resistance and load carrying capacity of 500 SN base oil were increased and the friction coefficient was decreased when 10–70 nm-sized titanium borate was added to the oil. This improvement of antiwear resistance was reported to be due to the deposition of nanometer-sized titanium borate on the wear scar. Duan and Lei (2001) studied the effect of particle size on lubricating properties of colloidal polystyrene in a water-based fluid and reported that overall there was improvement in antiwear and load carrying capabilities but the difference due to particle size on antiwear ability is small, even though better load carrying capability was shown by larger particles. Chen et al. (1998) studied the effect of the effects of the addition of DDP (dialkyl-dithiophosphate)-coated PbS nanoparticles in liquid paraffin and found that the antiwear ability improved even at very low concentrations with the optimum value at 0.05 percentage by weight of DDP-PbS. This improvement was attributed to the formation of a boundary film on the steel surface. Xue et al. (1997) carried out studies on friction and wear properties of a 2-ethylhexoic acid (EHA) surface modified using TiO_2 as an additive in liquid paraffin and found good improvements in these properties. It was suggested that an organic-coated nanoparticle will be effective as a lubricant for steel in both high and low loads, due to the formation of a boundary film consisting of the organic acid at low loads, while at higher loads the inorganic core of TiO_2 nanoparticles will form the boundary film with the latter more effective. Zhou et al. (2001) studied the effect of organic surface-coated LaF_3 nanoparticles in liquid paraffin on its antiwear and load bearing capacity and found that it resulted in improvement of these properties better than when zinc di-iso-amyl-octyl-dithiophosphate (ZDDP) alone is added to liquid paraffin. Boundary film on the worn surface consisted of LaF_3 nanoparticles depositing film and tribochemical reaction film of the elements S and P from the organic compound. The tribological properties of 10 nm average-sized Ni particles as additives in oil was studied and found to offer considerable improvement in wear, friction, and load characteristics over the base oil up to a concentration of 1%. Further addition degraded the tribological properties and this was reported as due to the formation of larger particles during tribotesting with the maximum size reaching 50 nm. Hu and Dong (1998b) studied the tribological properties of 500 SN oil with 20 nm particles of titanium oxide, and found that the wear resistance and load carrying capacity were increased and the friction coefficient decreased. Time dependent tests on base oil and oil with additives showed that the wear scar diameter and friction coefficient did not improve during the initial stages, but after a small time delay. The improvement shown later was explained, due to the deposition of nanoparticles on the rubbing surface.

Experimental investigations have been reported in the literature by Sajith et al. (2006) on the influence of the addition of cerium oxide nanoparticles on the viscosity of diesel, which indicated an increase in viscosity with an increase of the concentration of nanoparticles in the range 20 to 50 ppm.

In the convective heat transfer in nanofluids, apart from the thermal conductivity of the base fluid, the heat transfer coefficient depends highly on the viscosity, density, and specific heat. The kinematic viscosity is a property given by the ratio of the dynamic viscosity and density, and thus represents the combined effects of these two individual properties. The kinematic viscosity of nanofluids also might depend on the methods used for dispersion and stabilization of nanoparticles in the suspension, apart from the material and the volume fraction of the nanoparticles.

Results of experimental studies conducted at the Nanotechnology Research Laboratory, National Institute of Technology, Calicut, India, to measure the variations in kinematic viscosity due to nanoparticle addition in lubricating oils, as part of ongoing research activities on hydrocarbon liquids, will be presented here. The additives used in the lubricating oil were aluminum oxide and copper oxide in the nanoparticle form. The average size of the nanoparticles had a range of 30–40 nm and the bulk density was 0.260 g/cm^3. The lubricating oil used was Servo SAE 20W-40, which is commonly used for two stroke engines. The required quantity of the nanoparticles was accurately weighed using a precision electronic balance and mixed with the lubricating oil. An ultrasonic shaker was used for the preparation of the nanofluid by mixing the nanoparticle additives in the lubricating oil. The time of agitation was fixed as 30 min, based on past experience in producing a stable suspension with sufficient time for sedimentation to begin. The nanofluid sample was used in the experiments immediately after preparation, without providing considerable time for sedimentation to set in, in order to get good results. The kinematic viscosity of the oil was measured using a standard Redwood viscometer.

The experimental results showed that the kinematic viscosity of the oil is slightly influenced by the addition of the nanoparticles. Figure 6.17 shows the variation of the viscosity of the engine oil with different volume fractions of aluminum oxide nanoparticles, in a temperature range from 30°C to 65°C. Similar variations were obtained in the case of copper oxide nanoparticles also (Sajith et al. 2007). The kinematic viscosity was observed to increase with the addition of aluminum oxide nanoparticles up to a dosing level of 0.02% by volume and then to decrease with further nanoparticle addition. The same trend was observed at all of the temperatures, ranging from 30°C to 70°C. In the case of copper oxide a slight decrease in the kinematic viscosity was observed above 0.01% volume fraction at all temperatures ranging from 30°C to 70°C. One of the reasons for the reduction in

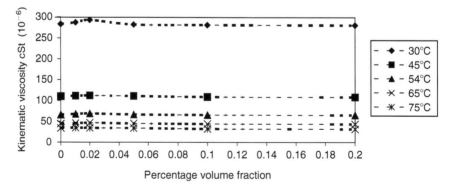

FIGURE 6.17 Variation of the kinematic viscosity with the volume fraction of aluminum oxide nanoparticles, at different temperatures.

kinematic viscosity at higher volume fractions of the nanoparticles could be the increase in the effective density of the suspension by the addition of the nanoparticles. When the relative increase in density overtakes the increase in absolute viscosity of the medium, there is a resulting fall in the kinematic viscosity, which is the ratio of the viscosity and the density; another possible reason could be the agglomeration of the nanoparticles at high dosing levels, which leads to the settling of the nanoparticles and subsequent decrease in the improved viscosity of the oil due to nanoparticle addition. Interestingly, either of these contradictory effects would lead to a decrease in the value of the kinematic viscosity of the fluid.

6.3.3 SURFACE TENSION

The rate of evaporation of a liquid is known to be inversely proportional to the surface tension. Experiments conducted by Armentadeu (1997) have established a relationship between the latent heat of evaporation and the surface tension and shown that in order to increase the rate of evaporation, the surface tension has to be decreased. As surface tension is one of the factors that affects evaporation rate, investigations on the effect of the nanoparticles on the surface tension is of importance. The experiments conducted by Das et al. (2003a) show that there is nominal surfactant effect with the addition of the particles to the fluid.

Surface tension measurements have been carried out in suspensions of aluminum oxide nanoparticles in water (average size 30–40 nm and bulk density 0.260 g/cm^3), as part of ongoing research at the Nanotechnology Research Laboratory, National Institute of Technology, Calicut, India. The experiment involved the measurement of capillary rise and contact angle in a liquid column contained in a vertically oriented capillary tube. The contact angle was measured from close-up photographs taken, using a high-resolution digital camera. The surface tension was calculated from the measured values as follows:

$$\sigma = \frac{hwd}{4\cos\theta} \tag{6.18}$$

where
  w is the specific weight of the liquid
  h is the capillary rise
  d is the diameter of the capillary tube
  θ is the contact angle

Measurements were carried out on suspensions with the aluminum oxide concentration varying from 1% to 4% (by volume). Table 6.1 gives the experimental results, which indicate that the surface tension is generally increased, due to the addition of nanoparticles, though no definite trend could be identified corresponding to the concentration of nanoparticles. The modification of the surface tension of fluids by the addition of metallic oxide nanoparticles could be attributed mainly to the changes in the bonds between the molecules of the surface, but will also be influenced by the surface geometry in the presence of the nanoparticles which makes the enhanced values the result of complex physical interactions in the fluid.

TABLE 6.1

Surface Tension of Nanofluids

Liquid	Surface Tension (N/m)
Distilled water	0.0733
1% Al_2O_3	0.0784
2% Al_2O_3	0.0762
3% Al_2O_3	0.0751
4% Al_2O_3	0.0784

6.3.4 ONSET OF NATURAL CONVECTION

Natural convection can find applications in various fields such as small energy conversion systems, thermosyphon circulating loops, and other situations where it is desirable to avoid moving parts. Such systems can also find use in electronics cooling where relatively low heat transfer rates are common. The natural convection heat exchange process is governed by the magnitude of various thermophysical properties of the fluid medium such as the coefficient of volumetric expansion, density, viscosity, specific heat, and thermal conductivity, other than the temperature potential and physical dimensions of the fluid domain (Holman 1989). The effects of all these are usually put together into a nondimensional group, the Rayleigh number, which decides the pattern of the natural convection field in the vicinity of a dissipating surface. The Rayleigh number is defined as

$$\mathrm{Ra} = \mathrm{Gr}\,\mathrm{Pr} = \left(\frac{g\beta\Delta T L^3}{\nu^2}\right)\left(\frac{C\mu}{k}\right) \qquad (6.19)$$

The key transition points in natural convection, namely the onset of natural convection and the transition to turbulence are determined by the magnitude of the Rayleigh number. In a transient process, the critical value of the instantaneous Rayleigh number can be taken as the criterion for these transitions.

The time required for natural convection to set in (the onset time) in a stagnant liquid domain, heated from an initial temperature level, provides important information necessary for the design of transient natural convection systems. One example is a transient operating device from which heat is dissipated to a surrounding fluid. In such a situation, it would be important that convection sets in as early as possible, as the mechanism of heat transfer before the onset will be pure conduction with limited heat dissipation rates due to relatively low thermal conductivities of the cooling fluid. It is of engineering interest to control the onset time, and one effective means to achieve this may be by introducing nanoparticles into the fluid, to form stabilized and well-dispersed nanofluids, which alters the thermophysical properties, and therefore, the temperature level at which the critical Rayleigh number is reached. However, as all thermophysical properties are affected independently, it is necessary to determine how the addition of nanoparticles and their concentration effects the convection onset time.

Experimental studies have been undertaken to determine the onset time for free convection in nanofluid samples with water as the base fluid, utilizing oxides of copper, aluminum, silicon, and cerium. These experiments were conducted at the Nanotechnology Research Laboratory, National Institute of Technology, Calicut, India. The nanoparticles used in the experimentation were of the size range 30–50 nm, and were supplied by M/s Sigma Aldrich. The bulk density of aluminum oxide, copper oxide, silicon dioxide, and cerium oxide nanoparticles used in the investigations were 0.260, 0.190, 0.150, and 0.340 g/cm^3, respectively.

The basic experimental setup was originally proposed by Venkateshan (1990) for determination of thermal conductivity and thermal diffusivity of liquids through the measurement of convection onset time, and has also been used to perform parametric studies on the onset of natural convection in liquid cells (Srinivasulu et al. 1997). This essentially consists of a liquid cell bounded by a thin copper foil at the bottom and heated using an infrared lamp, thermocouples and a PC-based Agilent Benchlink data logger to obtain the transient temperature variation of the foil and the liquid cell. The schematic of the experimental arrangement is shown in Figure 6.18. The onset time is obtained by monitoring the transient variation of the temperature difference between the thin copper foil and the fluid temperature at a small distance away from the foil, which is taken as the reference temperature (temperature excess of the foil with respect to the fluid). The onset condition is indicated by a sudden change in the slope of the temperature–time plot (corresponding to the instantaneous dip as convection sets in), as shown in Figure 6.19, which shows the temperature response of the metal foil in a liquid cell with pure distilled water, under a particular heating level, giving the convection onset time for that case equal to 36 s. The measured temperature difference in the experiments can be used to obtain the onset Rayleigh number if the thermophysical properties of the fluid are

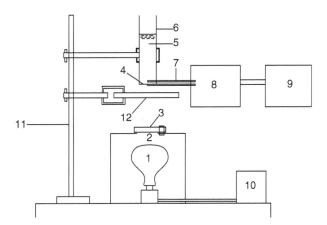

FIGURE 6.18 Schematic diagram of the test facility to obtain the convection onset time. 1. Infrared lamp. 2. Aperture. 3. Shutter. 4. Thin copper foil. 5. Liquid column. 6. Glass tube (test cell). 7. Thermocouples. 8. Data logger. 9. Computer. 10. Variac. 11. Stand. 12. Glass plate.

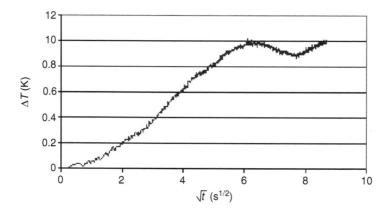

FIGURE 6.19 Typical transient response of the temperature excess of the foil showing the convection onset corresponding to a set of experimental conditions.

known. Such a calculation was used to test the accuracy of the setup by using pure water as the test fluid before conducting tests on the nanofluids.

The experiments on nanofluids with different nanoparticle materials have shown some interesting results. Figure 6.20 shows the variation of the natural convection onset time as a function of the volume fraction, under identical heating conditions, for the four different materials used. As shown in the graph, the nanoparticle material has a significant influence in the trends of variation of the onset time. It is found that in the case of copper oxide, aluminum oxide, and silicon dioxide, earlier onset is

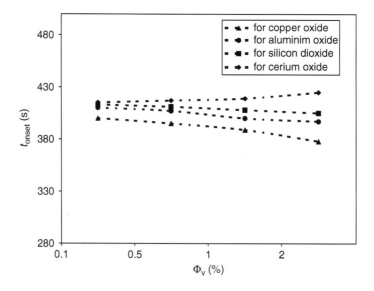

FIGURE 6.20 Variation of the natural convection onset time as a function of the volume fraction, under identical heating conditions, for the four different materials.

observed with the addition of nanoparticles compared to the base fluid (the onset time for pure water in the experimental cell was found to be 415 s), and this becomes progressively earlier as the volume fraction of nanoparticles is increased, as shown in the graph. In the case of cerium oxide, the addition of the nanoparticles is found to progressively delay the convection onset time.

These experimental results indicate that the nanoparticle material has a significant role in deciding the convection onset time, as expected from the definition of the Rayleigh number. Onset of natural convection depends upon the Rayleigh number which is greatly affected by the fluid properties and is a function of the temperature drop in the fluid, which is in turn determined by the thermal conductivity. Thus, due to the individual changes in thermophysical properties (thermal conductivity, viscosity, specific heat, and density) and the influence of the thermal conductivity on the temperature drop, the nanofluid Rayleigh number is altered in a natural convection situation, and the onset time is affected due to the complex interaction of all of the altered phenomena, making the convection onset either early or delayed, depending on the nanoparticle material among other parameters.

One of the problems associated with nanofluids is the effect of sedimentation of nanoparticles, which alters the properties of the suspension as time progresses. In the experiments, samples of nanofluids were tested after allowing various time periods after preparation, apart from varying the percentage concentration of the nanoparticles. The effects of these parameters on the convection onset time were studied for different nanoparticles, and a typical set of results for copper oxide are shown in Figure 6.21. The influence of sedimentation on the convection onset is clearly seen from the graph presented.

Additional discussions of the various experimental studies of natural convection and its onset under various heating conditions in polymer nanofluids, particularly where the convection onset is delayed due to the addition of nanoparticles, are presented in later sections.

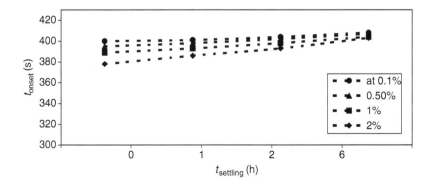

FIGURE 6.21 Variation of the onset time with respect to the volume fraction of copper oxide nanoparticles, and the time allowed for sedimentation after preparation.

6.3.5 Forced Convection in Heat Exchangers

Conventional enhancement methods employed to enhance heat exchanger perform-ance usually lead to alterations in design and operating conditions. A means of overcoming this limitation may be the enhancement of the thermal properties of heat exchange fluids. Suspensions of solid particles, including nanoparticles, in engineer-ing fluids have been used, which have shown improved thermal conductivity com-pared to base fluids. Heat transfer coefficient in forced convection has been reported as increasing with the addition of nanoparticles (Choi 1995a,b). Unlike conduction or forced convection, some natural convection experiments have shown decrease in heat transfer with the addition of nanoparticles (Keblinski et al. 2005).

The addition of nanoparticles in the working fluid influences its different ther-mophysical properties to different rates, which interact together and bring about a change in the performance of a heat exchanger. Discussions are presented here on an experimental study conducted at the Nanotechnology Research Laboratory, National Institute of Technology, Calicut, India, in which the overall enhancement of the heat transfer in the heat exchanger, due to the addition of nanoparticles in water has been investigated. The addition of nanoparticles is expected to increase the effectiveness of the heat exchanger as the suspended particles increase the surface area and the heat capacity of the fluid, apart from introducing an enhancement in thermal con-ductivity. A concentric pipe heat exchanger has been studied, aiming at exploring the use of nanofluids as the working fluid to achieve enhanced heat transfer performance.

Commercially available nanoparticles of aluminum oxide with sizes ranging from 40 to 47 nm and having a density of 0.26 g/cc, and 33 nm copper oxide particles with a density of 1.25 g/cc were used to prepare the nanofluids. Standard ultrasonic agitation methods were used, with an agitation time of 20 min. The prepared fluid was used immediately, so that sedimentation did not occur. Nanofluid samples of various concentrations were used in the experiments.

A double pipe counterflow mini heat exchanger with the hot fluid flowing through the inner tube and the cold fluid through the annulus was used in the experiments to determine the effectiveness. The experimental setup consisted of a pump for circulat-ing the nanofluid, a rotameter for measuring the mass flow rate of the hot fluid, a heating element for maintaining constant inlet temperature, and a valve for controlling the mass flow rate. A heating coil was wound over the copper tube for heating the fluid flowing through the circuit. The input power to the heating element and hence the inlet temperature of the hot water was varied by means of a variable transformer. A centrifugal pump was used for circulating the hot fluid in the circuit. The mass flow rate of the hot fluid was measured by means of a rotameter and that of the cold fluid by measuring the time of collection of water for a specific period. The specifications of the heat exchanger are given in Table 6.2 and a schematic of the setup is shown in Figure 6.22.

Base line tests on the heat exchanger were done with water as the base fluid at different inlet temperatures of the hot fluid. Then, the experiments were conducted with the nanofluid as the hot fluid inside the inner tube. Experiments were performed with two concentrations in each of the nanofluids—0.13% and 0.26% by weight of the aluminum oxide and copper oxide nanoparticles, respectively, suspended in water.

TABLE 6.2
Description of the Heat Exchanger
Experimental Setup

Inner tube diameter	8.5 mm
Outer tube diameter	30 mm
Length of the heat exchanger	69.6 cm
Pump capacity	50 W, 1000 LPH
Heating coil capacity	1000 W
Thermocouples	T-Type

The mass flow rates of the two fluids were kept constant. The hot fluid inlet temperature was varied from 50°C to 70°C by varying the input power to the heater element. The inlet and outlet temperatures of the hot and cold fluids were measured using calibrated thermocouples. The effectiveness of the heat exchanger for the different concentrations of the nanoparticles at different temperatures was determined as follows:

$$\text{€} = C\text{c}(T_{\text{co}} - T_{\text{ci}})/C_{\text{min}}(T_{\text{hi}} - T_{\text{ci}}) \tag{6.20}$$

where
T_{hi} is the hot fluid inlet temperature
T_{ci} is the cold fluid inlet temperature
T_{co} is the cold fluid outlet temperature
Cc is the heat capacity of cold fluid
C_{min} is the minimum heat capacity

The inlet and exit temperatures of the fluids were measured using T-type thermocouples. The calculated effectiveness of the heat exchanger is subject to the error in the temperature measurements which amounts to ±0.5°C for each thermocouple (Eckert and Goldstein 1976). This was found to produce an overall

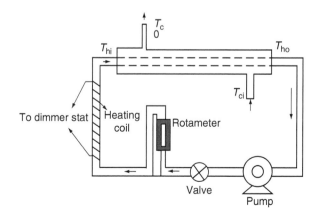

FIGURE 6.22 Schematic of the setup for experimental study of heat exchanger.

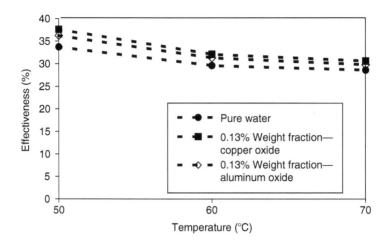

FIGURE 6.23 Effectiveness of the heat exchanger with the hot fluid containing aluminum oxide and copper oxide, compared with that of pure water as the hot fluid.

error of ± 0.03 in the estimation of the effectiveness of the heat exchanger as indicated by an error calculation, in the temperature range used.

The effectiveness of the heat exchanger with the nanoparticle suspension as the hot fluid was compared with the value using pure water as the hot fluid, for otherwise identical conditions. Experiments were conducted in a range of temperatures between 50°C and 70°C, with the two nanofluids, containing various concentrations of aluminum oxide and copper oxide nanoparticles. The effectiveness of the heat exchanger was observed to increase with increases in the concentration of the nanoparticles in the suspension. Typical results are shown in Figure 6.23 and Figure 6.24. Similar trends were obtained for aluminum oxide and copper oxide nanofluids (Rao et al. 2007).

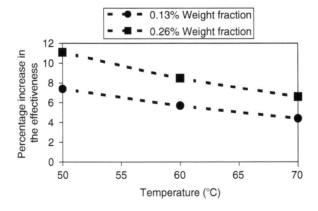

FIGURE 6.24 Variation of percentage increase in effectiveness of the heat exchanger with temperature for aluminum oxide nanoparticles in the hot fluid.

The convective heat transfer in heat exchangers is dependent on different thermophysical properties of the fluid, and their local variations in the passages— the density, thermal conductivity, specific heat, and viscosity of the fluid. Various reasons have been identified for the enhancement of the thermal conductivity in fluids. These include increases in surface area with the addition of nanoparticles, intensification of the interaction among particles, fluid and fluid passage, local turbulence, and molecular level interactions. In general, the enhancement in the effectiveness of the heat exchanger is understood to be an interactive effect of the individual changes in the various thermophysical properties, due to the addition of nanoparticles in the base fluid.

6.3.6 PHASE CHANGE HEAT TRANSFER

Evaporation is basically a surface phenomenon, which takes place at any temperature, when the surface molecules of the liquid possess enough energy to overcome the work function of the surface. The factors which influence evaporation include the temperature of the fluid, ambient temperature, relative dampness, composition and concentration of the surfactants, interfacial and surface tension, and heat exchange by convective processes (Al-Shammiri 2002, Armentadeu 1997).

An experimental study aimed at investigating the effects of inclusion of nanoparticles in suspended form, on the rate of evaporation and surface tension of water, has been carried out at the Nanotechnology Research Laboratory, National Institute of Technology, Calicut, India. The rate of evaporation was determined by measuring the reduction of the weight of a beaker carrying the nanofluid, due to evaporation. A high precision electronic balance with a sensitivity of 0.0001 g was used for determining the weight measurement. A lightweight tripod stand made of zinc and tin was kept on the electronic balance and a glass beaker containing 50 ml of water, alumina nanofluid was placed over it. A 100 W nichrome heating coil was used for heating the fluid. A T-type thermocouple, inserted into the beaker and connected to a data logger was used for continuous measurement of temperature of the fluid. Ambient conditions during the experimentation were 25°C and 60% RH. Experiments were conducted using pure water and water–alumina nanofluid with 1%, 2%, 3%, and 4% volume fraction of alumina nanoparticles. Figure 6.25 shows the variation of the loss of weight of the nanofluid with time at different temperatures and concentration levels. It is evident that the addition of nanoparticles to water decreases the evaporation rate compared to that of water. Experimental results show that especially at lower temperature levels, the addition of nanoparticles drastically reduces the evaporation rate compared to higher temperature levels. However, no clear functional relation is observed in the rate of evaporation with the dosing level of nanoparticles.

Distillation experiments can also be used to obtain information about evaporation rates in fluids, and such experiments also have been conducted. The ASTM distillation curve, which is a plot between temperature and percentage recovery of the condensate, indicates the nature of evaporation of the fluid under consideration. When pure water is distilled in this apparatus, most of the condensate recovery takes place at the boiling point of water itself. A standard ASTM apparatus was used for

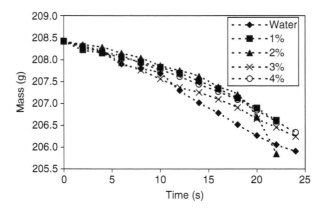

FIGURE 6.25 Loss of weight in an evaporating mass of fluid with time for pure water and for nanofluids with different volume fractions of aluminum oxide (values representing the concentrations at the beginning of evaporation). The graph indicates that the rate of evaporation is reduced by the addition of nanoparticles.

performing the distillation of 100 ml of the test fluid. The test fluid was gradually heated by means of a 1 kW heater and the temperature corresponding to every 10% condensate recovery was noted and plotted to obtain the distillation curve. Mass flow rate of cooling water in the condenser was set as 80 LPH.

Figure 6.26 shows the standard ASTM distillation curves for pure water and nanofluids with various percentage concentration (by volume) of suspended aluminum oxide nanoparticles. The graphs show the variation of the temperature of the fluid with the percentage of condensate collected. The results show that with the addition of nanoparticles, the percentage of condensate collected at a particular temperature decreases, which indicates a decrease in the evaporation rate.

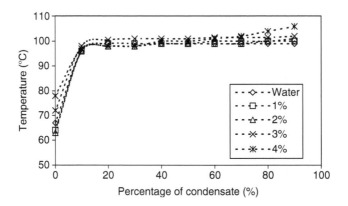

FIGURE 6.26 Percentage of condensate collected plotted against the temperature (ASTM distillation curve) for pure water and for nanofluids with various concentrations of aluminum oxide nanoparticles.

It should be noted that as the evaporation proceeds, the concentration of nano-particles in the base fluid increases and the concentration values shown in the graphs are those at the beginning of the evaporation of the nanofluid at room temperature.

6.3.6.1 Boiling

Investigations have been reported in the literature to characterize the boiling perform-ance of liquids in the presence of nanoparticles. Pool boiling experiments conducted in water–alumina nanofluids by Das et al. (2003a) have shown deterioration in boiling performance, at all levels of nanoparticle concentration. This observation was against the expectations, because the only unfavorable change in fluid property with the presence of nanoparticle is an increase in the viscosity. Latent heat and surface tension are almost unaffected and thermal conductivity enhances considerably due to the addition of nanoparticles, and all these factors are expected to enhance the boiling performance. The authors attributed the anomalous deterioration of boiling perform-ance to the change in surface characteristics due to the deposition of nanoparticles.

Pool boiling experiments by Bang and Chang (2005) have also confirmed the deterioration in boiling performance in nanofluids. The maximum deterioration was noted for 2% and 4% volume fractions. Longer natural convection stage and delayed nucleate boiling have been cited as major possible reasons. Nucleate boiling activity was found to be less intensive in water-based nanofluids, compared to that in water. The effect was also attributed to the Brownian motion of particles and subsequent agglomeration and attachment to the heater surface, which produces a situation similar to fouling and a change in surface roughness. Accordingly, if the original surface roughness is smaller than the nanoparticle size, it will increase further with the addition of nanoparticles, thereby providing more nucleation sites and intensified nucleate boiling. If the original surface roughness is larger than the nanoparticle size, it will decrease upon the addition of the nanoparticles, which is unfavorable for boiling heat transfer.

Vasallo et al. (2003) have performed experiments in silica–water nanofluid which showed no enhancement in nucleate boiling regime, but gave a significant enhancement in the critical heat flux. They also attributed the increase in the surface roughness of the heater wire, due to the coating of nanoparticles as a possible reason for critical heat flux enhancement. Enhancement in critical heat flux in nanofluids has been widely accepted, but efforts to quantify this enhancement have not given conclusive results.

Experiments were conducted at the Nanotechnology Research Laboratory, National Institute of Technology, Calicut, India to identify the boiling characteristics of nano-fluids and determine the variation of the heat flux as a function of the superheat (temperature excess of the heater above the saturation temperature). These experiments consisted of measurement of a nichrome-wire heater temperature and the fluid tempera-ture utilizing the power dissipation as the control variable. A 200 ml glass beaker was used as the test cell, with galvanized iron electrodes. A 4 cm length of 40 SWG nichrome wire was used as the heating element. The power input for heating was measured by a programmable digital multimeter. An autotransformer was used to vary the electric current through the nichrome wire in order to control the power dissipation. Using this setup, the pool boiling characteristic of pure water at saturation temperature was first obtained. Further experiments were conducted on nanofluids, in order to obtain the results presented, based on comparison with the results on water.

Experiments were conducted with nanofluids containing two materials (aluminum oxide and copper oxide) with various volume fractions, and corresponding boiling characteristics were plotted. As mentioned previously, these experiments used independent control of the input power to the heating element (power-controlled experimental method) which is easy to implement. The boiling characteristics were also used to identify the burnout heat flux of the heating element, which gives a fairly good indication of the critical heat flux, and to obtain the addition of the effects of the nanoparticle.

In order to understand the effects of the preparation time (sonication/agitation time) in making the nanoparticle suspension on pool boiling, a few experiments were carried out with various times used for the ultrasonic shaking process. Another feature studied in the experimental work was the effect of sedimentation of nanoparticles. For this, samples of nanofluids were tested after definite periods after preparation, to obtain the boiling characteristics. The effects of these parameters on the critical heat flux were also deduced by observing the burnout heat flux in these cases. Representative results of the experiments are presented and discussed below.

Figure 6.27 illustrates the boiling characteristics of the nanofluid with aluminum oxide nanoparticles, at various volume fractions. Similar trends have also been observed for nanofluids with copper oxide nanoparticles also. From the figure, it is clear that the presence of nanoparticles brings in differences in the pool boiling performance of the base fluid. A rightward shift of the boiling curve was also observed with the addition of the nanoparticles, which was noticed in both the nanofluids. This rightward shift of the boiling curve indicates that for a particular temperature level of the heating element above the saturation temperature, the boiling heat transfer is reduced on addition of nanoparticles. This is not a favorable effect, which indicates a deterioration of the pool boiling performance.

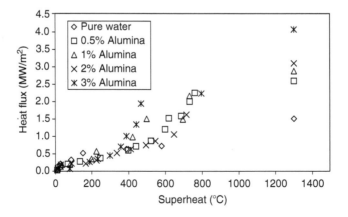

FIGURE 6.27 Boiling characteristics of aluminum oxide(Al_2O_3)/water suspension, obtained through power-controlled method.

6.3.6.2 Burnout Heat Flux

The burnout heat flux is an important factor in heater design. The heat flux corresponding to the burnout point of the heater wire can also be used to approximately understand the critical heat flux condition, though the actual critical heat flux occurs at an earlier superheat temperature on the boiling curve. Experiments performed using water, with different concentrations of nanoparticles, have shown a significant enhancement in the burnout heat flux, for both aluminum oxide and copper oxide nanofluids, as shown in Figure 6.28. Enhancement in the burnout heat flux was found to increase steadily with increasing volume fractions of nanoparticles.

6.3.6.3 Effects of Agitation Time and Sedimentation on Burnout

Two important aspects worth investigating in the study of nanofluids are the duration of the ultrasonic agitation process in preparing the nanofluid (termed agitation time here) and the time elapsed after preparation, when the nanofluid is actually used (termed sedimentation time). These two are found to affect the performance of the nanofluid. The mixing process determines the degree of dispersion of the suspended particles, and the level of agglomeration. Conversely, the concentration of the nanoparticles changes continuously, due to sedimentation, when the suspension is kept idle. The agitation process, in turn affects the effect of sedimentation also; the more well dispersed and nonagglomerated the particles are, it can be expected that the sedimentation process will be relatively slower.

In the investigations of the phase change phenomena in nanofluids described previously, the two factors described above have also been included as parameters in the study of the burnout point. Nanofluids prepared using ultrasonic agitation applied for various time periods have been tested. Also, samples of the prepared nanofluids have been tested after different time periods have elapsed after preparation. Typical experimental observations are presented here.

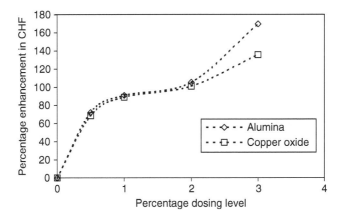

FIGURE 6.28 Variation of percentage enhancement of burnout heat flux with volume fraction of nanoparticles.

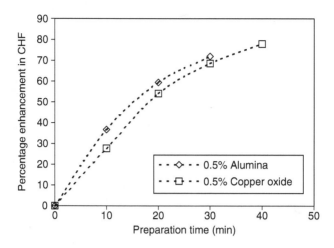

FIGURE 6.29 Variation of percentage enhancement in the burnout heat flux with preparation time for 0.5% volume fraction of nanoparticles.

The influence of the agitation time in preparing nanofluids on the burnout heat flux is interesting to note. The nanofluid was prepared using a standard ultrasonic shaker and the preparation time was varied in steps of 10 min. The experimental results show a strong dependence of enhancement of burnout heat flux on the agitation time which suggests that the critical heat flux also may increase steadily with increasing agitation time. Figure 6.29 illustrates the percentage enhancement of the burnout, in comparison with the base fluid. In water–aluminum oxide suspension, the enhancement in burnout heat flux was 25% when preparation time was 10 min, which increased to around 100% when the agitation time is increased to 40 min, signifying the influence of dispersion of the nanoparticles on the performance. Water–copper oxide suspension showed a similar trend. It is found that the suspension gets stabilized when the agitation time is increased, and so the relative improvement with respect to an increase in agitation time becomes insignificant beyond a certain level, as is seen from the plots. Thus, depending on the application, it is possible to experimentally determine an optimum agitation time to obtain a stable dispersion of the nanoparticles in the fluid.

Figure 6.30 illustrates the effect of the time elapsed after preparation, on the percentage enhancement in the burnout heat flux for various volume fractions of the copper oxide nanofluid. Similar trends were also observed for aluminum oxide. It is clear that the percentage enhancement in the burnout heat flux decreases as time elapses, for both nanofluids. Thus, even though the burnout heat flux is enhanced to a larger extent when nanoparticle concentration in water is increased, the effect is reduced when the prepared nanofluid is kept idle, due to sedimentation. Also, as the volume fraction of nanoparticles is increased, the drop in percentage enhancement was observed to be increasing, which makes the initial concentration of the nanoparticles rather insignificant, if the nanofluid is not used immediately after preparation.

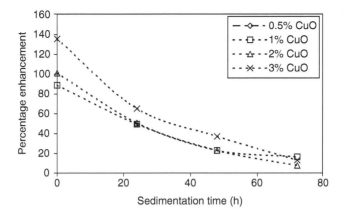

FIGURE 6.30 Variation of percentage enhancement of burnout heat flux with sedimentation time for different volume fractions of copper oxide nanoparticles.

6.4 THEORETICAL INVESTIGATIONS

A number of theoretical investigations leading to observations, hypotheses, practical expressions, and correlations have been reported in the literature pertaining to the addition of nanoparticles to base fluids. While some of them dealt with basic models for effective thermal conductivities, others aimed at a more refined treatment, by incorporating effects such as Brownian motion and thermophoresis.

More than a century ago, Maxwell predicted that adding a high thermal conductivity particle to a liquid would enhance the effective thermal conductivity and result in a composite material with a high effective thermal conductivity (Maxwell 1892). Maxwell understood that the addition of particles to a base fluid to create a suspension could result in an enhanced effective thermal conductivity, due to a change in the combined bulk property of the base fluid and the particles. Using this approach, Maxwell developed an analytical expression that has served as the basis for numerous theoretical investigations of the effective thermal conductivity of particles in suspension. This expression defined the effective thermal conductivity in terms of the thermal conductivity of the two components and the volume fraction, and was expressed as

$$\frac{k_{\mathrm{eff}} - k_{\mathrm{f}}}{k_{\mathrm{f}}} = \frac{3(\alpha - 1)\phi}{(\alpha + 2) - (\alpha - 1)\phi} \tag{6.21}$$

where
  k_{eff} is the effective thermal conductivity of the suspension
  k_{f} is the thermal conductivity of the base fluid
  α is the ratio of the thermal conductivities of the nanoparticle and base fluid
  ϕ is the volume fraction of the nanoparticles in the suspension

This model, while it accurately predicts the effective thermal conductivity of particle suspensions when the particles are relatively large, does not produce results which compare well with the experimental data for suspensions with very small particles. In fact, previous investigations indicate that as the particle size decreases the ability of this expression to accurately predict the experimentally measured results decreases.

Because many engineering heat transfer applications could benefit from a working material that could combine the comparably high thermal conductivity of a solid with the flow properties of liquids, there have been a considerable number of experimental investigations conducted with millimeter and micrometer-sized particle suspensions, since that initial hypothesis. These investigations have, for the most part, supported the initial theoretical prediction. More recently, these investigations have been directed at the evaluation of the effects of variations in the size of the particles and the relative response time.

The original ideas that form the basis of the field of nanotechnology are believed to have started with two articles by Richard Feynman: "There's Plenty of Room at the Bottom" delivered in 1959, which was republished 33 years later in the *Journal of Microelectromechanical Systems* (1992), and "Infinitesimal Machinery" republished again in 1993. The study of the thermal behavior of nanomaterials is rapidly increasing in importance and is currently one of the most exciting and significant fields of study within thermal science. At present, the study of the thermal behavior in nanometer-sized particles is primarily focused on experimental research: Tests on the specific heat, thermal conductivity, thermal expansion, etc.; the study of individual nanoparticles, nanometer membranes, nanotubes, and nanoparticle suspensions; and the thermal behaviors of many compositions or structures formed by nanometer-sized materials. Meanwhile, theoretical methods that utilize molecular simulation and the Boltzmann transport equation, have been devised and are still under development to try to understand the fundamental thermophysical phenomenon that govern the behavior of these materials.

The study of heat transfer in nanoparticle suspensions has increased in importance and achievements in this arena, allied with achievements in other areas of nanotechnology, offer tremendous promise for engineering applications. In order to understand the fundamental reasons why nanoparticle suspensions could cause dramatic increase in thermal conductivity, a number of experimental and theoretical investigations have been conducted. While informative, none of these have provided information that can clearly reveal the nature of this seemingly anomalous thermal behavior.

Table 6.3 gives a summary of the important observations and results of both theoretical and experimental investigations. For a more detailed review and discussion on the topic, the extensive paper by Li and Peterson (2005) is suggested as reading material.

6.4.1 MOLECULAR DYNAMICS SIMULATION

Molecular dynamics simulation is a versatile, discrete modeling technique which can provide a powerful tool for the analysis of nanoscale thermal phenomena. On the basis of the determination of the velocities and potential energies of molecules in the computational domain (simulation box), this method tackles the thermal energy transport problem as a dynamics problem. Thus, essentially, the governing equations solved in molecular dynamics are the Newton's laws of motion, pertaining to the individual

TABLE 6.3

Summary of Theoretical Investigations on Effective Thermal Conductivity and Associated Phenomena in Nanofluids

Investigations on Models of Effective Thermal Conductivity

Source	Nature and Objectives	Highlights, Findings, and Correlations
Fricke (1924)	Theoretical work. Effective thermal conductivity based on the work of Maxwell	Influence of geometric factors of the suspended particles was considered through a parameter x in the equation: $(k/k_1 - 1)/(k/k_1 + x) = \phi(k_2/k_1 - 1)/(k_2/k_1 + x)$
Fricke (1924)	Prediction of effective thermal conductivity of suspensions based on a review of previous research	$\dfrac{\frac{k_e}{k_1} - 1}{\frac{k_e}{k_2} + 2} = \phi \dfrac{\frac{k_2}{k_1 - 1}}{\frac{k_2}{k_1} + 2}$ Equation for suspensions of spherical particles $\dfrac{\frac{k_e}{k_1} - 1}{\frac{k_e}{k_2} + x} = \phi \dfrac{\frac{k_2}{k_1 - 1}}{\frac{k_2}{k_1} + x}$ Effective thermal conductivity of suspensions of nonspherical particles
Meredith and Tobias (1961)	Theoretical expression for effective thermal conductivity	$\dfrac{k_e}{k_f} = \left(\dfrac{2 + 2X\phi}{2 - X\phi}\right)\left(\dfrac{2 + (2X - 1)\phi}{2 - (X + 1)\phi}\right)$ Volume fraction of surrounding material around fluid is given by $1 - \phi = \dfrac{k_e/k_f - k_p/k_f}{(k_e/k_f)^{1/3}(1 - k_p/k_f)}$
Hamilton and Crosser (1962)	Theoretical expression for effective thermal conductivity of heterogenous system consisting of continuous phase with discontinuous phase dispersed in it	$k = k_1\left[\dfrac{k_2 + (n - 1)k_1 - (n - 1)\phi(k_1 - k_2)}{k_2 + (n - 1)k_1 + \phi(k_1 - k_2)}\right]$ Introduction of n and ψ ψ is the ratio of the surface area of the sphere to the surface area of a nonspherical particle having the same volume n depends on the shape of the dispersed particle and the ratio of thermal conductivities of discontinuous and continuous phase $n = \frac{3}{\psi^k}$ $k = 1, 2, 1.5$ depending upon the shape of particle
Keller (1963)	Theoretical investigation for determining the expression for effective thermal conductivity	$k/k_1 = -(\pi/2)\log[(\pi/6) - \phi] + \cdots (\pi/6) - \phi \ll 1$ $R = \frac{1}{k_e c}$, R is the resistance of the gap between two adjacent cylinders and c is the distance between the axes of the two adjacent cylinders. k_e is the effective conductivity Found that expressions of effective conductivity of nonconducting and perfectly conducting cylinders are inversely related
Leal (1973)	Theoretical analysis of effective conductivity of a dilute suspension in a simple shear flow	$k = k_1\left[1 + \phi\left\{\frac{3(k_2 - k_1)}{k_2 + 2k_1} + \left(\frac{1.176(k_2 - k_1)^2}{(k_2 + 2k_1)^2} + \frac{2\mu_1 + 5\mu_2}{\mu_1 + \mu_2}\right.\right.\right.$ $\left.\left.\left.\left(0.12\frac{2\mu_1 + 5\mu_2}{\mu_1 + \mu_2} - 0.028\frac{k_2 - k_1}{k_2 + 2k_1}\right)\right)Pe_1^{3/2} + 0(Pe_1^2)\right\}\right]$ Effective conductivity influenced by the thermal conductivity of suspended particle

(continued)

TABLE 6.3 (continued)

Summary of Theoretical Investigations on Effective Thermal Conductivity and Associated Phenomena in Nanofluids

Investigations on Models of Effective Thermal Conductivity

Source	Nature and Objectives	Highlights, Findings, and Correlations
Jeffrey (1973)	Made use of probability approach	$k/k_1 = 1 + 3\beta\phi + 3\beta\phi^2\left(\beta + \sum\limits_{p=6}^{\infty} \frac{B_p - 3A_p}{(p-3)2^{p-3}}\right)$
Rocha and Acrivos (1973)	Theoretical expression for effective conductivity of suspension	$\dfrac{k_{e,ij}}{k_1} = \delta_{ij} + 2[(\alpha - 1)/(\alpha + 1)]\phi$ $\times\left\{\delta_{ij} + V_2^{-1}\sum\limits_{m=1}^{\infty}(2\pi)^m[(\alpha-1)/(\alpha+1)]^m \int_{\Omega}\right\}$ On the basis of the expression for the heat conduction in a statistically homogenous suspension using Fourier's law
Batchelor (1974)	Transport properties of two-phase heterogeneous mixtures	$k_e = k_1\{1 + [2(\alpha - 1)/(\alpha + 2)]\phi + [3(\alpha - 1)^2/(\alpha + 2)^2]\phi^2\}$ Theoretical effective conductivity expression for noninteraction between particles in dilute suspension
Willis (1977)	Studied the bounds and self-consistent estimates for the effective thermal conductivity of a composite material comprising a matrix with a random volume fraction of dispersed particles with bounds of the Hashin–Shtrikman type and self-consistent estimate	Assuming the thermal conductivity of dispersed particle to be zero, expression for effective thermal conductivity is given by $k_e \leq \dfrac{2(1-\phi)}{(2+\phi)}k_1$ for the self-consistent estimates of effective thermal conductivity is given as $k_e = \left(1 - \dfrac{3}{2}\phi\right)k_1$ Assuming dispersed particle to be perfectly conducting, the effective thermal conductivity is given by $k_e \geq \dfrac{(1+2\phi)}{(1-\phi)}k_1$ and the self-consistent estimates of the effective thermal conductivity is given as $k_e = \dfrac{k_1}{1-3\phi}$
O'Brien (1979)	Theoretical analysis for determining expression for effective conductivity	Probability theory based expression for effective conductivity of suspension
Hatta and Taya (1985)	Analysis of randomly oriented short fibers in a continuous matrix	Equation for the effective conductivity of two-dimensional and three-dimensional cases, based on the equivalent inclusion method
Davis (1986)	Analytically developed an expression for effective conductivity	$k/k_1 = 1 + \dfrac{3(\alpha - 1)}{[\alpha + 2 - (\alpha - 1)\phi]}\{\phi + f(\alpha)\phi^2 + 0(\phi^3)\}$ $f(\alpha) = \sum\limits_{p=6}^{\infty}\dfrac{B_p - 3A_p}{(p-3)2^{p-3}}$

TABLE 6.3 (continued)
Summary of Theoretical Investigations on Effective Thermal Conductivity and Associated Phenomena in Nanofluids

Investigations on Models of Effective Thermal Conductivity

Source	Nature and Objectives	Highlights, Findings, and Correlations
Hasselman and Johnson (1987)	Theoretical expression for effective thermal conductivity based on the experimental results of spherical nickel dispersions in a sodium borosilicate glass matrix and carbon fiber reinforced glass ceramic	$k_e = k_1 \dfrac{\left[2\left(\frac{k_2}{k_1} - \frac{k_2}{ah_c} - 1\right)\phi + \frac{k_2}{k_1} + \frac{2k_2}{ah_c} + 2\right]}{\left[\left(1 - \frac{k_2}{k_1} + \frac{k_2}{ah_c}\right)\phi + \frac{k_2}{k_1} + \frac{2k_2}{ah_c} + 2\right]}$ Found that thermal barrier resistance at the interface between the dispersed particle and the continuous matrix could have an important influence on the effective conductivity
Brady et al. (1988)	Developed a method for calculating the hydrodynamic interaction among particles in an infinite suspension, with infinitely small Reynolds number	$k_e = k_1 - n(S/G^E)$
Lu and Kim (1990)	Investigated the relationship between the microstructure of the dispersed particles and the thermal conductivity of anisotropic composites	$k_e = \hat{I} + \left(-\dfrac{M}{V_2 r_1}\right)\phi + \left(\dfrac{-\Xi + \Omega}{V_2 r}\right)\phi^2$
Kim and Torquato (1990, 1991, 1993)	Utilized a computer simulated first passenger time algorithm	Method principally concerned with Brownian motion of dispersed particle
Lu and Lin (1996)	Expression for effective conductivity tensor to calculate effective conductivity in the form of virial expansion in the dispersed volume fraction, ϕ	$\dfrac{k_{e,ij}}{k_1} = 1 + \dfrac{\alpha - 1}{(\alpha - 1)h_i + 1}\phi$ $+ \dfrac{(\alpha - 1)^2 h_i}{[(\alpha - 1)h_i + 1]^2}\phi^2 + \dfrac{(\alpha - 1)^3 h_i}{[(\alpha - 1)h_i + 1]^3}\phi^3 + \cdots$ Particle shape considered was ellipsoid
Chen (1996)	Expression for effective conductivity with a single nanoparticle immersed in fluid	$k/k_1 = \dfrac{3\tau_1/4}{3\tau_1/4 + 1}$ Dependence of particle size of effective thermal conductivity was investigated

(continued)

TABLE 6.3 (continued)
Summary of Theoretical Investigations on Effective Thermal Conductivity and Associated Phenomena in Nanofluids

Investigations on Particle-Related Phenomena in Suspensions

Source	Nature and Objectives	Highlights, Findings, and Correlations
Xue (2003)	Expression for effective conductivity by utilizing polarization theory and by considering the interface effects between solid particles and the base fluid	$9\left(1 - \dfrac{v}{\lambda}\right)\dfrac{k - k_1}{sk + k_1} + \dfrac{v}{\lambda}\left[\dfrac{k - k_{c,x}}{k + B_{2,x}(k_{c,x} - k)}\right.$ $\left. + 4\dfrac{k - k_{c,y}}{sk + (1 - B_{2,x})(k_{c,y} - k)}\right] = 0$
Brown (1827) Cited in Haw (2002), Mazo (2002)	Immersed clarkia pulchella pollen particles in water and observed that many of the particles were in continual motion	Observed and explained Brownian motion Strong temperature dependence of Brownian motion
Perrin (1909) Cited in Haw (2002), Mazo (2002)	Experimental work to indirectly prove quantitatively that Brownian motion is, in fact, the reflection of molecular kinetic theory	Proved Brownian motion is the reflection of molecular kinetic theory Sedimentation speed of the particle was measured for different heights inside the suspension Expression for diffusion constant of particles in suspension obtained by Einstein (Mazo 2002) was proven to be accurate by the experiments of Perrin (Haw 2002, Mazo 2002) Expression for diffusion constant $D = kT/6\pi\eta a$
Grasselli and Bossis (1995)	Applied a three-dimensional micrometer-size particle-tracking technique that utilized microscopy and computer digitized imaging	Diffusion coefficients in X and Y direction were determined Measurements of both Brownian motion and sedimentation made along with particle hydrodynamic diameter and density
Gaspard et al. (1998)	Verified the hypothesis of microscopic chaos experimentally	Diffusion coefficient was calculated
Goodson and Kraft (2002)	Developed stochastic algorithm for the modeling of nanoparticle dynamics in the free molecular regime	Introduction of new simulation technique Simulation technique more efficient compared to direct simulation Monte Carlo and other algorithm used earlier
Batchelor (1972,1982) Torquato (1985)	Theoretical study of sedimentation. Formula obtained for arbitrary dimensional electrical conductivity in a composite	Derived a formula to predict the mean sediment velocity of various particles $(\beta_{ij}\phi_i)^2\left(\frac{k_e + (d-1)k_j}{k_i - k_j}\right) = \phi_i\beta_{ij} - \sum_{n=3}^{\infty} A_n^{(i)}\beta_{ij}^n$ Use of Green's function to describe Maxwell electric field and induced polarization field

TABLE 6.3 (continued)
Summary of Theoretical Investigations on Effective Thermal Conductivity and Associated Phenomena in Nanofluids

Investigations on Particle-Related Phenomena in Suspensions		
Source	Nature and Objectives	Highlights, Findings, and Correlations
Torquato and Kim (1989)	Simulation of effective transport properties of disordered heterogenous media	$k_e = \dfrac{X^2}{2d\left(\sum_i R_i^2/2dk_i\right)}$ Simulation included the concept of the first passage time probability distribution and grid method
Langevin (1908)	Expression for the Brownian motion velocity	$\frac{d}{dt}y(t) = -B(y(t)) + f(t)$ Applied stochastic equation of motion to derive the expression
Uhlenbeck and Ornstein (1930)	Analysis of Brownian motion	Calculated the mean value of all the powers of the velocity and displacement of Brownian Motion
Mazur and Bedeaux (1992)	Studied the stochastic properties of the random force in Brownian motion	$\dfrac{dp(t)}{dt} = -\beta(p(t))p(t) + f(t)$ Application of Langevin equation
Batchelor (1972)	Analysis of particles including Brownian motion	$U = U_0[1 + S\phi + O(\phi^2)]$ Method to determine the velocity of particles of small Re, under the influence of gravity, other particles, and Brownian motion
Brady et al. (1988)	Dynamics simulation	Developed a dynamics simulation of hydrodynamically interacting single-species particle suspension
Gupte et al. (1995)	Numerical investigation of microconvection	Microconvection induced by the sedimentation of particles greatly enhance the overall heat transfer at a macroscopic scale, in a stationary fluid
Waldmann (1961)	Studies on particles in gas mixture	Studied the forces on a particle in a static binary diffusing gas mixture with a temperature gradient and particle number density gradient
Brock (1962)	Theoretical studies on thermal force acting on particles in a slip flow regime	Equation for thermal force acting on particle was derived $F = -9\pi \dfrac{\mu^2}{\rho T}a\left(\dfrac{1}{1+3c_m\frac{l}{a}}\right)\left(\dfrac{\frac{k_f}{k_p}+c_t\frac{l}{a}}{1+2\frac{k_f}{k_p}+2c_t\frac{l}{a}}\right)\nabla T_\infty$ Explained the discrepancies between the experimental observations and theoretical predictions
Rosner et al. (1992)	Dynamics and transport of small particles suspended in gases exposed to a temperature gradient and undergoing heat transfer	$\text{Kn} > 1$ $(\alpha_T D)_p = \left(\dfrac{3}{4}\right)\nu_g\left\{1 + \left(\dfrac{\pi}{8}\right)\alpha_{\text{mom}}\right\}^{-1}$ $\text{Kn} < 1$ $(\alpha_T D)_p = \dfrac{2c_s\nu\left(\frac{k_g}{k_p}+c_t\frac{l_g}{a}\right)\left[1+\frac{l_g}{a}\left(A+be^{-ca/l_g}\right)\right]}{\left(1+3c_m\frac{l_g}{a}\right)\left(1+2\frac{k_g}{k_p}+2c_t\frac{l_g}{a}\right)}$

(continued)

TABLE 6.3 (continued)

Summary of Theoretical Investigations on Effective Thermal Conductivity and Associated Phenomena in Nanofluids

Investigations on Particle-Related Phenomena in Suspensions

Source	Nature and Objectives	Highlights, Findings, and Correlations
Huisken and Stelzer (2002)	Experimental investigation of photophoresis force	$F_z = -\int\limits_S P\cos\theta \, dA$
Buevich et al. (1990)	Diffusion process in flows for particle suspension	Evaluated self-diffusion and gradient diffusion coefficient Utilized the momentum conservation equation and introduced the thermodynamic forces
Jang and Choi (2004)	Effect of Brownian motion of nanoparticles on the effective thermal conductivity of nanofluids	$k_{\text{eff}} = k_{\text{BF}}(1-f) + k_{\text{nano}}f + 3C_1 \dfrac{d_{\text{BF}}}{d_{\text{nano}}} k_{\text{BF}} \text{Re}_{d\text{nano}}^2 \text{Pr} f$
Prasher et al. (2005)	Effect of Brownian motion of nanoparticles on the effective thermal conductivity of nanofluids	$\dfrac{k}{k_{\text{f}}} = (1 + A\text{Re}^m \text{Pr}^{0.333}\phi) \left[\dfrac{(1+2\alpha) + 2\phi(1-\alpha)}{(1+2\alpha) - \phi(1-\alpha)} \right]$
Yu et al. (1999)	Observed a thin film of tetrakis silane film that had spread on a silicon substrate which was cleaned in strong oxidizer, using a specular x-ray beam	Density of the liquid film changed along the liquid film thickness
Wang et al. (2002)	Applied Gibbs energy analysis method	Found that Gibbs energy of the system reaches a minimum when the nanoparticle and liquid surrounding it reaches equilibrium
Keblinski et al. (2002) Xu et al. (2003)	Applied nonequilibrium molecular dynamics to the study of thermal resistance of liquid–solid interfaces	Liquid with a weak atomic bond at the particle–matrix interface exhibited high thermal resistance, while for wetting systems, interfacial resistance was small
Ahuja (1975a,b)	Experimental investigation of the effective thermal conductivity of suspension	$h = (\dot{m}c/\pi d_m L)[(t_{\text{sus},e} - t_{\text{sus},i})/(\Delta t_i - \Delta t_e)] \ln(\Delta t_i/\Delta t_e)$ $\dfrac{1}{h} = \dfrac{1}{h_1} + \dfrac{b}{k_{\text{tube}}} + \dfrac{1}{h_2}$ $h_1 D = 3.65 k_{\text{sus}} + \dfrac{0.27\dot{m}c/\pi L}{1 + 0.04(4\dot{m}c/\pi L k_{\text{sus}})^{2/3}}$ where $b = \dfrac{1}{2}d_m \ln(D_0/D)$ and c the specific heat Equations were used to calculate effective thermal conductivity

molecules. In obtaining the equilibrium states of the molecules, the intermolecular potentials are used, the gradients of which, in turn, provide the intermolecular forces. A brief description of the basics of the method has been provided in Chapter 1, and the practical implementation of the method to the cases of nanofluids will be discussed here.

A few cases of molecular dynamics simulations pertaining to nanofluids have been undertaken at the Nanotechnology Research Laboratory, National Institute of Technology, Calicut, India, which are described in the following sections. The molecular dynamics method can be applied for the evaluation of thermophysical properties of media, as well as for analyzing thermal processes, both of which will be described in the case of nanofluids. Application of the method for obtaining the thermal conductivity of nanofluids and quantifying its enhancement due to nanoparticle addition will be dealt with. Such an approach will also be presented for viscosity. In the analysis of thermal phenomena, a practical application to the determination of the heat transfer coefficients in forced convection, using the computed properties of nanofluids will be demonstrated. Further, molecular dynamics simulation of boiling heat transfer in nanofluids will be presented as an example. Analysis of a few fluid flow problems in nanochannels using molecular dynamics simulation will also be discussed.

6.4.1.1 Molecular Dynamics for Nanofluids

As has been discussed extensively so far, it has been understood through experimental observations that the introduction of nanoparticles produces anomalous enhancement effects on the thermophysical properties of the fluids. Though there are several experimental results, generalization of such experimental observations over a wider range of problems involving different classes of materials is difficult to achieve. Moreover, while experimentation gives direction about the enhancement of thermophysical properties alone, the real physical reasons for such anomalous behavior will have to be interpreted intuitively in many practical situations. Such limitations of experimental research and the need for a complete understanding of the original physical processes endorse the inevitability of discrete modeling techniques as molecular dynamics, which is wholly based on the fundamental principles of the mechanics at the microscopic level. The application of molecular dynamics becomes extremely useful when microscopic phenomena are dealt with, or while analysis is done on domains where continuum hypothesis and usual computational methods are inadequate, such as in the case of the analysis of nanofluids containing dispersed nanoparticles in suspended form.

The problems discussed here are of two classes—one is to simulate nanofluids and calculate their thermophysical properties, which can later be used for studying the heat transfer phenomena in them. The second class of problems model the thermal phenomena as such, so as to observe how the molecular level dynamics manifest their effects on a larger macroscale. A few cases of practical interest are dealt with and the background research explained briefly in the following sections. In the calculations, the initial procedures of formulation are algorithmically similar for most of the physical problems, and the major differences arise out of the boundary conditions specified by the problem and the physical state of the matter simulated.

The major limitation of the calculations reported here is that the effect of the nanoparticles is introduced in the calculation by choosing the number of solid molecules in the simulation box and applying a "cluster potential" to represent the overall effect. However, the procedure followed is found to give good representative results, which can be improved upon using more realistic modeling.

The failure in predicting thermophysical properties in nanoscale domains using the methods for macroscale domains is mainly because of the molecular basis of these properties which are size affected. Since molecular dynamics deals directly with the molecular level motions, it helps to visualize and calculate thermophysical properties in the light of molecular interactions.

The values of the thermophysical properties are of direct use in approaching physical problems and predicting the overall behavior of physical systems, but these properties have been defined on a phenomenological basis, utilizing macroscopic observations and laws (the thermal conductivity is defined based on the Fourier's law of heat conduction, and the viscosity based on Newton's law of shear stress). In the size-affected domain, thus, the validity of the definition is lost, and so also, is the application of continuum modeling. For instance, it is not possible to computationally simulate the convective heat transfer process in a nanofluid using conventional macroscale modeling techniques of the continuum, in order to quantify the results in terms of the heat transfer coefficient. In such a case, as a useful practical approach, the properties of the nanofluid constituting the heat transfer coefficient, namely, the thermal conductivity, viscosity, density, and specific heat, may be simulated and estimated to calculate the overall convective heat transfer coefficient using appropriate correlations. Similarly a phase change heat transfer problem can be approached by simulating and estimating the values of the latent heat, using discrete computations in molecular dynamics.

Nevertheless, it should be noted that the effects of nanoparticle inclusions are quite small and might not be significantly reflected on the thermodynamic properties (such as density and specific heat) of nanofluids, unlike thermophysical properties, due to the very small volume fractions of nanoparticles introduced in nanofluids. This means, in most of the practical situations, evaluation of properties such as the density and specific heat using the effective medium theories will be sufficient, but it is definitely not so in the case of the size-affected thermophysical properties (like thermal conductivity and viscosity).

In a molecular dynamics simulation, estimation of thermophysical properties is done (at the final stage of the computation) harnessing the information from the essential procedural steps to obtain the position vectors and velocities of the individual molecules in the system considered. The thermophysical properties are usually evaluated by integrating the appropriate autocorrelation functions, in the equations known as Green–Kubo relations (Allen and Tildesely 1989, Rappaport 2004). The approach is based on statistical mechanics.

Some of the results obtained from the simulations of nanofluids and general fluid flow in nanoscale conduits, performed at the author's laboratory (Nanotechnology Research Laboratory, National Institute of Technology, Calicut, India) are explained in the subsequent sections. The application of molecular dynamics simulation of nanofluids has been demonstrated in water–platinum systems (representing a metallic particle suspension), essentially because of the availability of data on molecular potentials of these materials. It is expected that the simulation results will help the readers to appreciate the applicability of the method to predictive computations in nanofluids.

6.4.1.2   General Methodology

Before going into the application of molecular dynamics to specific problems of interest, the general methodology of simulation will be briefly explained here. However, to obtain an in-depth knowledge of molecular dynamics simulation and its application to thermal problems, the reader is advised to refer to the excellent publication by Maruyama (2000) and the book by Allen and Tildesely (1989).

Molecular dynamics is essentially a process which allows a system to evolve in time, based on the laws of classical mechanics. Like any evolutionary process, molecular dynamics, too, requires an initial configuration. Generally the molecules are assorted over a known crystal structure like Face Centered Cubic (FCC) or Body Centered Cubic (BCC) which gives the initial positions. It must be noted that the initial state does not have much influence over the final results of the simulation, and more careful attempts to construct a typical initial state is of little benefit. Random velocities are assigned to the molecules based on the temperature, and the size of the cell is chosen based on the density of the initial state. The crux of the method, termed a "time-evolution method," is to obtain the positions and velocities of the molecules at the next state of the system, which is obtained by stepping the system, with respect to time using a time integration algorithm, based on the finite difference method. There are several time integration algorithms like the Gear predictor–corrector algorithm and the Verlet algorithm (Allen and Tildesely 1989). The suitability and selection of an appropriate algorithm are primarily based on physically and computationally important factors such as the computational speed, the scope for the use of large time steps, and memory allocation.

The force between the molecules is obtained as the negative spatial gradient of the interatomic/intermolecular pair potentials. The potential is the most important data required for the molecular dynamics simulation. Without a realistic potential correctly defining the behavior of the physical interaction between two molecules of the same or different materials, the simulation will not give any result of physical significance. Various interaction potential functions have been suggested for implementing molecular dynamics simulations of fluid and solid phase molecules. The most common potential used in developing and benchmarking systems is the Lennard Jones potential, which was originally proposed to describe the interaction between the molecules of noble gases. This is given by

$$\Phi_{I,J}(r) = 4\varepsilon \left[\left(\frac{\sigma}{r}\right)^{12} - \left(\frac{\sigma}{r}\right)^{6} \right] \tag{6.22}$$

where
   σ is the distance below which the force becomes repulsive
   r is the magnitude of the distance vector between molecules
   ε is the energy (Rahman and Stillinger 1971)

Various augmentations have been incorporated to this to describe the intermolecular potential for real materials. Examples are the SPC/E (Single Point Charge/Extended) potential which considers the oxygen and hydrogen interaction within the water molecule by incorporating the van der Waals and Coloumbic (charge) terms and the Morse potential which describes most of the metal molecular interactions (Maruyama et al. 2000). Specific potential functions are also used for particular pairs of molecules such as the Spohr and Heinzinger potential (Spohr and Heinzinger 1988) used for describing water–platinum interactions. These potential functions are used in the molecular dynamics simulations to be described later in this book, for water–platinum systems.

Once the system starts evolving in time, the task is to observe when the system attains equilibrium so that the "measurements" (observations of the velocities and positions) can be made. There are several methods to benchmark and accelerate the equilibration process. For example one method is to see whether the velocity distribution of the molecules matches a standard distribution, in order to determine the equilibrium state of the system.

Considering a particular molecule in a system having N number of molecules to be interacting with all other molecules of the system, the computation time scales as N^2. There are several efficient techniques such as engaging a cutoff distance, which can be used to reduce the computation time scaling with respect to N. While using the cutoff distance method, the interactions of the atoms and molecules are limited to those inside a spherical space of a particular radius.

The boundary conditions for the system depend on the problem under analysis. Generally "periodic boundary conditions" are used in molecular dynamics, according to which, corresponding to an atom or molecule that leaves the system across a boundary, one is brought back to the system through another boundary. This takes care of the conservation of molecules in the simulation box.

Generally, a normalized unit system is used in molecular dynamics simulations. These units are called reduced units. The most common standards for the basic units are σ, the unit of length, ε, the unit of energy, and m, the unit of mass (mass of the atom or molecule in the system). The primary advantage of using the reduced units is that it gives a possibility to study a number of systems in a generalized way. Also the use of reduced units is of much practical convenience, while working with very large or small values for the system parameters, since the reduced values will generally lie around a numerical value of 1. In practical computation, this gives a means of identifying errors, as whenever a particularly large or small value is encountered, it can indicate a significant error. The reduced form of some of the common properties and parameters of the system are shown in Table 6.4.

6.4.1.3 Simulations for Thermophysical Properties

Molecular dynamics simulations have been utilized to predict the effect of nanoparticle addition to water, on the thermal conductivity and viscosity of the base fluid. The method has also been utilized to calculate the variations of the enhancement of these thermophysical properties with respect to the temperature and volume fraction of nanoparticles in the nanofluid, taking the case of a water–platinum nanofluid as an example. The method and the results will be presented and briefly explained in the following section.

TABLE 6.4

Reduced Unit System Used in Molecular Dynamics Simulations

Property	Notation	Reduced Form
Length	$R*$	r/σ
Time	$t*$	$t/\tau = t(\varepsilon/m\sigma^2)$
Temperature	$T*$	$k_B T/\varepsilon$
Force	$f*$	$f\sigma/\varepsilon$
Energy	$\phi*$	ϕ/ε
Pressure	$P*$	$P\sigma^3/\varepsilon$
Number density	$N*$	$N\sigma^3$
Density	$\rho*$	$\sigma^3\rho/m$
Surface tension	$\gamma*$	$\gamma\sigma^2/\varepsilon$

Note: r, intermolecular distance; σ, characteristic distance; ε, molecular energy; and m, molecular mass.

6.4.1.3.1 Thermal Conductivity

In the modeling of thermal conductivity, some standard interaction potential functions have been utilized to describe the water–water, platinum–platinum, and water–platinum interactions. The Lennard Jones potential has been used for water as a simplification, and the Morse potential has been used for platinum–platinum interactions. The water–platinum interaction is modeled using the Spohr and Heinzinger potential.

Usually, the thermal conductivity is evaluated through a Green–Kubo formula based on the time autocorrelation function (heat current autocorrelation function) that integrates the microscopic heat flux (Allen and Tildesely 1989, Rappaport 2004). The thermal conductivity is described through a function in the following form:

$$k = \frac{1}{3Vk_B T^2} \int_0^\infty \langle J(t) \cdot J(0) \rangle \, dt \qquad (6.23)$$

where J is the microscopic heat flux given by

$$J = (1/V)\left[\sum_j (1/2)e_j v_j + (1/2)\sum_{i\neq j}(r_{ij}:F_{ij})\cdot v_j\right] \qquad (6.24)$$

and e_j is the instantaneous excess energy of the atom j, represented as

$$e_j = \sum_j (1/2)m_j v_j^2 + (1/2)\sum_{i\neq j}\Phi_{ij} \qquad (6.25)$$

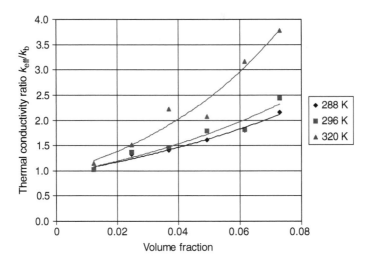

FIGURE 6.31 Molecular dynamics predictions of thermal conductivity enhancement in nanofluids.

F_{ij} in Equation 6.24 is the force experienced by a particle (molecule or atom as the case may be) when it interacts with a neighboring particle. The value of F_{ij} is equal to the spatial derivative of the interaction potential, which is a function of the distances between the simulated particles. Φ is the appropriate intermolecular potential, m is the molecular mass, v is the velocity, and r_{ij} is the distance vector.

A typical result for the thermal conductivity enhancement values obtained from the simulation of water–platinum nanofluids is as shown in Figure 6.31 (Sobhan et al. 2006). Usually a simulation based on an NVT (constant Number, constant volume, and constant temperature) ensemble is used to estimate the values of thermal conductivity. It can be seen in Figure 6.31 that the thermal conductivity of the nanofluid increases with the volume fraction or the volume fraction of the nanoparticles and with the temperature. Also, the values obtained are much higher than the ones predicted by the Maxwell's effective medium theory, as shown in Figure 6.32, which proves that it cannot be applied to explain nanoscale phenomena.

6.4.1.3.2 Viscosity

Similar to the thermal conductivity, viscosity of the nanofluid can also be evaluated from a Green–Kubo relation based on integrated autocorrelation function of the pressure tensor as follows:

$$\eta = \frac{V}{3T} \int_0^\infty \left\langle \sum_{x<y} P_{xy}(t) P_{xy}(0) \right\rangle dt \qquad (6.26)$$

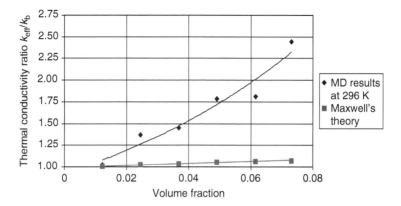

FIGURE 6.32 Comparison of predicted effective thermal conductivity with Maxwell's model.

where

$$P_{xy} = \frac{1}{V} \left[\sum_j m_j v_{jx} v_{jy} + \frac{1}{2} \sum_{i \neq j} r_{ijx} F_{ijy} \right] \tag{6.27}$$

P_{xy} is the component of the pressure tensor (the negative of which gives the stress tensor). In this, averaged values in x and y are used for the vector components, as indicated by the suffixes, so as to improve the statistics involved in the calculation (Rappaport 2004).

A typical result for the viscosity enhancement in the water–platinum system analyzed is as shown in Figure 6.33. It can be observed that the viscosity of the nanofluid increases with the dosing level and decreases with the temperature.

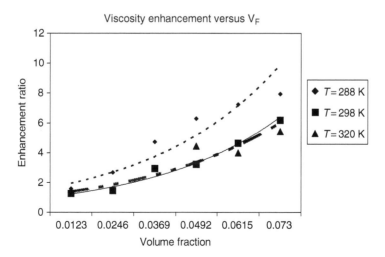

FIGURE 6.33 The enhancement of viscosity due to nanoparticle addition, predicted in a water–platinum nanofluid system using molecular dynamics simulation.

6.4.1.4   Modeling of Thermal Phenomena

A method to quantify the effects of thermal phenomena in nanofluids is to use existing correlations, using the computed values of enhanced thermophysical properties, though the interaction between the various properties may not be well understood to ensure the validity of conventional correlations in nanofluids unless direct experimental studies are done on thermal phenomena. These calculations can still give practical results applicable to design predictions. In this method, the values of the thermophysical properties obtained through simulation in the size-affected particle domain are utilized in conventional correlations, to study the overall thermal phenomenon. Such an approach is used in the evaluation of the heat transfer coefficients in forced convection from computed property values, in the section to follow.

Another approach is to simulate the physical phenomenon directly. As we know that the bulk thermophysical and thermodynamic properties of the nanofluids manifest out of the nanoscale physical interactions between the fluid molecules and the nanoparticle molecules, they can be evaluated to a very good degree of accuracy using a method which gives freedom to suitably manipulate the molecular motions of a hypothetical molecular system, representative of the actual thermophysical system. Molecular dynamics is such a tool which provides tremendous scope for accurate system modeling. This approach is used in modeling the phase change heat transfer problem in the presence of nanoparticles (where the state of the fluid, whether liquid or vapor, is tracked using the energy content of individual particles obtained through molecular dynamics), the results of which will be presented later. The direct approach has also been used to analyze some common flow problems in nanoscale dimension channels, which also will be presented in a subsequent section.

6.4.1.4.1   Convective Heat Transfer

Convective heat transfer coefficients in nanofluids can be obtained, with fairly good degree of accuracy, using the thermophysical property values (thermal conductivity and viscosity) estimated from molecular dynamics, in conventional heat transfer correlations, along with weighed average values of density and specific heat. To illustrate this, the results of calculation to estimate the heat transfer coefficients (from the Nusselt number) for laminar forced convection are presented here. The correlation used in this calculation is given below (Peng et al. 1994).

$$\text{Nu} = 1.86 \left(\frac{D}{L}\right)^{\frac{1}{3}} (\text{RePr})^{\frac{1}{3}} \left(\frac{\mu}{\mu_{\text{w}}}\right)^{0.14} \tag{6.28}$$

Nu is the Nusselt number
D is the diameter of the channel
L is the length
μ is the viscosity of the fluid
μ_{w} is the fluid viscosity at the wall

Figure 6.34 depicts the variation of the enhancement ratio (with respect to the values for water) of the heat transfer coefficient (for laminar flow of water–platinum nanofluid through a circular channel), with the volume fraction of nanoparticles in

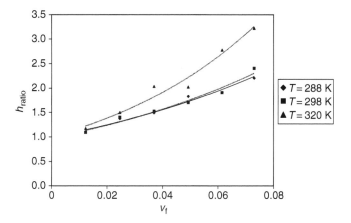

FIGURE 6.34 Effect of nanoparticle addition on the enhancement of forced convection heat transfer coefficient, estimated using properties evaluated with molecular dynamics modeling.

the nanofluid. It can be observed that the enhancement of the convective heat transfer coefficient for laminar flow increases with an increase in the volume fraction of the nanofluid. This variation of the convective heat transfer coefficient is an interactive effect of the individual variations of thermal conductivity, viscosity, specific heat, and density, the first two of which have been calculated using the molecular dynamics simulation method.

6.4.1.4.2   Phase Change

The pool boiling phenomenon has been simulated using molecular dynamics, primarily to obtain the latent heat variations due to nanoparticle inclusion. The simulation has been done on an NPT ensemble (constant number of molecules, constant pressure, and constant temperature) for the platinum nanofluid, comprising water and platinum, in a hypothetical two-dimensional box, with periodic boundary conditions. The initial procedures, the system initialization and equilibration, are similar to that for evaluating the thermal conductivity. The potentials used for the simulation are as described previously, for thermal conductivity and viscosity. The specified system temperature and pressure are attained and their constancy is maintained with the help of "thermostats (Sadus 1999) and barostats (Berendsen et al. 1984)," till the system attains equilibrium. (The thermostat is an artificial means by which a system is brought to a desired temperature, by suitably rescaling the velocities during simulation. Similarly, a barostat is used to maintain a constant pressure, by rescaling the length of the simulation box by a factor determined based upon the pressure at any computational stage.) The value of the latent heat of vaporization is obtained by calculating the difference observed in the total energy (which consists of the potential energy alone since the total kinetic energy of the molecules essentially remains constant during the phase change phenomenon, a constant temperature process) of the molecules while they change their state from liquid to vapor, at the boiling point. A liquid molecule is differentiated from a vapor molecule by the standard procedure of calculating the number of neighboring atoms. If the thermal conductivity is also

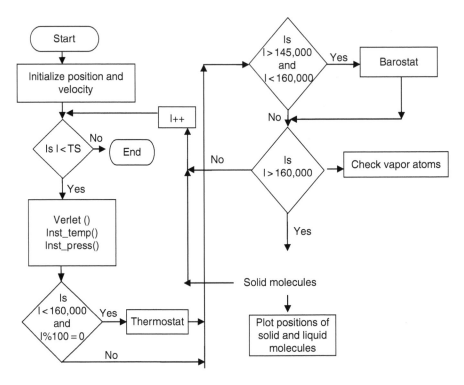

FIGURE 6.35 Flow chart showing the computation procedure for the molecular dynamics simulation of pool boiling in nanofluids.

obtained, a boiling heat transfer coefficient also can be calculated using the standard definition. The above procedure is represented in the flowchart given in Figure 6.35.

Figure 6.36 shows the variation of the latent heat of vaporization, calculated for the platinum nanofluid. The variation with the volume fraction of the nanoparticles is also given in Figure 6.36. Experimentation has been conducted with nanoparticles of aluminum oxide of average size in the range 40–50 nm suspended in water, and the

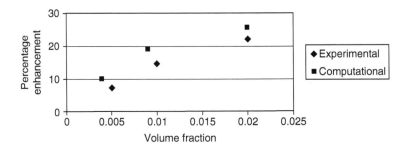

FIGURE 6.36 Result of molecular dynamics modeling of pool boiling of nanofluid at atmospheric pressure, showing the variation of the percentage enhancement of latent heat due to nanoparticle addition.

results have been shown in the graph, for qualitative comparison. The trends obtained the simulation and experiments are found to match, though different nanoparticles are used in the simulation and experiments.

6.4.1.4.3   Analysis of Fluid Flow in Nanoscale Domains

Miniaturization of engineering systems and devices essentially lead to passages of dimensions of order of nanometers. In such nanoscale domains, the need for a direct simulation capable of incorporating the effects emanating from the molecular and atomic level physical interaction is quite inevitable. Flow and thermal phenomena in nanoscale domains can be simulated directly with reasonable accuracy using various discrete modeling methods such as the Monte Carlo simulation, the Lattice–Boltzmann method, and molecular dynamics.

Discrete modeling of the dynamics of the molecules involves an approach strictly according to the basic physical phenomena (for example fluid flow), and accounting for geometrical features of the nanoscale domain such as the overall dimensions and surface roughness, among other surface effects. These effects are remarkably predominant in nanoscale domains. For example, for very small domains of nanoscale dimensions, the surface to volume ratio will be very high. An interesting finding from the analysis on fluid flow in nanochannels, signifying the prominence of the nanoscale behavior is shown in Figure 6.37 which depicts the density profile of a fluid confined in a nanochannel.

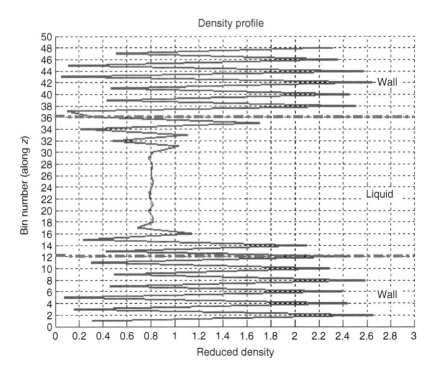

FIGURE 6.37 Distribution of the reduced density in a channel of nanoscale cross-sectional dimension, as obtained from a molecular dynamics simulation.

In this, the bulk liquid away from the walls shows an almost constant density, and the density profile of the solid walls shows peaks. This behavior is expected, since the solid molecules oscillate about their mean position and do not show much diffusive motion. The liquid region adjacent to the solid walls shows a density profile which has a behavior close to the solid, with peaks, although of a reduced magnitude. Thus, the liquid particle appears to experience some ordering due to the presence of the solid layer, and behaves similar to the solid. This could be one of the reasons why films of nanometer dimensions exhibit a higher thermal conductivity than bulk liquid, looking at it from the point of view of individual particles.

To demonstrate the application of molecular dynamics computations in nanoscale domains, results of the analysis of a few problems of liquid flow in nanochannels are discussed below.

Fluid flow and the convective heat transfer in nanochannels are of great research interest and practical value, in fields such as MEMS and NEMS. Molecular dynamics simulations of two flow problems, the Couette flow and the Poiseuille flow have been undertaken, and are presented here.

To simulate Couette flow, a representative case of liquid argon flowing through a platinum nanochannel is considered to demonstrate the use of a well-established interaction potential between the two materials. Both the liquid and solid states in the nanochannel are initialized as having FCC lattices for the molecules. The dynamics of the molecules are guided by a modified Lennard Jones Argon–Platinum interatomic potential (Maruyama 2000). The platinum atoms constituting the solid walls are given a constant velocity along the channel axis, which induces velocities in the liquid molecules through the direct solid–liquid molecule interaction. The velocity profile of the liquid flow can be plotted as shown in Figure 6.38.

It can be clearly observed from Figure 6.38 that in the nanoscale, the velocity profile of the liquid flow deviates from the linear profile observed in Couette flow in

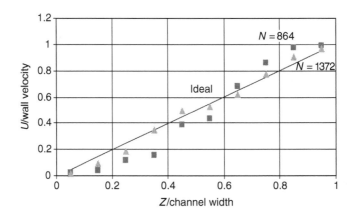

FIGURE 6.38 Normalized velocity distribution in the direction of the channel width for Couette flow, which approaches the standard linear profile when the channel becomes larger in size, showing the size effect in the solution.

the macroscale domain. As the size of the nanochannel increases (which is implemented in computation by appropriately increasing the number of liquid molecules) the flow profile tends toward linear, which is characteristic of Couette flow.

It is also interesting to note that, if a linear profile is fitted to the velocity profile obtained from the simulation of nanoscale domains, the fluid velocity at the fixed wall is nonzero (which results in a y-intercept in the graph). This means that slip-boundary condition prevails in nanoscale domains, as opposed to the no-slip condition traditionally employed in solving the Navier–Stokes equation in macroscale domains. This establishes the need for considering the slip condition, in the design of nanochannels, if continuum modeling is to be used in nanoscale domains.

In a way similar to the above, the Poiseuille flow has also been simulated. Here, the flow is induced by imposing a constant force on the liquid molecules along the flow direction, in addition to those arising from interatomic interactions. Figure 6.39 shows that as the channel size decreases the surface–fluid interaction becomes dominant. The velocity profile deviates from a parabolic profile of a macroscopic flow domain. This observation underlines the specialties in the flow profiles and characteristics in nanoscale fluid domains, which essentially brings in differences in physical effects such as frictional shear and convective heat transfer in nanochannel domains.

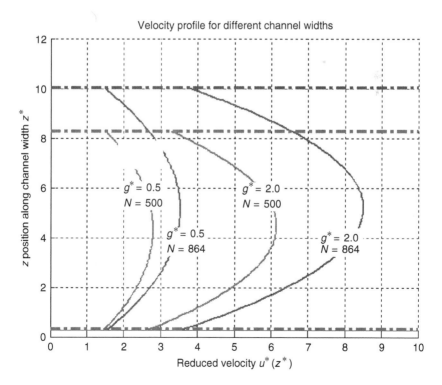

FIGURE 6.39 Velocity profiles in Poiseuille flow in nanochannels obtained by molecular dynamics simulation.

6.5 SPECIAL TOPICS IN THERMAL PHENOMENA

Investigations on certain special topics related to thermal phenomena in nanofluids, and discussions based on observations are presented in this section.

6.5.1 NATURAL CONVECTION UNDER VARIOUS HEATING CONDITIONS

Two mechanisms have been proposed to explain the enhancement of the effective thermal conductivity in nanofluids, the first one being the contribution from the higher thermal conductivity of the nanoparticles and the second one, the contribution from the Brownian motion of the nanoparticles. There is considerable controversy on the significance of the latter of these two, and some would take the position that it has no impact whatsoever. Presented here, however, is the evidence that this contribution is significant and must be included in order to accurately represent the physical phenomena in nanosuspensions. The significance of the two contributions are closely related to the bulk temperature of the nanoparticles suspension, the size of the nanoparticles, and the volume fraction of the nanoparticles in the nanofluid, as well as the thermophysical properties of the nanoparticles and the base fluid materials. The mechanisms that govern the contribution from the higher thermal conductivity of the particles can be calculated using the equations found in Maxwell (1892) or Hamilton and Crosser (1962), and has been shown to accurately predict the effective thermal conductivity for fluids with particles with mean diameters of around 1 mm or larger. The contribution from the Brownian motion of the nanoparticles in a base fluid still remains somewhat controversial and is not well understood, in spite of a considerable number of investigations (Jang and Choi 2004, Li and Peterson 2005, Prasher et al. 2005). Investigations on other heat transfer models such as natural and forced convective heat transfer and boiling heat transfer of nanofluids have not yet resulted in any definitive and widely accepted conclusions.

Okada and Suzuki (1997) experimentally measured the change of layering and concentration of soda glass/water suspensions in a rectangular cell heated at the central portion of the horizontal bottom wall and with isothermal vertical walls, and later the situation for 2.97 μm SiO_2/water suspensions in the rectangular vessel heated and cooled from two opposing vertical walls with adiabatic horizontal walls. Khanfar et al. (2003) and Jou and Tzeng (2006) numerically studied the natural convection heat transfer of nanofluids in a two-dimensional enclosure. On the basis of ideal assumptions such as thermal and flow equilibrium between the fluid phase and the nanoparticles and the uniform shape and size of the nanoparticles, several conclusions were developed. These include the hypothesis that at various volume fractions the nanofluid will substantially increase the heat transfer rate regardless of the Grashof number, and that the enhancement will increase with volume fraction. This later point was in direct contrast to the results of the experimental investigations. For instance, Putra et al. (2003) tested Al_2O_3 and CuO nanoparticle/water nanofluids in a horizontal cylinder heated at one end and cooled at the other, and observed a systematic and definite deterioration in natural convection heat transfer compared to the case of pure water. The same phenomena were found in the experimental investigation of various volume fractions of TiO_2/water nanofluids

resulting from natural convection heat transfer (Wen and Ding 2005). Visual observation of micron-sized particle suspensions under intermittently heated natural
convection were reported and used to explain the phenomena in them (Chen et al.
2005). Similar work on nanometer-sized particle suspensions under different heating
situations is necessary, however, in order to fully understand these phenomena.

The work reported here, performed at the Two Phase Heat Transfer Laboratory,
University of Colorado at Boulder, USA, utilized a microscope and a digital recording system to visually observe the onset and flow patterns of natural convection for
the situation in which a platinum ribbon was used to both heat and measure
the average bulk temperature. The investigation was useful in understanding the
phenomena related to natural convective heat transfer in nanoparticles suspensions.
The investigation was designed to determine the contribution of the Brownian
motion mixing effect on the natural convective heat transfer in nanoparticles suspensions. A set of experimental investigations were conducted to examine the
nanoparticles movement and the resulting mass and energy mixing effects under
different geometrical orientations of the heated surface, i.e., vertical and horizontal
heating modes.

Calibrated platinum ribbons of width 2.5 mm and 1 mm were used for the
application of heat as well as the measurement of temperature in the experiments.
From the calibration experiments, the following resistance–temperature relationships
were established (Equations 6.29 and 6.30) for the 2.5 mm and 1 mm ribbons,
respectively:

$$R = 3.398 \ [1 + 0.00752(T - 278.2)] \tag{6.29}$$

$$R = 3.345 \ [1 + 0.00838(T - 296.7)] \tag{6.30}$$

For the case of bottom heating, a 2.5 mm wide, 0.5 mm deep rectangular channel was
cut using standard MEMS techniques on a glass slide. The bottom was covered by a
0.05 mm thick, 2.5 mm wide platinum ribbon, and then covered with a 0.1 mm thick
glass slide. The test cell channel was charged with an 850 nm diameter polystyrene/
water suspension at 0.1% volume fraction, diluted from an original 2.5% volume
fraction suspension, and sealed with an inert epoxy. For the side heating case, a
1.0 mm wide, 1.0 mm deep rectangular cross-sectional channel was cut using the
same MEMS technique. In this case, however, one side wall of the channel was covered
with a 0.05 mm thick, 1 mm wide platinum ribbon. The remainder of the preparations
and configuration was identical to that of the bottom heating case.

Figure 6.40 shows the cells for both the bottom heating and the side heating test
cases. After the cells were fabricated, the electric power connections and thermocouples were connected and attached to the two cells. The temperature of the upper
surfaces of the test cells was measured with the thermocouples and the heating
surface temperatures were calculated through the relationships described above.
These experimental data were also used to calculate the natural convective heat
transfer coefficient of the nanoparticle suspensions.

The test procedure for both the side heating and bottom heating cases was the
same. Initially, the data acquisition was activated to begin the transient temperature

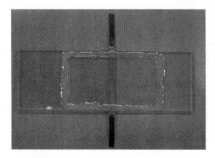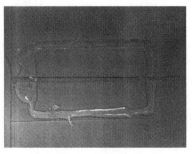

FIGURE 6.40 The test cells of the bottom heating (left) and the side heating (right).

measurement and record the information obtained from the three thermocouples and the electric resistance test wires. A small voltage was then applied to the platinum ribbon. The initial input power applied was approximately 2.7 W. The movement of nanoparticles in suspension was observed through a high-power microscope with a digital imaging system attached to a high-speed CCD digital recording system. The resulting time synchronized images were analyzed in combination with the transient temperature data.

The high-speed visual observations of the particle movement resulted in some interesting findings. For the bottom heating case, once the power was applied to the platinum ribbon, the particles initially remained in the same location, exhibiting typical Brownian motion. As the thermal gradient was established, the vast majority of the particles exhibited a very slow drift in the direction of the temperature gradient, while still exhibiting the characteristics typical of Brownian motion. A few particles, however, moved against the direction of the drift, in part due to the Brownian motion. As time passed, the drift became stronger and the magnitude was much more dominant when compared to the Brownian motion. At this point, the onset of natural convection was observed, and a vertical layer of particle concentration was found in the middle of the channel. These phenomena are similar to those reported by Okada and Suzuki (1997), but in the present case, the layer is believed to be caused by natural convection as opposed to the double diffusive convection, due to temperature and concentration gradients in their work. In the first several minutes after the power was applied and the heating initiated, there was no sign of convection in the channel. Then, at approximately 210 s after the power had been applied, the onset of convection was observed. This phenomenon continued for approximately 30 to 60 s, and then full natural convection was observed and continued through the remainder of the tests. The temperature changes of both the bottom heated surface and the upper cooled surface were recorded as shown in Figure 6.41. Prior to the onset of natural convection, the temperature of the bottom heated surface increased relatively sharply, while at the point of the onset of natural convection, the temperature dropped suddenly nearly 5°C and then slowly returned to the previous value. Following the onset period, the temperature increased more slowly than it did prior to the onset of natural convection. The temperature of the upper cooling surface increased continuously without showing the change brought about by the onset of

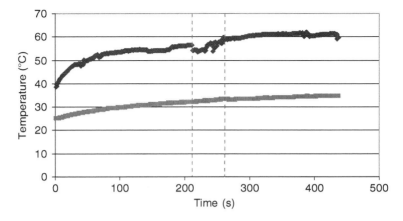

FIGURE 6.41 Temperature change of bottom heating surface (upper line) and upper cool surface (lower line).

natural convection, but it generally illustrated the same trends as the temperature of the bottom surface.

For the side heating case, once power was applied to the platinum ribbon, the particles followed the flow almost immediately and, while doing so, continued to exhibit typical Brownian motion, albeit smaller in magnitude than the drift. It was again apparent that a few particles moved against the direction of the drift, due to the Brownian motion of the individual particles. The velocity of the drift remained relatively constant, but as was the case for the bottom heating, the magnitude of this drift was much greater than the velocities observed due to the Brownian motion. In this case, no onset of natural convection was observed. Nor was a particle concentration layer found anywhere in the channel. The observations indicate that, since no onset is observed, the natural convection occurs, in all likelihood, very soon after the heat is applied. The temperatures recorded during the side heating case were found to increase much faster than the temperature for the bottom heating surface case, which increased to approximately 35°C, and remained relatively constant with only a very slight increase. The temperature of the upper cooling surface, however, continued to increase continuously throughout the tests.

Previous investigations have resulted in two relatively independent conclusions: One is that the presence of the nanofluid deteriorates the natural convective heat transfer coefficient when compared to water at the same condition (Das et al. 2003b, Wen and Ding 2005); the second is that the nanofluid enhances the overall heat transfer coefficient (Jou et al. 2006, Khanfar et al. 2003). However, there are a number of factors that should be taken into consideration before any final conclusion can be drawn.

In the investigation presented here, under the same heating conditions, both the bottom heating and side wall heating experimental results indicated a smaller initial value for the Nusselt number for the nanofluids than pure water. In addition, the rate of change of the Nusselt number (decrease) was more rapid for the nanofluid than for water. Specifically, for the side wall heating case, the experimental observations and corresponding temperature record indicate no evidence of instability or stratification of concentration layers, which confirms the results of the experimental study of Putra et al. (2003). For the bottom heating case, the experimental results illustrate a flow instability phenomenon, while no such phenomenon has previously been reported for the case of bottom heating (Wen and Ding 2005).

In stark contrast to the experimental observations made here and in other investigations, several numerical studies have all presented results that indicate that nanofluids have a higher natural convective heat transfer coefficient than water (Khanfar et al. 2003, Jou et al. 2006). The reasons for this disparity may be the assumptions used in the simulations which may not accurately reflect the actual situation. For instance, effects such as the nonuniform particle size, the particle–fluid hydrodynamic interaction, the sedimentation of particles, and the aggregation of particles, may all impact the experimental results and are not typically incorporated in the numerical models.

During the entire process of heating for both cases, the Brownian motion of particles exhibited a strong influence on the natural convection flow and heat transfer. Before the onset of natural convection in the bottom heating case, it is possible that the Brownian movement restrained the onset of natural convection slightly by increasing the effective thermal conductivity of the suspension through the mixing and stirring effect of the Brownian motion. This reduces the temperature gradient and, in turn, restrains the onset of natural convection for the bottom heating case.

The results of the experiments were compared with the results from previous numerical studies and the resulting conclusions contrasted. The results confirmed the previous statements on the deterioration of natural convective heat transfer coefficient of nanofluids. Possible explanation for this trend could be due to a number of reasons, including the high thermal conductivity of the nanofluids; the high viscosity of the nanofluids; or the deformation of flow and thermal boundaries resulting from the mixing and stirring effect of the Brownian motion of the individual particles. Other fluid and hydrodynamic interactions such as aggregation, sedimentation, or nonuniform particle sizes could also be possible reasons. Such complex interactions might produce material dependent effects, as has been observed in the early onset of natural convection in oxide nanofluids, as discussed earlier. Additional investigations are required to clarify and differentiate the contribution of each of these potential factors.

6.5.2 MIXING EFFECT DUE TO BROWNIAN MOTION

The earliest explanation of the mechanisms contributing to the enhanced effective thermal conductivity of particle suspensions was proposed by Maxwell in 1873, which is for noninteracting particles and nonmoving particles, and consisted of an approximation based upon the relative thermal conductivity of the combination of the particle and base fluid Maxwell (1892). The resulting equation has been successfully applied to predict the effective thermal conductivity of a wide range of particle and fluid/solid matrices, primarily for particles in the millimeter or micron range. While the Maxwell approximation techniques yield accurate results for suspensions in which the particle sizes are in the range of a millimeter or larger, the accuracy of the models diminishes as the particle size decreases.

The Maxwell model for the effective thermal conductivity of particle suspensions, and the equations developed by various investigators based on Maxwell's original work, are found to be inadequate to predict the nature of nanoparticle suspensions. This is expected, because most of these models, with the exception of the Maxwell-Garnett model (Hasselman and Johnson 1987, Nan et al. 1997), do not consider the effect of the particle size, and none of them consider Brownian motion of the particles, and the changes in the rheological and thermal properties of the suspension, which are characteristics of nanofluids. Aggregation and sedimentation effects which also are important in nanofluids are not considered in the models.

On the basis of the mechanisms mentioned above, a number of new models have been developed. Many of these newer models, which incorporate the effect of the Brownian motion, are based on a combination of Maxwell's equation and the effect of the Brownian motion (Kumar et al. 2004, Li and Peterson 2005, Prasher et al. 2005) or based solely on the Brownian motion effect. Koo and Kleinstreuer (2004) and still others have based findings on other possible mechanisms, such as a combination of adsorption, Maxwell's equations, and the Brownian motion effect (Jang and Choi 2004).

In order to better understand the governing phenomena and to determine the contribution of the Brownian motion, a numerical simulation was conducted to determine the validity of the previously stated conclusions, which is presented here.

Visual observations of the Brownian motion indicate that each nanoparticle can be modeled as having a local periodic motion within the suspension, as shown in Figure 6.42. As illustrated, point A and C can be used to represent the farthest points of local periodic motion, and point B is the location at which the local periodic motion has the highest velocity. Based upon this diagram, the velocity and range of influence can be determined from the following expressions:

$$m\frac{\mathrm{d}^2x}{\mathrm{d}t^2} + bx = 0 \tag{6.31}$$

or

$$\frac{\mathrm{d}^2x}{\mathrm{d}t^2} + \omega^2x = 0, \quad \omega = \sqrt{\frac{b}{m}} \tag{6.32}$$

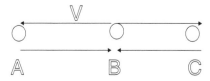

FIGURE 6.42 The local periodic motion of a single nanoparticle.

$$x(t) = x_0 \cos \omega t + \frac{v_0}{\omega} \sin \omega t \qquad (6.33)$$

where
   m is the mass
   x is the displacement
   t is the time
   b is a constant
   ω is the angular velocity
   v_0 is the initial velocity
   x_0 is the initial location

The range of influence for this local periodic motion model can be calculated and the governing equations for the convection caused by the motion of the nanoparticles in three-dimensional form can be solved within the domain. The governing equations are the continuity, momentum, and energy equations. Details of the formulation are presented in the literature by Li and Peterson (2007b).

The model was generated using CFX 5.5.1 software (computational fluid dynamics software of British AEA Technology Inc.). The size of the particle was enlarged from the nanometer to the micron scale, due to the limitations of the software, and a similitude method was used to ensure the validity of the simulation results, which includes the geometrical similarity, the spherical geometry, kinematic similarity, Reynolds number (Re), and thermodynamic similarity, Prandl number (Pr) (Schuring 1977). Properties of matter used were selected from standard figures available from the CFX software. Initially a single nanoparticle case was studied along with the isothermal pressure and the velocity fields were imaged. Then, the difference between a single nanoparticle and two adjacent nanoparticles was studied, and finally the multiparticle case was investigated.

The corresponding temperature, pressure, and velocity fields were simulated using CFX 5.5.1 software and a finite-volume algorithm. The simulations for single, adjacent, and multiple nanoparticles are discussed in detail in the literature (Li and Peterson 2007b). The results showed that microconvection/mixing induced by the Brownian motion of the nanoparticles could significantly affect the macro heat transfer capability of the nanofluids. It also clearly indicated the effect of the microconvection/mixing caused by the Brownian motion of the nanoparticles in these types of suspensions, and demonstrated that the Brownian motion is one of the key factors behind the observed high effective thermal conductivity of the nanofluids.

Further investigation of the transport phenomena should be directed at experimental validation and theoretical quantification of the heat transfer enhancement caused by this microconvection/mixing effect, and other possible mechanisms, such as the impact of the adsorbed fluid layer, cluster formation, and thermophysical property changes resulting from the size effect and clustering.

6.5.3 Microconvection in Nanofluids

Examination of the experimental results indicates that the expression originally developed by Maxwell in 1873 and presented in Maxwell (1892) for effective thermal conductivity does not accurately predict the increases observed for these nanoparticle solutions, particularly as the particle size gets very small. None of the early theoretical models is capable of accurately predicting the experimental data obtained for nanoparticle suspensions over a wide range of temperature, volume fractions, and particle sizes.

Jang and Choi (2004) and Prasher et al. (2005) have developed two different theoretical models for effective thermal conductivity, which can be found in Table 6.3. These two models are in reasonably good agreement with existing experimental data, provided that the appropriate constant, C_1, is selected for the model of Jang and Choi (2004) or the two constants, A and m, are selected for the model of Prasher et al. (2005). In these expressions, the enhanced effective thermal conductivity is presumed to be due to a combination of the contribution resulting from the thermal conductivity of the nanoparticle, which is typically higher than that of the fluid, and the convective heat transfer contribution, due to the transport of energy by the particles which absorb heat from the fluid in one location and reject it in another, due to the temperature difference between the fluid flowing past the nanoparticles, referred to as "microconvection."

To better understand the reasons for the improved thermal conductivity of nanoparticle suspensions, it is important to understand the physical situation that exists in these suspensions. Envision a nanoparticle suspension enclosed in a rectangular container with one wall heated and the opposite wall cooled. As noted by Keblinski et al. (2002), the thermal diffusion in the base fluid is typically much faster than the Brownian diffusion of the nanoparticles in the suspension. As a result, heat transferred through the base liquid by thermal diffusion will be much faster than the heat transfer resulting from the movement of the nanoparticles. This implies that when heat is conducted from the heated surface into the nanoparticle suspension, the temperature of the base fluid will be higher than, or at least the same as, the temperature of the nanoparticles at any given location. Further, as the bulk temperature increases, the thermal conductivity of most base fluids, e.g., water, increases, while the thermal conductivity of most nanoparticles, e.g., aluminum or copper oxide, decreases. Hence, if the thermal conductivity of the nanoparticle is initially higher than that of the surrounding fluid, the thermal conductivity difference will decrease with increasing temperature, and as a result, the contribution resulting from the combined thermal conductivity will decrease with increasing temperature. In addition, the temperature difference between nanoparticle and the fluid molecules around it will be decreasing and, in turn the contribution from the microconvection heat transfer between the nanoparticles and the base fluid will decrease. This is quite different from what has been observed experimentally, where the higher the bulk temperature, the greater the enhancement of the effective thermal conductivity. Finally, based on the computational simulation of Wang et al. (2004), it appears that the presence of the nanoparticles will enhance the effective thermal conductivity of the suspension, even if the nanoparticle has a very low thermal conductivity, indicating that it is the movement of the particle alone, that enhances the overall effective thermal conductivity in nanoparticle suspensions. For these reasons, it seems unlikely that the enhancement observed in a number of experimental investigations, is due to the heat transfer relationships between the nanoparticles and the base fluid alone, as described in several previous investigations (Jang and Choi 2004, Prasher et al. 2005).

As predicted and described by Maxwell, the presence of particles with a greater thermal conductivity than that of the base fluid will enhance the effective thermal conductivity of these suspensions, however, in order to more fully understand why these correlations work for suspensions with relatively large particles, but not for nanoparticle suspensions, one must carefully examine how the Brownian motion of the nanoparticles could contribute to the effective thermal conductivity of nanoparticle suspensions.

When the particles are very small, the particles may contribute to the effective thermal conductivity in a variety of ways and hence, the enhancement for nanoparticles suspended in a fluid is the result of a combination of at least two different mechanisms, i.e.,

$$\frac{k_{eff}}{k_f} = \frac{k_{eff,1}}{k_f} + \frac{k_{eff,2}}{k_f} \qquad (6.34)$$

where the first term on the right-hand side of Equation 6.34 is the contribution resulting from the thermal conductivity of the nanoparticle, which may be greater or in some situations, less than that of the base fluid, and the second term on the right-hand side is the contribution due to the movement of the nanoparticles.

Understanding of the contribution due to the movement of the nanoparticle requires a fundamental understanding of the basic mechanisms that occur as a result of the motion of the nanoparticles and how these mechanisms contribute to an increase in the effective thermal conductivity of nanoparticle suspensions.

Very small particles, such as those used in nanoparticle suspensions, experience a thermally induced random motion. This Brownian motion provides the mechanism that gives rise to the anomalously high effective thermal conductivity reported in recent experiments by stirring the base fluid which results in a locally ordered mass exchange, while at the same time, causes a local energy exchange, due to this mass exchange. With millions and millions of nanoparticles effectively stirring the base fluid, these local energy exchanges result in a macro energy exchange, and hence a higher effective thermal conductivity in these suspensions than would normally be predicted by the difference in the thermal conductivity of nanoparticles alone. This enhanced heat transfer is the result of the instantaneous velocity of the particle, which pushes the base fluid molecules in front of the particle in the direction of its movement. Physically, the fluid right in front of the nanoparticle will experience a positive pressure as well as the same velocity as the particle, and the fluid immediately behind the nanoparticle will experience a reduced pressure and be dragged along at the same velocity as the nanoparticle.

Because the nanoparticle moves randomly in all directions, some of the movements will be in the direction of the bulk heat transfer. This will cause the molecules immediately in front of the particle to be pushed along the direction of the temperature gradient and move orthogonally across the temperature contour lines. The net result is to enhance the heat transfer along the temperature gradient by means of mass transfer along the temperature gradient through a stirring action. In this way a kind of

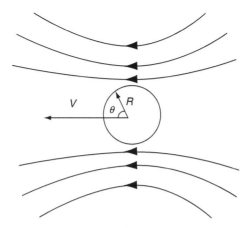

FIGURE 6.43 Flow profile of microconvection around nanoparticle due to the Brownian motion (higher temperature on the right side).

microconvection or perhaps better termed micromixing occurs around the nanoparticle, which can be represented as Stokes flow, as illustrated in Figure 6.43.

The contribution of this microconvection on the enhancement of the effective thermal conductivity of nanoparticle suspensions can be described by evaluating the two-dimensional continuity, momentum, and energy equations expressed in spherical polar coordinates, with usual notations, as

$$\frac{1}{r}\frac{\partial}{\partial r}(r^2 v_r) + \frac{1}{\sin\theta}\frac{\partial}{\partial\theta}(v_\theta \sin\theta) = 0 \tag{6.35}$$

$$\rho\left(\frac{Dv_r}{Dt} - \frac{v_\theta^2}{r}\right) = -\frac{\partial P}{\partial r} + \mu\left(\nabla^2 v_r - \frac{2v_r}{r^2} - \frac{2}{r^2}\frac{\partial v_\theta}{\partial\theta} - \frac{2v_\theta \cot\theta}{r^2}\right) \tag{6.36}$$

$$\rho c_p\left(v_r\frac{\partial T}{\partial r} + \frac{v_\theta}{r}\frac{\partial T}{\partial\theta}\right) = k\left[\frac{1}{r^2}\frac{\partial}{\partial r}\left(r^2\frac{\partial T}{\partial r}\right) + \frac{1}{r^2 \sin\theta}\frac{\partial}{\partial\theta}\left(\sin\theta\frac{\partial T}{\partial\theta}\right)\right] \tag{6.37}$$

with the boundary conditions:

$$r = R : v_r = U\cos\theta : v_\theta = -U\sin\theta$$

$$r \rightarrow \infty : v_r = v_\theta = 0, \ p = p_\infty$$

Applying a scale analysis to the energy equation for the direction where $\phi = 0$ ($v_f = 0$) with the velocity of the nanoparticle, U, and the diameter, $d = 2r$, the energy equation can be modified from the form:

$$\rho c_p\left(v_r\frac{\partial T}{\partial r}\right) = k\frac{1}{r^2}\left(2r\frac{\partial T}{\partial r} + r^2\frac{\partial^2 T}{\partial r^2}\right) \tag{6.38}$$

to

$$\frac{Ud}{\alpha}\frac{\Delta T}{d^2} \sim \frac{1}{d^2}\left(2d\frac{\Delta T}{d} + d^2\frac{\Delta T}{d^2}\right) \tag{6.39}$$

where $\frac{Ud}{\alpha} = \text{Pe}$ is the ratio of the heat transfer, due to the microconvection or micromixing of the nanoparticle, to the heat transferred in the same direction, resulting from the thermal conductivity of the base fluid. This can be rewritten as $\frac{k_{\text{eff},2}}{k_f}$.

In a spherical unit cell with a radius of $2r$, the velocity of the displaced liquid in the region between the outer surface of the particle and a distance R from the surface, i.e., $r < R < 2r$ is assumed to be the same as the velocity of the nanoparticle, which yields a critical nanoparticle volume fraction of 0.060777. The velocity components calculated with the two momentum equations can be expressed as follows:

$$\begin{cases} v_r = \frac{1}{2}U\left(3\frac{R}{r} - \frac{R^3}{r^3}\right)\cos\theta \\ v_\theta = -\frac{1}{4}U\left(3\frac{R}{r} + \frac{R^3}{r^3}\right)\sin\theta \end{cases} \tag{6.40}$$

Figure 6.44 illustrates the relative velocity of the microconvection to that of the nanoparticle, i.e., the velocity of the fluid surrounding the particle and the particle itself. Here the x-axis is the ratio between the distance from the center of the nanoparticle and a point in the fluid, and the radius of the nanoparticle; the y-axis is the arc degree, ϕ, and the z-axis represents the ratio between the molecular velocity of the fluid and the velocity of the nanoparticle.

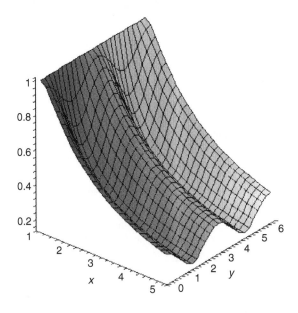

FIGURE 6.44 Relationship between the local fluid velocity and the particle position.

The velocity of a nanoparticle can be obtained from the equipartition theorem, which states that when a nanoparticle reaches thermal equilibrium with the environment, the velocity can be described as

$$U = \sqrt{\frac{k_B T}{m}},$$

where

m is the mass of the nanoparticle

$k_B = 1.38054 \times 10^{-23}$ J/K, i.e., the Boltzmann constant

T is the bulk temperature

Examination of the expression $\frac{k_{eff,2}}{k_f} = $ Pe indicates that the contribution due to this microconvection is of the same order of magnitude as that resulting from the thermal conductivity enhancement of the nanoparticle, $\frac{k_{eff,1}}{k_f}$.

This expression for the microconvection contribution, $\frac{k_{eff,2}}{k_f} = $ Pe, also assumes that the contribution of the microconvection at a given volume fraction will have a linear relationship with respect to the ratio between the volume fraction and the critical volume fraction, i.e., $\phi = 0.060777$. This implies that the microconvection contribution will increase or decrease linearly with increases or decreases in the nanoparticle volume fraction, respectively. This yields the following relationship for the effective thermal conductivity as a function of the volume fraction:

$$\frac{k_{eff}(\phi)}{k_f} = \frac{k_{eff,1}(\phi)}{k_f} + \frac{k_{eff,2}(\phi)}{k_f} \tag{6.41}$$

where

$$\frac{k_{eff,2}(\phi)}{k_f} = \frac{\phi}{0.060777} \text{Pe}(0.060777) = \frac{\phi}{0.060777} \frac{Ud}{\alpha} \tag{6.42}$$

Figure 6.45 presents a comparison of the theoretical predictions obtained using this model with $U = 0.301$ m/s, $T = 300$ K, and $\alpha = k_p/k_f = 140.625$, and the experimental results obtained for Al_2O_3/ethylene glycol suspensions at the same conditions. Here the contribution of the critical volume fraction is Pe$(0.060777) = 0.0802$. As illustrated, the smaller diameter nanoparticles lead to a greater enhancement in the effective thermal conductivity.

Figure 6.46 illustrates a similar comparison for Al_2O_3/water suspensions, where $\alpha = k_p/k_f = 58.625$. Here, the contribution of the critical volume fraction is Pe $(0.060777) = 0.0575$. As shown, in Figure 6.46, this new theoretical model matches quite well with the experimental data of several investigators.

In the new model presented above, the dependence of the effective thermal conductivity on the nanoparticle size and temperature are also represented. As indicated, the smaller the nanoparticle, the greater the velocity and hence, the greater the contribution of the microconvection to the overall effective thermal conductivity. In addition, the higher the temperature, the greater the velocity, which in turn, increases the contribution of this microconvection.

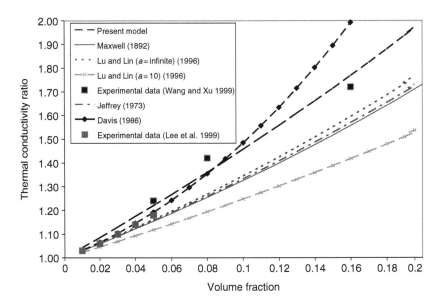

FIGURE 6.45 Comparison of the theoretical predictions and the experimental results for Al$_2$O$_3$/ethylene glycol suspensions. (Diameter of nanoparticle: 38 nm [Lee et al. 1999] and 28 nm [Wang et al. 1999].)

Careful examination of Equation 6.42 indicates that as the mass of the nanoparticle increases, the velocity, U, decreases and hence, the influence of the microconvection component decreases. Further evaluation indicates that at a particle size over 100 nm, Equation 12 will approximate the results of Maxwell's expression quite well.

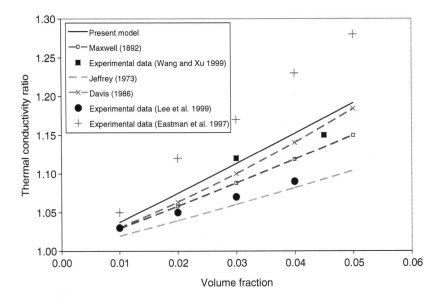

FIGURE 6.46 Comparison of the theoretical predictions and the experimental results for Al$_2$O$_3$/water suspensions. (Diameter of nanoparticle: 33 nm [Eastman et al. 1997], 38 nm [Lee et al. 1999], and 28 nm [Wang et al. 1999].)

This raises some interesting points about the experimental data illustrated in Figure 6.46. As indicated in Figure 6.46, the experimental data of Wang et al. (1999), Eastman et al. (1997), and Lee et al. (1999) were all obtained for Al_2O_3/water suspensions with particle diameters of 28, 33, and 38 nm, respectively. While the data obtained by Wang et al. (1999) and Eastman et al. (1997), for the particle diameters of 28 and 33 nm, respectively, appear to be consistent with the theory, i.e., the impact of the microconvection increases with respect to both particle size and volume fraction, the data for a particle size of 38 nm (Lee et al. 1999) would appear to be somewhat inconsistent. Furthermore, it is not clear why or how nanoparticles, particularly when the thermal conductivity is higher than that of the base fluid could result in an effective thermal conductivity that is lower than that predicted by Maxwell's original equation. One possible solution is that as indicated by Vadasz et al. (2005), thermal wave effects via hyperbolic heat conduction could play some role that may result in a high measurement uncertainty of the transient hot-wire method for these suspensions.

The model developed presents a fundamental explanation of the mechanisms that contribute to the microconvection occurring in nanoparticle suspensions, and the relative relationship between this term and the contribution resulting from the change in the bulk properties, due to the individual contributions of the particle and base fluid thermal conductivities as explained by the initial formulation developed by Maxwell. On the basis of this new analysis, it is clear that the effective thermal conductivity of nanoparticle suspensions is not only dependent upon the thermal conductivity of the suspensions, but also on a microconvection or micromixing contribution resulting from the Brownian motion of the nanoparticles. This explanation gives rise to the concept that the particle motion provides a significant contribution to the overall effective thermal conductivity and increases with the size of the particle, and hence, the level of Brownian motion that occurs. This model also demonstrates that the volume fraction is an important parameter in the calculation of the effective thermal conductivity, not only due to the contribution resulting from the change in the bulk properties, but also due to the role it plays in controlling the microconvection contribution.

This is further supported by the concept that higher order terms of ϕ are necessary to accurately predict the resulting effective thermal conductivity (Jeffrey 1973, O'Brien 1979, Davis 1986, Lu and Lin 1996). Moreover, this model provides an explanation of why, when the particle size is large, the models of Maxwell, Hamilton and Crosser, and others yield reasonably accurate predictions of the effective thermal conductivity of suspensions, yet as the particle size decreases the accuracy of the models decreases proportionally. Finally, using this new model the relative contributions of the microconvection can be determined as a function of the particle size and volume fraction.

6.6 CONCLUDING REMARKS

The recent trends in miniaturization and the advent of MEMS have made nanoscale heat transfer a very challenging and innovative area of study, research, and application. Methods of achieving the transfer of high heat fluxes from micro- and nanoscale systems, the problems associated with fabrication, including molecular manufacture, and the need for development of measurement techniques to characterize such processes are some of the areas of importance in modern technology. The possibility of modifying and controlling thermophysical properties and thermal phenomena by the use of nanoparticles is also an emerging area of great significance.

Starting from a brief introduction of nanoscale heat transfer and considerations in its analysis, this chapter focused on one particular area of interest—nanoparticle suspensions. Studies and analyses related to nanofluids and their modified properties were discussed in detail, with thoughts on the possible fundamental mechanisms involved. Experimental and theoretical investigations, both reported and ongoing, were presented including those of the authors. It is expected that the chapter would give an introductory idea about the excellent possibilities of the application of nanofluids for enhancing and controlling various heat transfer processes relevant to a large spectrum of engineering applications.

REFERENCES

Ahuja, A.S. 1975a. Augmentation of heat transport in laminar flow of polystyrene suspensions. I. Experiments and results. *Journal of Applied Physics* 46: 3408–3416.

Ahuja, A.S. 1975b. Augmentation of heat transport in laminar flow of polystyrene suspensions. II. Analysis of data. *Journal of Applied Physics* 46: 3417–3425.

Allen, M.P. and D.J. Tildesely. 1989. *Computer Simulation of Liquids.* Oxford: Clarendon Press.

Al-Shammiri, M. 2002. Evaporation rate as a function of water salinity. *Desalination* 150: 189–200.

Armentadeu, C. 1997. Increasing evaporation rate for fresh water production-Application to energy savings in renewable energy sources. *Renewable Energy* 11: 197–209.

Ball, P. and L. Garwin. 1992. Science at the atomic scale. *Nature* 355: 761–766.

Bang, I.C. and S.H. Chang. 2005. Boiling heat transfer performance and phenomena of Al_2O_3–water nano-fluids from a plain surface in a pool. *International Journal of Heat and Mass Transfer* 48: 2407–2419.

Batchelor, G.K. 1972. Sedimentation in a dilute dispersion of spheres. *Journal of Fluid Mechanics* 52: 245–268.

Batchelor, G.K. 1974. Transport properties of two-phase materials with random structure. *Annual Review of Fluid Mechanics* 6: 227–255.

Batchelor, G.K. 1982. Sedimentation in a dilute polydisperse system of interacting spheres, part 1. General theory. *Journal of Fluid Mechanics* 119: 379–408.

Berendsen, H.J.C., J.P.M. Postma, W.F. Van Gunsteren, A. Dinola, and J.R. Haak. 1984. Molecular dynamics with coupling to an external bath. *Journal of Chemical Physics* 81(8): 3684–3690.

Brady, J.F., R.J. Phillips, J.C. Lester, and G. Bossis. 1988. Dynamic simulation of hydrodynamically interacting suspensions. *Journal of Fluid Mechanics* 195: 257–280.

Brock, J.R. 1962. On the theory of thermal forces acting on aerosol particles. *Journal of Colloid Science* 17: 768–780.

Buevich, Y.A., A.Y. Zubarev, and A. Isaev. 1990. The hydromechanics of suspensions. *Journal of Engineering Physics* 57(3): 1030–1038.

Buhrman, R.A. and W.P. Halperin. 1973. Fluctuation diamagnetism in a "zero-dimensional" superconductor. *Physical Review Letters* 30: 692–695.

Challoner, A.R. and R.W. Powell. 1956. Thermal conductivities of liquids: New determinations for seven liquids and appraisal of existing values. *Proceedings of the Royal Society of London, Series A (Mathematical and Physical Sciences)* 238: 90–106.

Che, J., T. Cagin, and W.A. Goddard. 2000. Thermal conductivity of carbon nanotubes. *Nanotechnology* 11: 65–69.

Chen, B., F. Mikami, and N. Nishikawa. 2005. Experimental studies on transient features of natural convection in particles suspensions. *International Journal of Heat and Mass Transfer* 48: 2933–2942.

Chen, G. 1996. Nonlocal and nonequilibrium heat conduction in the vicinity of nanoparticles. *Journal of Heat Transfer* 118: 539–545.

Chen, G., D. Borca-Tasciuc, and R.G. Yang. 2004. Nanoscale heat transfer. In *Encyclopedia of Nanoscience and Nanotechnology* H.S. Nalwa (Ed.). California: American Scientific Publishers, pp. 1–30.

Chen, S., W. Liu, and L. Yu. 1998. Preparation of DDP-coated PbS nanoparticles and investigation of the antiwear ability of the prepared nanoparticles as additive in liquid paraffin. *Journal of Wear* 218: 153–158.

Choi, S.U.S. 1995a. Enhancing thermal conductivity of fluids with nanoparticles. In D.A. Siginer and H.P. Wang (Eds.), *Developments and Applications of Non-Newtonian Flows*. New York: American Society of Mechanical Engineers.

Choi, S.U.S. 1995b. Enhancing thermal conductivity of fluids with nano particles. *ASME Fluids Engineering Division* 231: 99–103.

Choi, S.U.S. 1998. Nanofluid technology: Current status and future research. Korea–U.S. Technical Conference on Strategic Technologies, Oct. 22–24, Vienna, Virginia.

Czarnetzki, W. and W. Roetzel. 1995. Temperature oscillation techniques for simultaneous measurement of thermal diffusivity and conductivity. *International Journal of Thermophysics* 16: 413–422.

Das, S.K., N. Putra, and W. Roetzel. 2003a. Pool boiling characteristics of nano-fluids. *International Journal of Heat and Mass Transfer* 46: 851–862.

Das, S.K., N. Putra, P. Thiesen, and W. Roetzel. 2003b. Temperature dependence of thermal conductivity enhancement for nanofluids. *Journal of Heat Transfer* 125: 567–574.

Davis, R.H. 1986. The effective thermal conductivity of a composite material with spherical inclusions. *International Journal of Thermophysics* 7: 609–620.

Duan, B. and H. Lei. 2001. The effect of particle size on the lubricating properties of colloidal polystyrene used as water based lubrication additive. *Journal of Wear* 249: 528–532.

Duncan, A.B., G.P. Peterson, and L.S. Fletcher. 1989. Effective thermal conductivity within packed beds of spherical particles. *Journal of Heat Transfer* 111: 830–836.

Eastman, J.A., S.U.S. Choi, S. Li, L.J. Thompson, and S. Lee. 1997. Enhanced thermal conductivity through the development of nanofluids. *Nanophase and Nanocomposite Material II*: 3–11.

Eastman, J.A., S.U.S. Choi, S. Li, W. Yu, and L.J. Thompson. 2001. Anomalously increased effective thermal conductivities of ethylene glycol-based nanofluids containing copper nanoparticles. *Applied Physics Letters* 78: 718–720.

Eckert, E.R.G. and R.J. Goldstein. 1976. *Measurements in Heat Transfer*. New York: Hemisphere/McGraw Hill.

Fricke, H.A. 1924. Mathematical treatment of the electric conductivity and capacity of disperse systems. *Physical Review* 24: 575–587.

Gaspard, P., M.E. Briggs, M.K. Francis, et al. 1998. Experimental evidence for microscopic chaos. *Nature* 394: 865–868.

Gleiter, H. 1969. Theory of grain boundary migration rate. *Acta metallurgica* 17: 853–862.

Goodson, M. and M. Kraft. 2002. An efficient stochastic algorithm for simulating nano-particle dynamics. *Journal of Computational Physics* 183: 210–232.

Granqvist, C.G. and P.A. Buhrman. 1976. Ultrafine metal particles. *Journal of Applied Physics* 47: 2200–2219.

Grasselli, Y. and G. Bossis. 1995. Three-dimensional particle tracking for the characterization of micrometer-size colloidal particles. *Journal of Colloid and Interface Science* 170: 269–274.

Guczi, L. and D. Horvath. 2000. Modeling gold nanoparticles: Morphology, electron structure, and catalytic activity in CO oxidation. *Journal of Physical Chemistry B* 104: 3183–3193.

Gupte, S.K., S.G. Advani, and P. Huq. 1995. Role of micro-convection due to non-affine motion of particles in a mono-disperse suspension. *International Journal of Heat and Mass Transfer* 38: 2945–2958.

Hamilton, R.L. and O.K. Crosser. 1962. Thermal conductivity of heterogeneous two-component systems. *Industrial & Engineering Chemistry Fundamentals* 1: 187–191

Hasselman, D.P.H. and L.F. Johnson. 1987. Effective thermal conductivity of composites with interfacial thermal barrier resistance. *Journal of Composite Materials* 21: 508–515.

Haw, M.D. 2002. Colloidal suspensions, Brownian motion, molecular reality; a short history. *Journal of Physics: Condensed Matter* 14: 7769–7779.

Holman, J.P. 1989. *Heat Transfer SI Metric Edition*. New York: McGraw Hill.

Hone, J., M. Whitney, and A. Zettl. 1999. Thermal conductivity of single-walled carbon nanotubes. *Synthetic Metals* 103: 2498–2499.

Hu, Z.S. and J.X. Dong. 1998a. Study on antiwear and reducing friction additive of nanometer titanium borate. *Journal of Wear* 216: 87–91.

Hu, Z.S. and J.X. Dong. 1998b. Study on antiwear and reducing friction additive of nanometer titanium oxide. *Journal of Wear* 216: 92–96.

Hu, Z.S., J.X. Dong, G.X. Chen, and J.Z. He. 2000. Preparation and tribological properties of nanoparticle lanthanum borate. *Journal of Wear* 243: 43–47.

Hu, Z.S., R. Lai, F. Lou, et al. 2002. Preparation and tribological properties of nanometer magnesium borate as lubricating oil additive. *Journal of Wear* 252: 370–374.

Huisken, J. and E.H.K. Stelzer. 2002. Optical levitation of absorbing particles with a nominally Gaussian laser beam. *Optics Letters* 27: 1223–1225.

Hummes, D., S. Neumann, F. Schmidt, M. Drotboom, and H. Fissan. 1996. Determination of the size distribution of nanometer-sized particles. *Journal of Aerosol Science* 27(1): S163–S164.

Jang, S.P. and S.U.S. Choi. 2004. Role of Brownian motion in the enhanced thermal conductivity of nanofluids. *Applied Physics Letters* 84: 4316–4318.

Jeffrey, D.J. 1973. Conduction through a random suspension of spheres. *Proceedings of the Royal Society of London, Series A (Mathematical and Physical Sciences)* 335: 355–367.

Jou, R. and S. Tzeng. 2006. Numerical research of natural convective heat transfer enhancement filled with nanofluids in rectangular enclosures. *International Communications in Heat Mass Transfer* 33: 727–736.

Kamat, P.V. and N.M. Dimitrijevic. 1990a. Colloidal semiconductors as photocatalysts for solar energy conversion. *Solar Energy* 44: 83–98.

Kamat, P.V. and N.M. Dimitrijevic. 1990b. Picosecond charge transfer processes in ultrasmall CdS and CdSe semiconductor particles. *Molecular Crystals and Liquid Crystals* 183: 439–445.

Kang, C., M. Okada, A. Hattori, and K. Oyama. 2001. Natural convection of water-fine particle suspension in a rectangular vessel heated and cooled from opposing vertical walls (classification of the natural convection in the case of suspension with a narrow-size distribution). *International Journal of Heat and Mass Transfer* 44: 2973–2982.

Keblinski, P., J.A. Eastman, and D.G. Cahill. 2005. Nanofluids for thermal transport. *Materials Today* 8: 36–44.

Keblinski, P., S.R. Phillpot, S.U.S. Choi, and J.A. Eastman. 2002. Mechanisms of heat flow in suspensions of nano-sized particles (Nanofluids). *International Journal of Heat and Mass Transfer* 45: 855–863.

Keller, J.B. 1963. Conductivity of a medium containing a dense array of perfectly conducting spheres or cylinders or nonconducting cylinders. *Journal of Applied Physics* 34: 991–993.

Khanfar, K., K. Vafai, and M. Lightstone. 2003. Buoyancy-driven heat transfer enhancement in a two-dimensional enclosure utilizing nanofluids. *International Journal of Heat and Mass Transfer* 46: 3639–3653.

Kim, I.C. and S. Torquato. 1990. Determination of the effective conductivity of heterogeneous media by Brownian motion simulation. *Journal of Applied Physics* 68: 3892–3903.

Kim, I.C. and S. Torquato. 1991. Effective conductivity of suspensions of hard spheres by Brownian motion simulation. *Journal of Applied Physics* 69: 2280–2289.

Kim, I.C. and S. Torquato. 1993. Effective conductivity of composites containing spheroidal inclusions: comparison of simulations with theory. *Journal of Applied Physics* 74: 1844–1854.

Kim, P., L. Shi, A. Majumdar, and P.L. McEuen. 2001. Thermal transport measurements of individual multiwalled carbon nanotubes. *Physical Review Letters* 87: 215502- 1–215502-4.

Kongsheng, W. and D. Yulong. 2004. Experimental investigation into convective heat transfer of nanofluids at the entrance region under laminar flow conditions. *International Journal of Heat and Mass Transfer* 47: 5181–5188.

Koo, J. and C. Kleinstreuer. 2004. A new thermal conductivity model for nanofluids. *Journal of Nanoparticle Research* 6: 577–588.

Kumar, D.H., H.E. Patel, V.R.R. Kumar, T. Sundararajan, T. Pradeep, and S.K. Das. 2004. Model for heat conduction in nanofluids, *Physical Review Letters* 93: 155301.

Langevin, P. 1908. Sur la théorie du mouvement Brownien, *C.R. Acad. Sci. (Paris)* 146: 530–533.

Leal, L.G. 1973. On the effective conductivity of a dilute suspension of spherical drops in the limit of low particle Peclet number. *Chemical Engineering Communication* 1: 21–31.

Lee, S.P. and S.U.S. Choi. 1996. Application of metallic nanoparticle suspensions in advanced cooling systems. In Y. Kwon, D. Davis, and H. Chung (Eds.), *Recent Advances in Solids/Structures and Application of Metallic Materials PVP-Vol. 342/MD-Vol. 72* (pp. 227–234). New York: The American Society of Mechanical Engineers.

Lee, S., S.U.S. Choi, S. Li, and J.A. Eastman. 1999. Measuring thermal conductivity of fluids containing oxide nanoparticles. *Transactions of the ASME* 121: 280–289.

Li, C.H. and G.P. Peterson. 2005. Dual role of nanoparticles in the thermal conductivity enhancement of nanoparticle suspensions. Proceedings of the 2005 ASME Congress, Orlando, Florida.

Li, C.H. and G.P. Peterson. 2006a. Experimental investigation of temperature and volume fraction variations on the effective thermal conductivity of nanoparticle suspensions (nanofluids). *Journal of Applied Physics* 99: 084314.

Li, C.H. and G.P. Peterson. 2006b. Transient and steady-state experimental comparison study of effective thermal conductivity of Al2O3/water nanofluid (Poster). Energy, Nanotechnology International Conference, MIT, Boston Massachusetts.

Li, C.H. and G.P. Peterson. 2006c. Transport Phenomena of Nanoparticle Suspensions (Nanofluids) Heated under Various Heating Conditions. Proceedings of the 2006 AIAA/ASME Summer Conference, San Francisco, California.

Li, C.H. and G.P. Peterson. 2007a. The effect of particle size on the effective thermal conductivity of Al_2O_3–water nanofluids. Journal of Applied Physics 101: 044312.

Li, C.H. and G.P. Peterson. 2007b. Mixing effect on the enhancement of thermal conductivity of nanoparticle suspensions (nanofluids). International Journal of Heat and Mass Transfer 50: 4668–4677.

Li, C.H., W. Williams, L. Hu, J. Buonqiorno, and G.P. Peterson. 2008. Transient and steady-state experimental comparison study of effective thermal conductivity of Al_2O_3/water nanofluid. Journal of Heat Transfer 130: in print.

Lu, S.Y. and H.C. Lin. 1996. Effective conductivity of composites containing aligned spheroidal inclusions of finite conductivity. Journal of Applied Physics 79:6761–6769.

Lu, S.Y. and S.T. Kim. 1990. Effective thermal conductivity of composites containing spheroidal inclusions. AIChE Journal 36: 927–938.

Mapa, L.B. and M. Sana. 2005. Heat transfer in mini heat exchanger using nano fluids. American Society for Engineering Education, Session BT4-4.

Maruyama, S. 2000. Molecular dynamics method for microscale heat transfer. Advances in Numerical Heat Transfer 2: 189–226.

Maxwell, J.C. 1892. A Treatise on Electricity and Magnetism. 3rd ed. New York: Oxford University Press.

Mazo, R.M. 2002. Brownian motion fluctuations, dynamics and applications. Oxford: Clarendon Press.

Mazur, P. and D. Bedeaux, 1992. Nature of the random force in Brownian motion. Langmuir 8: 2947–2951.

Meier, F. and P. Wyder. 1973. Magnetic moment of small indium particles in the quantum size-effect regime. Physical Review Letters 30: 181–184.

Meredith, R.E. and C.W. Tobias. 1961. Conductivities in emulsions. Journal of the Electro-chemical Society 108: 286–290.

Miller, R.G. and L.S. Fletcher. 1974. A facility for the measurement of thermal contact conductance. Proceedings of the 10th Southeastern Seminar on Thermal Sciences. New Orleans, Louisiana: 263–285

Nagasaka, Y. and A. Nagashima. 1981. Absolute measurement of the thermal conductivity of electrically conducting liquids by the transient hot-wire method. Journal of Physics E: Scientific Instrument 4: 1435–1439.

Nalwa, H.S. 2004. Encyclopedia of Nanoscience and Nanotechnology. California: American Scientific Publishers, pp. 1–30.

Nan, C.W., R. Birringer, D.R. Clarke, and H. Gleiter. 1997. Effective thermal conductivity of particulate composites with interfacial thermal resistance. Journal of Applied Physics 81: 6692–6699.

Novotny, V. and P.P.M. Meincke. 1973. Thermodynamic lattice and electronic properties of small particles. Physical Review B 8: 4186–4199.

O'Brien, R.W. 1979. A method for the calculation of the effective transport properties of suspensions of interacting particles. Journal of Fluid Mechanics 9: 17–39.

Okada, M. and T. Suzuki. 1997. Natural Convection of Water-fine Particle Suspension in a Rectangular Cell. International Journal of Heat and Mass Transfer 40: 3201–3208.

Patel, H.E., S.K. Das, and T. Sundararajan. 2003. Thermal conductivities of naked and monolayer protected metal nanoparticle based nanofluids: Manifestation of anomalous enhancement and chemical effects. Applied Physics Letters 83: 2931–2933.

Peng X.F., G.P. Peterson, and B.X. Wang. 1994. Heat Transfer characteristic of water flowing through microchannels. *Experimental Heat Transfer* 7: 265–283.

Peterson, G.P. and L.S. Fletcher, 1988. Thermal contact conductance of packed beds in contact with a flat surface. *Journal of Heat Transfer* 110: 38–41.

Peterson, G.P. and L.S. Fletcher. 1989. On the thermal conductivity of dispersed ceramics. *Journal of Heat Transfer* 111: 824–829.

Peterson, G.P. and L.S. Fletcher 1987. Effective thermal conductivity of sintered heat pipe wicks. *AIAA Journal of Thermophysics and Heat Transfer* 1: 343–347.

Peterson, G.P. and C.H. Li. 2006. An Overview of heat and mass transfer in fluids with nanoparticle suspensions. *Advances in Heat Transfer* 39: 261–292.

Prasher, R., P. Bhattacharya, and P.E. Phelan. 2005. Thermal conductivity of nanoscale colloidal solutions (Nanofluid). *Physical Review Letters* 94: 025901–025903.

Putra, N., W. Roetzel, and S.K. Das. 2003. Natural convection of nano-fluids. *Heat and Mass Transfer* 39: 775–784.

Rahman, A. and F.H. Stillinger. 1971. Molecular dynamics study of liquid water. *Journal of Chemical Physics* 55: 3336–3359.

Rao V.U., V. Sajith, T. Hanas, and C.B. Sobhan. 2007. An investigation of the effect of nanoparticles on the effectiveness of a heat exchanger. Proceedings of the IPACK. Vancouver, British Columbia, Canada.

Rappaport, D.C. 2004. *The Art of Molecular Dynamics Simulation*. 2nd ed. Cambridge: Cambridge University Press.

Rocha, A. and A. Acrivos. 1973. On the effective thermal conductivity of dilute dispersions. *Quarterly Journal of Mechanics and Applied Mathematics*. 26: 217–233.

Sadus, R.J. 1999. *Molecular Simulation of Fluids: Theory, Algorithm and Object Orientation*. New York: Elsevier Science.

Sajith, V., M. Sandhya, and C.B. Sobhan. 2006. An investigation into the effect of inclusion of Cerium oxide nanoparticles on the physicochemical properties of diesel oil. Proceedings of the IMECE2006, 2006 ASME International Mechanical Engineering Congress and Exposition, Chicago, Illinois.

Sajith, V., M. Mohiddeen, M.B. Sajanish, and C.B. Sobhan. 2007. An investigation of the effect of nanoparticle addition on lubricating oil properties, *ASME-JSME Thermal Engineering and Summer Heat Transfer Conference*. Vancouver, BC, Canada.

Schuring, D.J. 1977. *Scale Models in Engineering*. Oxford: Pergamon Press.

Shi, L., D. Li, C. Yu, et al. 2003. Measuring thermal and thermoelectric properties of one-dimensional nanostructures using a microfabricated device. *Journal of Heat Transfer* 125: 881–88.

Sobhan, C.B., N. Sankar, N. Mathew, and R. Ratnapal. 2006. Molecular Dynamics Modeling of Thermal Conductivity of Engineering Fluids and its Enhancement Using Nanoparticles, *CANEUS 2006 Micro-Nano Technology for Aerospace Applications*. Toulouse, France.

Spohr, E. and K. Heinzinger.1988. A molecular dynamics study on the water/metal interfacial potential, *Ber Bunsenges. Phys. Chem.* 92: 1358–1363.

Srinivasulu, M., C.B. Sobhan, and S.P. Venkateshan. 1997. Experimental studies on the onset of natural convection in a liquid cell. *Journal of Energy, Heat and Mass Transfer* ISSN 0970–9991 19:179–185.

Sunqing, Q., D. Junxiu, and C. Guoxu. 1999. Tribological properties of CeF_3 nanoparticles as additives in lubricating oils. *Journal of Wear* 230: 35–38.

Tanner, D.B. and A.J. Sievers. 1975. Far-infrared absorption in small metallic particles. *Physical Review B* 11: 1330–1341.

Tarasov, S., A. Kolubaev, S. Belyaev, M. Lerner, and F. Tepper. 2002. Study of friction reduction by nanocopper additives to motor oil. *Journal of Wear* 252: 63–69.

Torquato, S. 1985. Effective electrical conductivity of two-phase disordered composite media. *Journal of Applied Physics* 58: 3790–3797.

Torquato, S. and I.C. Kim. 1989. Efficient simulation technique to compute effective properties of heterogeneous media. *Applied Physics Letters* 55: 1847–1849.

Uhlenbeck, G.E. and L.S. Ornstein. 1930. On the theory of the Brownian motion. *Physical Review* 36: 823–841.

Vadasz, J.J. and S. Govender. 2005. Heat transfer enhancement in nano-fluids suspensions-possible mechanisms and explanations. *International Journal of Heat and Mass Transfer* 48: 2673–2683.

Vasallo, P., R. Kumar, and S. D'Amico. 2004. Pool boiling heat transfer experiments in silica-water nano fluids. *International Journal of Heat and Mass Transfer* 47: 407–411.

Venkateshan, S.P. 1990. Simultaneous determination of thermal conductivity and thermal diffusivity of liquids by a transient technique. Proceedings of the ASME Thermophysics and Heat Transfer Conference. Seattle, Washington.

Waldmann, L. 1961. On the motion of spherical particles in nonhomogeneous gases. In L. Talbot (Ed.), *Rarefied Gas Dynamics. Section 4: Applications of Kinetic Theory.* New York: Academic press.

Wang X., X. Xu, and S.U.S Choi. 1999. Thermal conductivity of nanoparticle–fluid mixture. *AIAA Journal of Thermophysics and Heat Transfer* 13: 474–480.

Wang, B.X., H. Li, and X.F. Peng, 2002. Research on the heat conduction enhancement for liquid with nanoparticle suspensions. *Journal of Thermal Science* 11: 214–219.

Wang, B.X., H. Li, and X.F. Peng. 2003. Research on the effective thermal conductivity of nano-particle colloids. The Sixth ASME–JSME Thermal Engineering Joint Conference, Hawai, March 2003.

Wang, B.X., H. Li, X.F. Peng, and G.P. Peterson. 2004. Numerical simulation for micro-convection around Brownian motion moving nanoparticles. Third International Symposium on Two-Phase Flow Modeling and Experimentation. Pisa, Italy, September 22–24.

Wang, X., and X. Xu. 1999. Thermal conductivity of nanoparticle–fluid mixture. *Journal of Thermophysics and Heat Transfer* 13: 474–480.

Wen, D. and Y. Ding. 2005. Formulation of nanofluids for natural convective heat transfer applications. *International Journal of Heat and Fluid Flow* 26: 855–864.

Williams, W.S. 2006. Experimental and Theoretical Investigation of Transport Phenomena in Nanoparticle Colloids (Nanofluids). PhD Dissertation, Massachusetts Institute of Technology.

Willis, J.R. 1977. Bounds and self-consistent estimates for the overall properties of anisotropic composites. *Journal of Mechanics and Physics of Solids* 25: 185–202.

Wu, Y.Y., W.C. Tsui, and T.C. Liu. 2007. Experimental analysis of tribological properties of lubricating oil with nanoparticle additives. *Wear* 262: 819–825.

Xie, H., H. Lee, W. Youn, and M. Choi. 2003. Nanofluids containing multiwalled carbon nanotubes and their enhanced thermal conductivities. *Journal of Applied Physics* 94: 4967–4971.

Xie, H., J. Wang, T. Xi, Y. Liu, and F. Ai. 2002a. Dependence of the thermal conductivity of nanoparticle–fluid mixture on the base fluid. *Journal of Materials Science Letters* 21: 1469–1471.

Xie, H., J. Wang, T. Xi, Y. Liu, and F. Ai. 2002b. Thermal conductivity enhancement of suspensions containing nanosized alumina particles. *Journal of Applied Physics* 91: 4568–4572.

Xu, Y.P., W.Y. Wang, D.F. Zhang and X.L. Chen. 2001. Dielectric properties of GaN nanoparticles. *Journal of Materials Science* 36: 4401–4403.

Xuan, Y. and Q. Li. 2000a. Heat transfer enhancement of nanofluids. *International Journal of Heat and Fluid Flow* 21: 58–64.

Xuan, Y. and Q. Li. 2000b. Heat transfer enhancement of nanofluids (Chinese). *Journal of Engineering Thermophysics* 20: 465–470.

Xuan, Y.M. and W. Roetzel, 2000. Conceptions for heat transfer correlation of nanofluids. *International Journal of Heat and Mass Transfer* 43: 3701–3777.

Xuan, Y. and Q. Li. 2003. Investigation on convective heat transfer and flow features of nanofluids. *Journal of Heat Transfer* 125: 151–155.

Xue, Q.W. Liu, and Z. Zhang. 1997. Friction and wear properties of a surface-modified TiO_2 nanoparticle as an additive in liquid paraffin. *Journal of Wear* 213: 29–32.

Xue, Q.Z. 2003. Model for effective thermal conductivity of nanofluids. *Physics Letters A* 307: 313–317.

Xue, L., P. Keblinski, S.R. Phillpot, S.U.S Choi, and J.A. Eastman. 2003. Two regimes of thermal resistance at a liquid–solid interface. *Journal of Chemical Physics* 118: 337–339.

Yaws, C.L. 1999. *Chemical Property Handbook, Physical, Thermodynamic, Environmental, Transport, Safety, and Health Related Properties for Organic and Inorganic Chenicals. 1st ed.* New York: McGraw-Hill.

Ye, P., X. Jiang, S. Li, and S. Li. 2002. Preparation of $NiMoO_2S_2$ nanoparticle and investigation of its tribological behavior as additive in lubricating oils. *Journal of Wear* 253: 572–575.

Yee, P. and W.D. Knight. 1975. Quantum size effect in copper: NMR in small particles. *Physical Review B* 11: 3261–3267.

You, S.M. and J.H. Kim. 2003. Effect of nanoparticles on critical heat flux of water in pool boiling heat transfer. *Applied Physics Letters* 83: 3374–3376.

You, S.M., F.E. Lockwood, and E.A. Brulkre. 2003. Anomalous thermal conductivity enhancement in nanotube suspensions. *Applied Physics Letters* 83: 33–37.

Yu, C.J., A.G. Richter, A. Datta, M.K. Durbin, and P. Dutta. 1999. Observation of molecular layering in thin liquid films using X ray reflectivity. *Physical Review Letters* 82: 2326–2329.

Zhou, J., J. Yang, J.Z. Zhang, W. Kliu, and Q. Xue. 1999. Study on the structure and tribological properties of surface modified Cu nanoparticles. *Materials Research Bulletin* 34: 1361–1367.

Zhou, J., Z. Wu, Z. Zhang, W. Liu, and H. Dang. 2001. Study on an antiwear and extreme pressure additive of surface coated LaF_3 nanoparticles in liquid paraffin. *Journal of Wear* 249: 333–337.

7 Numerical Examples

7.1 MICROSCALE CONDUCTION

PROBLEM 7.1

The temperature drop across a copper film is to be maintained at 5×10^{-4}°C. Assuming one-dimensional conduction, calculate the heat fluxes across the film, for a mean film temperature of 400 K, for films of thickness 1200, 1810, 3000, and 6450 Å. Plot the variation of the heat flux with the film thickness. Use Figure 2.1 for thermal conductivity data.

SOLUTION

The heat fluxes across the films can be calculated applying Fourier law in one dimension. The size-affected thermal conductivity values at 400 K for the different film thicknesses are taken from the graph (Figure 2.1) as follows:

For thickness = 1200 Å, $k = 0.55$ cal cm^{-1}K^{-1}s^{-1} = 230 W/m K

For thickness = 1810 Å, $k = 0.625$ cal cm^{-1}K^{-1}s^{-1} = 261 W/m K

For thickness = 3000 Å, $k = 0.75$ cal cm^{-1}K^{-1}s^{-1} = 313 W/m K

For thickness = 6450 Å, $k = 0.9$ cal cm^{-1}K^{-1}s^{-1} = 376 W/m K

Correspondingly, the heat flux values are obtained as follows:
Heat flux $q = -k\frac{\Delta T}{\Delta x}$, where $\Delta T = T_2 - T_1 = -0.0005$ K, and Δx is the film thickness (1 Å = 10^{-10} m). From this calculation,

For thickness = 1200 Å, $q = 95.8$ W/cm^2

For thickness = 1810 Å, $q = 72.1$ W/cm^2

For thickness = 3000 Å, $q = 52.2$ W/cm^2

For thickness = 6450 Å, $q = 29.1$ W/cm^2

Figure 7.1 shows the variation of the heat flux with the film thickness at a mean temperature of 400 K.

PROBLEM 7.2

In Problem 7.1, compare the heat flux values with those obtained using the bulk thermal conductivity values of copper at 400 K, as given in Figure 2.1.

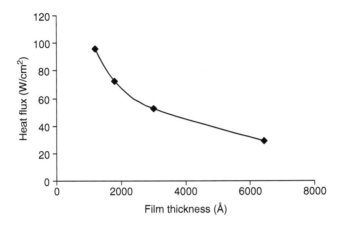

FIGURE 7.1 Variation of the heat flux with the film thickness at a mean temperature of 400 K.

SOLUTION

From Figure 2.1, corresponding to 400 K, the bulk thermal conductivity value $k = 397$ W/m K.

Correspondingly, for the various thicknesses, the heat fluxes are obtained as follows, applying Fourier's law:

$$\text{Film thickness} = 1200 \text{ Å}, \quad q = 165.4 \text{ W/cm}^2$$
$$\text{Film thickness} = 1810 \text{ Å}, \quad q = 109.7 \text{ W/cm}^2$$
$$\text{Film thickness} = 3000 \text{ Å}, \quad q = 66.2 \text{ W/cm}^2$$
$$\text{Film thickness} = 6450 \text{ Å}, \quad q = 30.8 \text{ W/cm}^2$$

Figure 7.2 shows the comparison of the size-affected values and the values obtained using the bulk thermal conductivity. The calculation shows the importance of the use of size-affected thermal conductivity values. It also indicates that the result approaches the macroscale result when the film thickness is increased.

PROBLEM 7.3

An experiment was conducted to determine the out-of-plane thermal conductivity of a thin dielectric film using periodic harmonic heating with a small bridge introduced in the film, for which the experimental observation is given in Figure 2.14a. Calculate the thermal conductivity for the data given in the figure, and verify with the value indicated. Assume $\psi = 1$.

SOLUTION

The film contribution, ΔT_f, is independent of the frequency (in the low frequency domain), and is given by

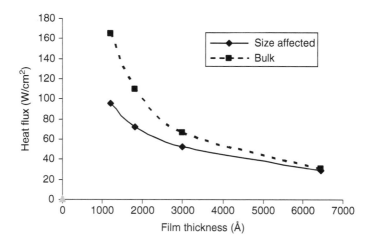

FIGURE 7.2 Comparison of heat flux calculated using size-affected values and bulk value of thermal conductivity.

$$\Delta T_f = \psi \frac{P'}{W} \frac{d_f}{k_f}$$

Width of the metal, $W = 50$ μm
Film thickness, $d_f = 1.6$ μm
$\psi = 1$
From Figure 2.14a corresponding to 1 Hz, $\Delta T_f/P' = 0.022$ m K/W

Thermal conductivity of the film, $k_f = d_f P' \psi / (W \Delta T_f)$
$$= (1.6 \times 10^{-6} \times 1)/(50 \times 10^{-6} \times 0.022)$$
$$= 1.45 \text{ W/m K (value is given as 1.5 W/m K}$$
$$\text{in Figure 2.14a)}$$

7.2 MICROSCALE CONVECTIVE HEAT TRANSFER

PROBLEM 7.4

A microchannel heat exchanger has been designed for cooling an electronic chip of size 20×20 mm, by fabricating an array of concealed longitudinal channels on it. Experiments reveal that in channels of the typical size used in the design, the transition Reynolds number is 700, and that the average Nusselt number is given by

$$\text{Nu} = 0.0384 \text{Re}^{0.62} \text{Pr}^{0.33} \text{ (for laminar flow)}$$
$$\text{Nu} = 0.00726 \text{Re}^{0.8} \text{Pr}^{0.33} \text{ (for turbulent flow)}$$

For a heat dissipation rate of 25 W/cm^2, calculate the average temperature difference between the substrate and cooling water. Also calculate the exit temperature of the cooling water, if water enters the channel at 30°C. Assuming a linear variation for water temperature, calculate the average substrate temperature. The process takes place at steady state.

Use the following data for calculations:
Permissible flow velocity $= 1.5$ m/s
Channel dimension $= 300$ μm side, square channels
Properties of water can be taken at 30°C

SOLUTION

Permissible flow velocity, $V = 1.5$ m/s

Channel size $\qquad = 300$ μm side

Hydraulic dimension, $D_h \qquad = 4 \times$ Area/wetted perimeter

$\qquad\qquad = 300$ μm

Properties of water at 30°C

Density, $\rho \qquad = 997.5$ kg/m^3,

Kinematic viscosity, $\upsilon \qquad = 0.8315 \times 10^{-6}$ m^2/s

Prandtl number, Pr $\qquad = 5.68$

Thermal conductivity, $k \qquad = 612.9 \times 10^{-3}$ W/m K

Specific heat Capacity, $C_p \qquad = 4.178$ kJ/kg K

Reynolds number, Re $\qquad = V\, D_h/\upsilon$

$\qquad\qquad = 1.5 \times 300 \times 10^{-6}/0.8315 \times 10^{-6}$

$\qquad\qquad = 541$

Since the Reynolds number is less than the transition Reynolds number the flow is in the laminar regime. Nusselt number correlation for laminar flow is given as

$$Nu = 0.0384 Re^{0.62} Pr^{0.33}$$

Substituting,

$$Nu = 0.0384 \times (541)^{0.62} \times (5.68)^{0.33} = 3.37$$

Average heat transfer coefficient, $h = Nu\, k/D_h$

$$= 3.37 \times 612.9 \times 10^{-3}/300 \times 10^{-6}$$

$$= 6885 \text{ W/m}^2 \text{ K}$$

Heat transfer by convection, $Q = hA_s\, (\Delta T)$ where A_s is the surface area

Temperature difference, $\Delta T = (Q/A_s)/h$

$$= 25 \times 10^4/6885 = 36.3°C$$

This is the average temperature difference between the substrate and the cooling water. Heat carried by water, $Q = m\, Cp\, (T_o - T_{in})$
Equating $h\, A_s\, (\Delta T) = \rho\, A_c\, V\, (T_o - T_{in})$

$$6885 \times (20 \times 10^{-3} \times 300 \times 10^{-6} \times 4) \times 36.3$$

$$= 997.5 \times (300 \times 10^{-6})^2 \times 1.5 \times 4178 \times (T_o - 30)$$

$T_o = 40.6°C$

Assuming a linear variation for water,

$$\text{Fluid mean temperature, } T_m = (T_{in} + T_{ex})/2$$
$$= (30 + 40.6)/2$$
$$= 35.3°C$$

$$\text{Average temperature of substrate, } T_{sub} = T_{bulk} + \Delta T$$
$$= 35.3 + 36.3$$
$$= 71.6°C$$

PROBLEM 7.5

Compare and plot the friction factor calculated using the conventional correlation and the microchannel correlation, for a water velocity of 0.75 m/s. The correlations for the friction factor can be taken as follows:

Conventional correlation:

$f = C/\text{Re}$

Microchannel channel correlation:

$f = C_1/\text{Re}^{1.9}$

where

D_h is the hydraulic diameter

Re is the Reynolds number based on the hydraulic diameter

f is the friction factor

C and C_1 are coefficients for laminar flow

The channel length is 50 mm. The hydraulic diameter and the coefficients are given in the table below. The properties of the water can be taken at a temperature of 30°C.

Test Specimen	Hydraulic Diameter (D_h) mm	C	C_1
1	0.4	56.46	42000
2	0.343	57.89	44800
3	0.3	56.91	30000
4	0.267	62.19	28600
5	0.24	59.32	42600

SOLUTION

Properties of water at 30°C

Kinematic viscosity, $v = 0.8315 \times 10^{-6}$ m^2/s

For hydraulic diameter 0.4 mm,

Reynolds number, $\text{Re} = VD_h/v = 0.75 \times 0.4 \times 10^{-3}/0.8315 \times 10^{-6} = 361$

According to conventional correlation,

$$f = C/\text{Re} = 56.46/361 = 0.16$$

By the microchannel channel correlation,

$$f = C_1/\mathrm{Re}^{1.9} = 42000/(361)^{1.9} = 0.58$$

Similarly the friction factor for the other cases are also determined and given below:

Test Specimen	Hydraulic Diameter (D_h) mm	Reynolds Number	f From Conventional Correlation	f From Microchannel Correlation
1	0.4	361	0.16	0.58
2	0.343	310	0.19	0.83
3	0.3	271	0.21	0.72
4	0.267	241	0.26	0.85
5	0.24	217	0.27	1.55

Figure 7.3 shows a comparison of the friction factor calculated using the conventional and the microchannel correlations. According to the experimental finding, the conventional correlation appears to under-predict the frictional performance of the channel.

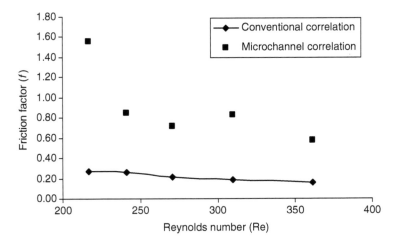

FIGURE 7.3 Comparison of the variation of the friction factor with the Reynolds number, predicted using the microchannel correlation and the conventional correlation.

PROBLEM 7.6

Water flows at a temperature of 30°C through microchannels of length 20 mm and hydraulic diameters as shown below. The Reynolds number for the flow is 600. Determine the effect of the hydraulic diameter of the channel on the pressure drop. Use the following correlation and the given coefficients for obtaining the friction factor:

$$f = C_1/\mathrm{Re}^{1.9}$$

Test Specimen	Hydraulic Diameter (D_h) mm	C_I
1	0.4	42000
2	0.343	44800
3	0.3	30000
4	0.267	28600
5	0.24	42600

SOLUTION

Properties of water at 30°C

Kinematic viscosity, $v = 0.8315 \times 10^{-6}$ m^2/s

Reynolds number, Re $= 600$

For hydraulic diameter, $D_h = 0.4$ mm

Friction factor, $f = C_I/\text{Re}^{1.9} = 42000/(600)^{1.9} = 0.221$

Velocity, $V = \text{Re} \times v/D_h = 600 \times 0.8315 \times 10^{-6}/0.40 \times 10^{-3} = 1.25$ m/s

Pressure drop, $\Delta P = \frac{fLV^2}{2gD} = 0.221 \times 20 \times 10^{-3} \times (1.25)^2/(2 \times 9.81 \times 0.40 \times 10^{-3})$

$= 0.88 = 0.88$ m of water $= 0.09$ bar

Similarly, the pressure drop for different hydraulic diameters are calculated and given below:

Test Specimen	Hydraulic Diameter (D_h) mm	C_I	Friction Factor (f)	Velocity m/s	Pressure Drop (m of Water)	Pressure Drop (Bar)
1	0.4	42000	0.221	1.25	0.88	0.09
2	0.343	44800	0.236	1.45	1.48	0.15
3	0.3	30000	0.158	1.66	1.48	0.15
4	0.267	28600	0.151	1.87	2.01	0.20
5	0.24	42600	0.325	2.08	5.96	0.58

The plot given in Figure 7.4 shows the increase in the pressure drop as the hydraulic diameter of the channel decreases.

PROBLEM 7.7

Water flows through a square channel of side 133 μm and length 20 mm at a temperature of 30°C. For a square channel, the hydraulic mean diameter will be equal to the side. Hence, use Figure 3.1 to obtain the pressure drop corresponding to Reynolds numbers equal to 350, 400, 450, 500, 550, and 600. Use the results to obtain the pumping power requirement for the microchannel, and plot it as a function of the Reynolds number.

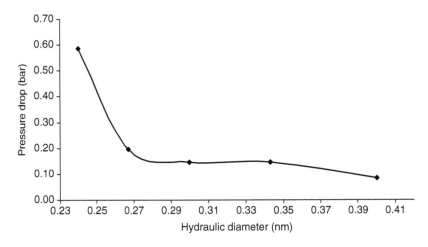

FIGURE 7.4 Variation of pressure drop with hydraulic diameter in the microchannel.

SOLUTION

Properties of water at 30°C
Kinematic viscosity, $v = 0.8315 \times 10^{-6}$ m²/s
Pressure drop in the channel,

$$\Delta P = \frac{fLV^2}{2gD}$$

For a Reynolds number 350, $V = \text{Re } v/D_h = 350 \times 0.8315 \times 10^{-6}/0.133 \times 10^{-3}$
$$= 2.19 \text{ m/s}$$
From the Figure 3.1, corresponding to the Reynolds number 350, $f = 0.8$

$$\Delta P = \frac{fLV^2}{2gD} = 0.8 \times 20 \times 10^{-3} \times (2.19)^2/(2 \times 9.81 \times 0.133 \times 10^{-3})$$
$$= 29.4 \text{ m} = 2.88 \text{ bar}$$

Similarly the pressure drop corresponding to the other Reynolds number values are calculated and given below:

Reynolds Number (Re)	Velocity m/s	Friction Factor (f)	Pressure Drop (Bar)
350	2.19	0.8	2.88
400	2.50	0.65	3.05
450	2.81	0.52	3.12
500	3.12	0.45	3.30
550	3.44	0.37	3.39
600	3.75	0.32	3.50

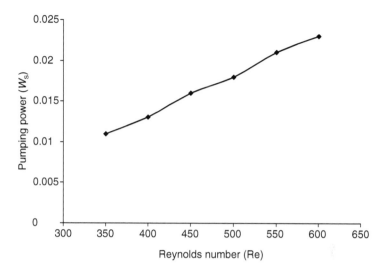

FIGURE 7.5 Variation of the pumping power with the Reynolds number.

For the sample calculation shown above, corresponding to the Reynolds number of 350,

$$\text{Volume flow rate, } Q = \text{Area of the channel} \times \text{velocity}$$
$$= (133 \times 10^{-6})^2 \times 2.19$$
$$= 3.87 \times 10^{-8}\,\text{m}^3/\text{s}$$

$$\text{Pumping power, } P = \text{Pressure drop} \times \text{Volume flow rate}$$
$$= 2.88 \times 10^5 \times 2.225 \times 10^{-8}$$
$$= 0.011 \text{ W}$$

Similarly, the increasing pumping power requirement with an increase in the flow velocity and hence the Reynolds number, varying from 350 to 600, are calculated as plotted in Figure 7.5.

PROBLEM 7.8

The experimental correlation for laminar flow and convective heat transfer in micro channels can be written in the following general form:

$$\text{Nu} = C_{\text{HI}}\,\text{Re}^{0.62}\text{Pr}^{0.33}$$

where C_{HI} is the coefficient for laminar flow, given as follows from experiments on various test specimen:

Test Specimen	D_h (mm)	C_{HI}
1	0.343	0.0580
2	0.3	0.0384
3	0.267	0.0426
4	0.24	0.0472
5	0.20	0.0468
6	0.15	0.0104

Compare the performance evaluated for the microchannel using the above correlation and the conventional correlation, by plotting the variation of $Nu/Pr^{0.33}$ with respect to Reynolds number for a water velocity of 3 m/s. The conventional correlation is as follows:

$$Nu = 1.86(D_h/L)^{0.33}(Re\ Pr)^{0.33}$$

where
    Nu is the Nusselt number
    D_h is the hydraulic diameter
    L is the length of the channel
    Re is the Reynolds number based on the hydraulic diameter
    Pr is the Prandtl number

Take the channel length as 50 mm. Take properties of water at 30°C.

SOLUTION

Properties of water at 30°C

Density	$= 985\ kg/m^3$
Kinematic viscosity, v	$= 0.8315 \times 10^{-6}\ m^2/s$
Prandtl number, Pr	$= 5.68$
Thermal conductivity, k	$= 651 \times 10^{-3}\ W/m\ K$
Specific heat Capacity, C_p	$= 4.183\ kJ/kg\ K$

For hydraulic diameter 0.343 mm,
Reynolds number, $Re = VD_h/v = 1238$

According to the conventional correlation,

$$Nu = 1.86(D_h/L)^{0.33}(PrRe)^{0.33}$$
$$= 1.86(0.343/50)^{0.33}(5.68 \times 1238)^{0.33} = 6.68$$

According to microchannel correlation,

$$Nu = C_{HL}(Re)^{0.62}(Pr)^{0.33}$$
$$= 0.058(1238)^{0.62}(5.68)^{0.33} = 8.51$$

The Nusselt number for all the cases are calculated on similar lines, and tabulated below. The value of the parameter $Nu/Pr^{0.33}$ are also determined.

Test Specimen	D_h (mm)	C_{HI}	Re	Microchannel Correlation		Conventional Correlation	
				Nu	$Nu/Pr^{0.33}$	Nu	$Nu/Pr_{0.33}$
1	0.343	0.058	1238	8.51	4.80	6.68	3.77
2	0.3	0.0384	1083	5.18	2.92	6.12	3.45
3	0.267	0.0426	964	5.35	3.02	5.67	3.19
4	0.24	0.0472	866	5.55	3.13	5.28	2.98
5	0.2	0.0468	722	4.91	2.77	4.68	2.64
6	0.15	0.0104	542	0.91	0.52	3.87	2.18

Figure 7.6 shows the plot of the performance parameter against the Reynolds number, using the conventional correlation and the microchannel correlation. As is seen, no clear trend of difference is depicted.

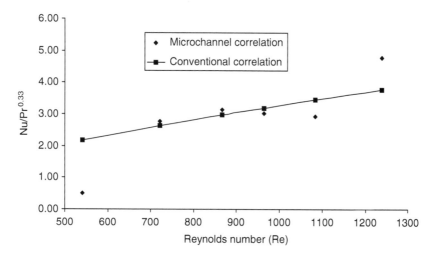

FIGURE 7.6 Plot of $Nu/Pr^{0.33}$ as a function of Reynolds number.

PROBLEM 7.9

Aqueous KCl solution of molar concentration 10^{-6} M flows through a parallel plate channel (Figure 3.3) made of silicon and having a separation height of 90 μm. The ions are completely dissociated in the solution. Width and length of the channel are much greater than the separation height. The flow is restricted to the length direction only. Assume fully developed velocity profile along the length direction. The bulk fluid temperature is 27°C and is constant throughout the flow. Find the characteristic thickness of the EDL.

The volume flow rate through a parallel plate channel without considering EDL effect is given by $Q_p = (2P_z a^3)/(3\mu)$. Where a is the half distance between the plates, and $P_z = -dP/dz$, the pressure gradient along the flow direction z. The actual flow rate considering EDL effect was found to be 1.5×10^{-6} m^3/s for a constant pressure drop of 35kPa. Find the apparent viscosity of the fluid flowing in the micro channel. What is the percentage change in the volume flow rate through the micro channel due to EDL effect?

Take the following data for calculations:

Dielectric constant of the solution, $\varepsilon = 80$

Permittivity of vacuum, $\varepsilon_o = 8.854 \times 10^{-12}$ Fm^{-1}

Boltzmann constant, $k_B = 1.3806 \times 10^{-23}$ JK^{-1}

Viscosity of the solution, $\mu_{solution} = 0.90 \times 10^{-3}$ kg m/s

Electron charge, $e = 1.6 \times 10^{-19}$ C

SOLUTION

Characteristic thickness of EDL $= (1/k)$ where k is the Debye–Huckel parameter

$$k = \left(\frac{2n_o z^2 e^2}{\varepsilon \varepsilon_o k_B T}\right)^{0.5}$$

n_o = Ionic concentration in the bulk fluid = 10^{-6} M = 6.022×10^{20}/m^3

z = valency of K$^+$ ion = 1

$T = 300$ K

Therefore, the characteristic thickness of EDL $1/k = 3.0846 \times 10^{-1}$ μm

$$Q_{p\ actual} = (2P_z a^3)/(3\mu_a) = 1.5 \times 10^{-6}\ \text{m}^3/\text{s}$$
$$P_z = 35\ \text{kPa}$$
$$A = 45\ \mu\text{m}$$

Apparent viscosity $\mu_a = 1.4175 \times 10^{-3}$ kg m s^{-1}

Theoretically,

$$Q_p = (2P_z a^3)/(3\mu) = 2.3625 \times 10^{-6}\ \text{m}^3/\text{s}$$

The percentage change of volume flow rate due to the EDL effect

$$((Q_p - Q_{p\ actual})/(Q_p)) \times 100 = 36.5\%$$

PROBLEM 7.10

It is required to compare the critical heat flux in flow boiling of water in square cross section mini channels for three cross-sectional dimensions—with side 1, 2, and

3 mm, the flow and heating conditions being kept constant. The pertinent data are given below:

$$\text{Length of the channels, } L \quad = 40 \text{ mm}$$
$$\text{Flow rate, } Q \quad\quad\quad\quad = 3 \text{ cm}^3/\text{min}$$
$$\text{Surface tension of water, } \sigma \quad = 65 \times 10^{-6} \text{ N/cm}$$
$$\text{Latent heat of vaporization, } h_{fg} = 2260 \text{ kJ/kg}$$

Calculate the critical heat flux values for the three cases. Plot the variation of the critical heat flux with the hydraulic mean diameter.

SOLUTION

The expression for critical heat flux for mini- and microchannels (Equation 4.9) is

$$\frac{q_{m,p}}{Gh_{fg}} = 0.16 \, We^{-0.19} \left(\frac{L}{D}\right)^{-0.54}$$

Mass velocity $G = \rho V = \rho(Q/A)$
For the channel with side $= 1$ mm, $G = 1000 \times [(3 \times 10^{-6}/60)]/(1 \times 10^{-3})^2 = 50 \text{ kg/m}^2 \text{ s}$
Weber number $We = \frac{G^2 L}{\rho\sigma} = [50^2 \times 40 \times 10^{-3}]/(65 \times 10^{-6} \times 10^2 \times 1000) = 15.29$

Critical heat flux, $q_{m,p} = G\,h_{fg}0.16 \, We^{-0.19} \left(\frac{L}{D}\right)^{-0.54}$

$$= 50 \times 2260 \times 0.16 \times 15.29^{-0.19} \, [(40 \times 10^{-3})/(1 \times 10^{-3})]^{-0.54}$$
$$= 1467.4 \text{ kW/m}^2$$

Proceeding similarly for the other cases, the results are obtained as follows:

$$\text{For the 2 mm channel, } We = 0.96 \quad q_{m,p} = 903.3 \text{ kW/m}^2$$
$$\text{For the 3 mm channel, } We = 0.19 \quad q_{m,p} = 680 \text{ kW/m}^2$$

7.3 MICROSCALE RADIATION

PROBLEM 7.11

A black body having a characteristic dimension of 7 μm emits a radiation of wavelength 90 μm. Find the temperature of the body if the monochromatic emissive power is 90 kW/m². Check whether the process is in the microscale regime.

SOLUTION

Characteristic dimension, $L = 7$ μm

Wave length, λ $= 30$ μm

Emissive power, $E_{b\lambda}$ $= 90$ kW/m^2

From data book, h $= 662 \times 10^{-36}$ J s

k_B $= 1.38066 \times 10^{-23}$ J/K

c $= 3 \times 10^8$ m/s

Monochromatic emissive power, $E_{b\lambda} = \dfrac{2\pi hc^2}{\lambda^5} \dfrac{1}{\exp\left(\dfrac{h\omega_{em}}{2\pi kBT}\right) - 1}$

Substituting the values and solving for T, temperature of the body, $T = 300$ K, the process can be assumed to be in the microscale regime if $L/l_c \leq O(1)$

Coherent length, $l_c = \dfrac{0.15 \, ch}{(k_B T)}$

$= \dfrac{0.15 \times (3 \times 10^8) \times (662 \times 10^{-36})}{(1.38066 \times 10^{-23} \times 300)}$

$= 7.2$ μm

$$L/l_c = 7/7.2 = 0.97$$

Since L/l_c is $O(1)$, the radiative transfer is in the microscale regime.

PROBLEM 7.12

Radiation of wavelength 10 μm is normally incident on a silicon wafer of 2.45 μm thickness. If the irradiation on the surface is 350 W/m^2 and the amount of radiation reflected from the surface is 125 W/m^2, find the optical density of the medium. The extinction coefficient of the film is 0.9. Also plot the variation of external transmittance when thickness of the medium is varied from 0.1 to 3 μm.

SOLUTION

Wavelength, λ $= 10$ μm

Thickness, d $= 2.45$ μm

Incident radiation $= 350$ W/m^2

Reflected radiation $= 125 \text{ W/m}^2$

Extinction coefficient, κ $= 0.9$

Reflectance, ρ $= 125/350 = 0.357$

For normal incidence, ρ $= \dfrac{(n-1)^2 + \kappa^2}{(n+1)^2 + \kappa^2}$

or, $\rho(n+1)^2 - (n-1)^2$ $= (1-\rho)\kappa^2$

$= 0.521$

By trial, refractive index, $n = 3.735$

Internal transmittance, τ $= \exp(-4\pi \times \kappa d/\lambda)$

$= \exp(-4\pi \times 0.9 \times 2.45 \times 10^{-6}/10^{-6})$

$= 0.0626$

External transmittance, T $= (1-\rho)^2 \tau = 0.026$

Optical density $= -\log_{10} T$

$= 1.585$

The external transmittance values calculated for different thicknesses are given below. The values are also shown plotted in Figure 7.7.

Thickness (μm)	External Transmittance (T)
0.1	0.369
0.2	0.33
0.3	0.29
0.5	0.235
0.7	0.187
0.9	0.15
1	0.133
1.3	0.095
1.5	0.076
1.7	0.06
2	0.043
2.5	0.0245
3	0.014

PROBLEM 7.13

The transmittance of a polyimide film of 2.45 μm thickness was measured using a Fourier transform infrared spectrometer for normal incidence. Find the refractive index of the medium if the reflectivity is 0.4 and the wave number is 5000 cm^{-1}.

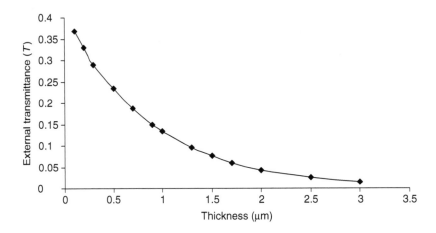

FIGURE 7.7 Variation of the external transmittance with film thickness.

SOLUTION

Thickness, d	$= 2.45\ \mu m$
Reflectivity, ρ	$= 0.4$
Wave number, v	$= 5000\ cm^{-1}$

$$\text{Transmittance, } T = \frac{(1-\rho)^2}{1+\rho^2 - 2\rho\cos(4\pi n v d)}$$

From Figure 5.12, for $v = 5000\ cm^{-1}$, $T = 0.81$

i.e., 0.81 $$= \frac{(1-0.4)^2}{1+0.4 - [2 \times 0.4 \times \cos(4\pi\,n \times 5000 \times 10^2 \times 2.4 \times 10^{-6})]}$$

which gives refractive index $n = 1.81$

Alternatively, in the low absorption region of the transmittance spectrum,

$$\Delta v = 1/(2nd)$$

From Figure 5.12, $\Delta v = 1150\ cm^{-1}$

$$n = 1/(2d\Delta v)$$
$$= 1/(1150 \times 10^2 \times 2 \times 2.4 \times 10^{-6})$$
$$= 1.81$$

7.4   NANOSCALE THERMAL PHENOMENA

PROBLEM 7.14

Calculate the effective thermal conductivity of a 2% by volume suspension of Al_2O_3 nanoparticles in water at 28°C. Compare the value with that obtained by Maxwell's predictive model. Find the percentage enhancement in thermal conductivity of the base fluid due to the addition of nanoparticles in this case. Also bring out the influence of the volume fraction by plotting the enhancement in thermal conductivity of water when 3%, 4%, 5%, and 6% by volume of Al_2O_3 nanoparticle is added to water at the same temperature.

SOLUTION

Volume fraction, $\Phi = 0.02$
Temperature of fluid, $t = 28°C$
Referring to standard tables, at $28°C$
Thermal conductivity of Al_2O_3, $k_{solid} = 36$ W/m K
Thermal conductivity of water, $k_f = 0.609$ W/m K
The expression for the effective thermal conductivity of Al_2O_3/water nanoparticle suspension is chosen as follows (Equation 6.16), where T is in degree Celsius:

$$(k_{eff} - k_f)/k_f = 0.7644\Phi + 0.01869t - 0.46215$$

i.e.,

$$(k_{eff} - k_f)/k_f = 0.7644 \times 0.02 + 0.01869 \times 28 - 0.46215$$
$$= 0.0765$$

or, percentage increase in thermal conductivity of base fluid $= 7.65\%$
The effective thermal conductivity of the nanofluid is

$$k_{eff} = k_f + 0.0765\, k_f$$
$$= 0.656 \text{ W/m K}$$

From Maxwell's model,

$$k_{eff}/k_f = 1 + \frac{3(\alpha - 1)\varphi}{(\alpha + 2) - (\alpha - 1)\varphi}$$
$$\alpha = k_{solid}/k_f = 36/0.609 = 59.03$$

The effective thermal conductivity based on Maxwell's model,

$$k_{eff} = 0.60988\left[1 + \frac{3(59.03 - 1) \times 0.02}{(59.03 + 2) - (59.03 - 1) \times 0.02}\right]$$
$$= 0.645 \text{ W/m K}$$

This shows that the Maxwell's model under-predicts the thermal conductivity of the nanoparticle suspension. The thermal conductivity enhancement has been calculated for various volume fractions of nanoparticles.

% Volume Fraction of Al$_2$O$_3$	Φ	$(k_{eff} - k_f)/k_f$	k_{eff}
2	0.02	0.076461	0.656
3	0.03	0.084106	0.661
4	0.04	0.09175	0.665
5	0.05	0.099395	0.670
6	0.06	0.10704	0.675

The enhancement has been plotted as a function of the volume fraction, as shown in Figure 7.8. The graph clearly shows an enhancement in thermal conductivity, which further increases with the volume fraction of the nanoparticle in the nanofluid.

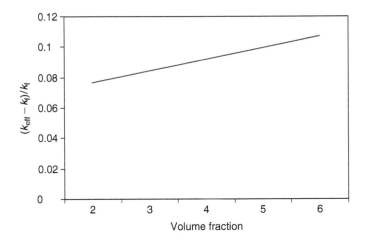

FIGURE 7.8 Variation of the increase in effective thermal conductivity with volume fraction for Al$_2$O$_3$/water nanoparticle suspension.

PROBLEM 7.15

A heat exchange tube of diameter 60 mm placed at the focal line of a parabolic reflector in a solar collector is subjected to a constant heat flux of 2000 W/m^2 on its surface. Water enters the tube at a rate 0.01 kg/s and a temperature of 24°C, and leaves the tube at 40°C. If the water is replaced by a 2% by volume suspension of water–copper nanofluid, with particles of 60 nm diameter as the working fluid, what will be the required diameter of the tube to maintain the same Nusselt number?

SOLUTION

Diameter of nanoparticles, $D_p = 60$ nm
Heat flux, $q'' = 2000$ W/m^2
Diameter of the tube, $D_t = 60$ mm
Mass flow rate of the fluid, $m = 0.01$ kg/s
Volume fraction of nanoparticle, $\varphi = 0.02$
It is assumed that the Prandtl number and viscosity of the nanofluid are the same as that of water (base fluid).
From standard data book, for water at a mean temperature of $(24 + 40)/2 = 32°C = 305$ K, thermal conductivity $k = 0.615$ W/m K, density $\rho = 997$ kg/m^3, kinematic viscosity $v = 0.7966 \times 10^{-06}$ m^2/s, Pr $= 5.412$.
For water, the conventional result for fully developed flow in the tube is
Nu $= 4.36$
For the copper–water nanoparticle suspension (Equation 6.12):
Nu $= \left(0.005991 + 7.6286 \ \varphi^{0.6886} \ Pe_{particle}^{0.001}\right) Re^{0.9238} Pr^{0.4}$
where $Pe_{particle} = Re_{particle} \cdot Pr$, and φ is the volume fraction.

$$\text{Velocity of fluid in the pipe, } v = 4m/(\pi D_p^2 \cdot \rho)$$
$$= 4 \times 0.01/[\pi \times (60 \times 10^{-9})^2 * 997]$$
$$= 3.5424 \times 10^{-3} \, \text{m/s}$$

$$Re_{particle} = v \cdot D_p/v = 3.5424 \times 10^{-03} \times 6 \times 10^{-09}/0.7966 \times 10^{-06}$$
$$= 2.668 \times 10^{-4}$$

$$Pe_{particle} = 2.668 \times 10^{-4} \times 5.412$$
$$= 1.446 \times 10^{-3}$$

Now, to maintain the same Nusselt number,
$4.36 = (0.005991 + 7.6286 \times 0.02^{0.6886} \times (1.446 \times 10^{-3})^{0.001}) \ Re^{0.9238} \times 5.412^{0.4}$
$Re_D = 4.824843$

$$\text{Diameter of the pipe, } D = v/(vRe_D)$$
$$= 0.7966 \times 10^{-6}/(3.5424 \times 10^{-3} \times 4.824843)$$
$$= 1.085 \times 10^{-3} \, \text{m} = 1.085 \, \text{mm}$$

This shows that by using a nanoparticle suspension as the working fluid, a system with much smaller dimensions can achieve the same amount of heat transfer.

PROBLEM 7.16

A cut-bar apparatus is used for measuring the thermal conductivity of a CuO–Water nanoparticle suspension. The copper bar is of diameter 2.54 cm. A heat flux of 42 W/m^2 is applied at the test cell. A rubber O-ring of 1.5 mm thickness encloses the test cell of height 1.5 mm. The temperatures at the top and bottom of the test cell are 35°C and 35.1°C, respectively. Calculate the volume fraction of the CuO nanoparticles in the suspension.

SOLUTION

Diameter of copper bar, $D_{bar} = 2.54$ cm
Thickness of O-ring, $t = 1.5$ mm
Height of test cell, $\Delta Z_{cell} = 1.5$ mm
Temperature difference across test cell, $\Delta T = 0.1°C$
Heat flux, $q'' = 42$ W/m^2
From standard data book, at 35°C,
Thermal conductivity of water, $k_f = 620.45 \times 10^{-3}$ W/m K
Thermal conductivity of copper, $k_{cu} = 386$ W/m K
Thermal conductivity of rubber, $k_{ring} = 162.8 \times 10^{-3}$ W/m K
Diameter of test cell, $D_{cell} = D_{bar} - 2t = 25.4 - 2 \times 1.5 = 2.24$ cm

$$\text{Cross-sectional area of O-ring, } A_{ring} = \pi(D_{bar}^2 - D_{cell}^2)/4$$
$$= \pi(0.0254^2 - 0.0224^2)/4$$
$$= 1.12626 \times 10^{-4} \text{ m}^2$$

$$\text{Cross-sectional area of bar, } A_{bar} = \pi D_{bar}^2/4$$
$$= \pi \times 0.0254^2/4$$
$$= 5.067 \times 10^{-4} \text{ m}^2$$

$$\text{Cross-sectional area of test cell, } A_{cell} = \pi D_{cell}^2/4$$
$$= \pi \times 0.0224^2/4$$
$$= 3.940 \times 10^{-4} \text{ m}^2$$

$$\text{Total heat flow through test cell, } q = q'' \times A_{bar}$$
$$= 45 \times 5.067 \times 10^{-4}$$
$$= 0.0228 \text{ W}$$

According to Equation 6.14,

$$k_{eff} = \left(q\frac{\Delta Z_{cell}}{\Delta T_{cell}} - k_{ring}A_{ring} \right)\bigg/A_{cell}$$

Effective thermal conductivity of the suspension,

$$k_{eff} = \left(0.0228 \times \frac{0.0015}{0.1} - 162.8 \times 10^{-3} \times 1.126 \times 10^{-4} \right)\bigg/3.940 \times 10^{-4}$$
$$= 821.39 \times 10^{-3} \text{ W/m K}$$

From Equation 6.17 for CuO–water nanoparticle suspension,

$$(k_{eff} - k_f)/k_f = 3.761088 \, \varphi + 0.017924 \, t - 0.30734$$
$$(k_{eff} - k_f)/k_f = (0.82139 - 0.62045)/0.62045$$
$$= 0.323862$$
$$= 3.761088 \, \varphi + 0.017924 \times 35 - 0.30734$$

$$\text{Hence, volume fraction, } \varphi = 0.001$$
$$= 0.1\%$$

Index